Mathematical Methods in Chemical Engineering

PRENTICE-HALL INTERNATIONAL SERIES
IN THE PHYSICAL AND CHEMICAL ENGINEERING SCIENCES

AMUNDSON *Mathematical Methods in Chemical Engineering*
ARIS *Introduction to the Analysis of Chemical Reactors*
ARIS *Vectors, Tensors, and the Basic Equations of Fluid Mechanics*
FREDRICKSON *Principles and Applications of Rheology*
HAPPEL AND BRENNER *Low Reynolds Number Hydrodynamics*
HOLLAND *Multicomponent Distillation*
LEVICH *Physicochemical Hydrodynamics*
LYKOV AND MIKHAYLOV *Theory of Energy and Mass Transfer*
PETERSEN *Chemical Reaction Analysis*
WILDE *Optimum Seeking Methods*

PRENTICE-HALL, INC.
PRENTICE-HALL INTERNATIONAL, UNITED KINGDOM AND EIRE
PRENTICE-HALL OF CANADA, LTD., CANADA

Mathematical Methods in Chemical Engineering

Matrices and Their Application

NEAL R. AMUNDSON

Professor and Head
Department of Chemical Engineering
University of Minnesota

PRENTICE-HALL, INC.

Englewood Cliffs, N. J.

PRENTICE-HALL INTERNATIONAL, INC., *London*
PRENTICE-HALL OF AUSTRALIA, PTY. LTD., *Sydney*
PRENTICE-HALL OF CANADA, LTD., *Toronto*
PRENTICE-HALL OF INDIA (PRIVATE) LTD., *New Delhi*
PRENTICE-HALL OF JAPAN, INC., *Tokyo*

Current printing (last digit):

10 9 8 7 6 5 4 3 2 1

Library of Congress Catalog Card Number 66–11768

Printed in the United States of America

C–56108

Preface

This book covers the material of the fall quarter of a course, Advanced Mathematics for Chemical Engineers, and has been offered for about fifteen years. Chapters 1, 2, and 3 are covered in their entirety, with the numerical and theoretical examples frequently left as exercises. In Chapter 4 the first eight sections are offered along with Sections 4.10 and 4.12. Practically all of the material of Chapter 5 is taught with a few minor deletions. Most of Chapter 6 is presented in the form of exercises and outside reading, although at times Section 6.4 is lectured on in detail since it presents a good example of the method of separation of variables and orthogonal sets of vectors as a prelude to later work on partial differential equations. In Chapter 7 usually a selection of material is presented such as reduction of a matrix to canonical form, quadratic forms, functions defined on matrices and differential equations, and reduction of a system of differential equations to normal form. Some of the remaining material is covered in exercises. The material in Chapter 8 has never been offered in a formal way, although it has been presented in outline form.

The chapter on linear programming has been included because of its importance in the chemical and petroleum industries. It can be taught without reference to matrices and vectors but the formalism of the latter makes the understanding of the computational procedure so simple that it was included. This book in retrospect appears not to be a text in the conventional sense, since it includes more detail and manipulation than one usually encounters. This was at least partially by design, for it was hoped that it might prove useful for self-study to engineers whose mathematical training had been obtained before the topics in this book had become so standard. It is felt that the material of the book should be appropriate for upper division and first year graduate students in chemical engineering. Engineering students who are serious about mathematics should take a proper course offered in a mathematics department and there are a number of fine texts for such, some of which are listed at the end of this book.

I am indebted to Dr. Victor Jenson of the University of Birmingham and to innumerable graduate students at Minnesota over the past years who have made comments, criticisms, and suggestions. Mr. Dan Luss aided with reading the proofs. Naturally, however, the author must ultimately bear the responsibility for both the good and the bad.

NEAL R. AMUNDSON

Contents

Mathematical Methods
in Chemical Engineering

Determinants and Their Properties 1

1.1 Introduction and Definition

Although the reader is probably familiar with determinants, an organized but brief exposition will be presented in this chapter. Determinants have few direct applications in physical science, but their use as a means of representation and in theoretical work is widespread. One's first exposure to determinants is usually through linear simultaneous algebraic equations where the solution may be obtained by Cramer's rule.

A determinant of the *n*th order is written in the form

$$D = \begin{vmatrix} a_{11} & a_{12} & \dots & a_{1n} \\ a_{21} & a_{22} & \dots & a_{2n} \\ \cdot & & & \\ \cdot & & & \\ \cdot & & & \\ a_{n1} & a_{n2} & \dots & a_{nn} \end{vmatrix} = |a_{ij}|$$

and is an array of n^2 elements, where the elements a_{ij} may be real or complex numbers or functions. A determinant itself is a function, and if the a_{ij} are thought of as variables, then D is a function of n^2 variables which by definition has the functional form

$$D = \sum (-1)^h (a_{1l_1} a_{2l_2} a_{3l_3} \dots a_{nl_n})$$

and where the summation is taken over all possible products of the a_{ij} in which each product has n elements, one and only one element arising from each row and each column. Because all possible combinations of elements are to be taken, one and only one from each row and each column, it follows that there are n choices for the first element. For the second element there are $(n - 1)$ choices each of which can be associated with the first choice. For the third element there are $n - 2$ choices, and so on, down to the last choice which is uniquely determined. Thus in all, there are $n(n - 1)(n - 2) \dots (3)(2)(1)$ choices to be made for a particular product and therefore the summation involves $n!$ terms. If in each product in the summation the elements are ordered by their first subscripts as shown; then, in general, the second subscripts will not be in their natural order, $1, 2, 3, \dots, n$, although all the numbers from 1 to n will appear. The value of h is defined as the number of transpositions required to transform the sequence of numbers $l_1, l_2, \dots, l_n$ into the order $1, 2, 3, \dots, n$, where a transposition is an interchange of two numbers l_i and l_j. Although the number h is certainly not unique, it may be shown to be always even or always odd for a given sequence. For example, consider the sequence 1 5 3 4 2. To put this sequence into its natural order, several alternate schemes are possible, as

1 5 3 2 4	1 5 3 2 4	1 2 3 4 5
1 3 5 2 4	1 2 3 5 4	
1 3 2 5 4	1 2 3 4 5	
1 2 3 5 4		
1 2 3 4 5		

In the first, only adjacent transpositions are made. The value of h is 5 or 3 or 1. With the third-order determinant

$$\begin{vmatrix} a_{11} & a_{12} & a_{13} \\ a_{21} & a_{22} & a_{23} \\ a_{31} & a_{32} & a_{33} \end{vmatrix} = \begin{matrix} a_{11}a_{22}a_{33} + a_{12}a_{23}a_{31} + a_{13}a_{21}a_{32} \\ -a_{13}a_{22}a_{31} - a_{12}a_{21}a_{33} - a_{11}a_{23}a_{32} \end{matrix}$$

one obtains the familiar expansion which the reader may verify.

Notice that it is immaterial whether the first or second subscripts are ordered and the number of transpositions counted in the sequence of second or first subscripts, respectively. For if one considers a particular term in the summation with first subscripts ordered

$$a_{1l_1} a_{2l_2} \dots a_{nl_n}$$

and considers the same term (that is the term with the same elements) but with second subscripts ordered

$$a_{k_1 1} a_{k_2 2} \dots a_{k_n n}$$

then obviously these two carry the same sign, since the number of transpositions of elements in the latter product required to put the first subscripts back into their natural order will disorder the second subscripts requiring the same number of transpositions of subscripts to reorder the second subscripts. Hence the definition of the sign of each term in the summation representing D could be based on either first or second subscripts, with the other having been previously ordered.

1.2 Elementary Operations

From the definition it follows that multiplication of a determinant by a scalar quantity k is equivalent to multiplication of the elements of any row or of any column by that scalar, since

$$kD = k\,|a_{ij}| = \sum (-1)^h (k a_{1l_1} a_{2l_2} \ldots a_{nl_n})$$

and, because every term in the summation contains an element from every row and every column, k may be associated with the elements from a particular row or column as

$$k\begin{vmatrix} a_{11} & a_{12} & a_{13} \\ a_{21} & a_{22} & a_{23} \\ a_{31} & a_{32} & a_{33} \end{vmatrix} = \begin{vmatrix} a_{11} & ka_{12} & a_{13} \\ a_{21} & ka_{22} & a_{23} \\ a_{31} & ka_{32} & a_{33} \end{vmatrix} = \begin{vmatrix} a_{11} & a_{12} & a_{13} \\ a_{21} & a_{22} & a_{23} \\ ka_{31} & ka_{32} & ka_{33} \end{vmatrix}$$

The effect on the value of a determinant of interchanging two rows or two columns may be determined immediately by examining a typical product in the definition. Consider the case first of two columns and note that the first subscript in a_{ij} stands for the number of the row. A typical term in the sum is

$$a_{1l_1} a_{2l_2} \ldots a_{il_i} \ldots a_{jl_j} \ldots a_{nl_n}$$

If the l_ith and l_jth columns are interchanged, then it is obvious from the definition that the same terms will appear in the expansion but the elements will have a different subscript designation. For example, the element originally in the ith row and l_ith column is designated by a_{il_i}. That same element when the l_ith and l_jth columns are interchanged is in a new position and is now designated by a_{il_j} and the sign is determined by the number of second subscript transpositions required in the term

$$a_{1l_1} a_{2l_2} \ldots a_{il_j} \ldots a_{jl_i} \ldots a_{nl_n}$$

since the first subscripts are not changed if columns are interchanged. Clearly one transposition is required to order the second subscripts, and the sign of D is changed if two columns have their positions interchanged. If one considers the determinant in which the columns and rows are interchanged so that the ith row becomes the ith column and the jth column becomes the jth row, the value of the determinant remains unchanged since a typical term in the summation will be such that wherever a_{ij} appeared it will now be designated as a_{ji} and thus the second subscripts will be ordered. From the preceding discussion on first or second subscript ordering, the sign of that term remains unchanged; hence the value of the determinant is unchanged. It follows that, if two rows are interchanged, the sign of the determinant changes, for the net effect is to interchange rows with columns and then interchange the two corresponding columns.

A determinant in which all elements in one column or row are zero is zero, since in the summation each term has a zero as a factor. If two rows or columns of a determinant are identical element by element then that determinant is zero, for if D is the determinant and D' is the determinant with these two rows or columns interchanged

$$D = D' = -D$$

since interchanging two identical rows or columns leaves the value of the determinant unchanged. Thus

$$D = 0$$

Addition of two determinants with the sum a single determinant is defined only for determinants of the same order. Two determinants may be added if they are identical except for one row (or one column) and the sum is defined as follows:

$$\begin{vmatrix} a_{11} & a_{12} & a_{13} \\ a_{21} & a_{22} & a_{23} \\ a_{31} & a_{32} & a_{33} \end{vmatrix} + \begin{vmatrix} a_{11} & b_{12} & a_{13} \\ a_{21} & b_{22} & a_{23} \\ a_{31} & b_{32} & a_{33} \end{vmatrix} = \begin{vmatrix} a_{11} & a_{12} + b_{12} & a_{13} \\ a_{21} & a_{22} + b_{22} & a_{23} \\ a_{31} & a_{32} + b_{32} & a_{33} \end{vmatrix}$$

and this may be verified using the definition of a determinant. In general, if all but the ith rows of two determinants of the same order are the same, then the determinants may be added and the ith row of the sum is the sum of the ith rows in each of the original determinants as

$$\sum (-1)^h (a_{1l_1} a_{2l_2} \ldots a_{il_i} \ldots a_{nl_n}) + \sum (-1)^h (a_{1l_1} a_{2l_1} \ldots b_{il_i} \ldots a_{nl_n})$$
$$= \sum (-1)^h [a_{1l_1} a_{2l_2} \ldots (a_{il_i} + b_{il_i}) \ldots a_{nl_n}]$$

The most important property of a determinant involves multiplication of one row (or column) by a constant and addition to another row (or column) element by element. For simplicity consider the following determinant

$$\begin{vmatrix} a_{11} & a_{12} + ka_{11} & a_{13} \\ a_{21} & a_{22} + ka_{21} & a_{23} \\ a_{31} & a_{32} + ka_{31} & a_{33} \end{vmatrix}$$

and by the law for addition of determinants this can be separated into

$$\begin{vmatrix} a_{11} & a_{12} & a_{13} \\ a_{21} & a_{22} & a_{23} \\ a_{31} & a_{32} & a_{33} \end{vmatrix} + \begin{vmatrix} a_{11} & ka_{11} & a_{13} \\ a_{21} & ka_{21} & a_{23} \\ a_{31} & ka_{31} & a_{33} \end{vmatrix}$$

In the second determinant, the k may be removed to give

$$k \begin{vmatrix} a_{11} & a_{11} & a_{13} \\ a_{21} & a_{21} & a_{23} \\ a_{31} & a_{31} & a_{33} \end{vmatrix}$$

which is zero since it has two identical columns. Thus

$$\begin{vmatrix} a_{11} & a_{12} & a_{13} \\ a_{21} & a_{22} & a_{23} \\ a_{31} & a_{32} & a_{33} \end{vmatrix} = \begin{vmatrix} a_{11} & a_{12} + ka_{11} & a_{13} \\ a_{21} & a_{22} + ka_{21} & a_{23} \\ a_{31} & a_{32} + ka_{31} & a_{33} \end{vmatrix}$$

and this may be generalized to arbitrary determinants. Therefore the value of a determinant is unchanged if a row or column is multiplied by a constant and added to another row or column element by element. This property is a useful one if one desires to evaluate a determinant from its definition since it may be possible to produce a large number of zeros by appropriate manipulation of the rows and columns.

1.3 Minors and Cofactors

Given a determinant of order n, other determinants called *minors* may be formed by striking out an equal number of whole rows and whole columns. If, for example, in an nth-order determinant m rows and m columns are removed, what is left is a determinant of order $n - m$. The elements at the intersection of the rows and columns which have been struck also form a determinant of order m, and this determinant and the determinant of order $n - m$ are said to be *complementary minors* and one is the complement of the other. For example, in the following fifth-order determinant,

$$\begin{vmatrix} a_{11} & a_{12} & a_{13} & a_{14} & a_{15} \\ a_{21} & a_{22} & a_{23} & a_{24} & a_{25} \\ a_{31} & a_{32} & a_{33} & a_{34} & a_{35} \\ a_{41} & a_{42} & a_{43} & a_{44} & a_{45} \\ a_{51} & a_{52} & a_{53} & a_{54} & a_{55} \end{vmatrix}$$

the two minors obtained by striking out the second and fourth columns and the third and fourth rows are

$$\begin{vmatrix} a_{11} & a_{13} & a_{15} \\ a_{21} & a_{23} & a_{25} \\ a_{51} & a_{53} & a_{55} \end{vmatrix} \quad \text{and} \quad \begin{vmatrix} a_{32} & a_{34} \\ a_{42} & a_{44} \end{vmatrix}$$

The concept of algebraic complement occurs more frequently and is defined in the following way: If D is a determinant of order n and M is an m-rowed minor of D in which the rows of D numbered $k_1, k_2, \ldots, k_m$ and the columns numbered $l_1, l_2, \ldots, l_m$ are represented in M, then the algebraic complement of M is

$$\text{alg comp } M = (-1)^{\sum_1^m l_j + \sum_1^m k_j} (\text{comp } M)$$

In the preceding example

$$\text{alg comp} \begin{vmatrix} a_{11} & a_{13} & a_{15} \\ a_{21} & a_{23} & a_{25} \\ a_{51} & a_{53} & a_{55} \end{vmatrix} = (-1)^{1+3+5+1+2+5} \begin{vmatrix} a_{32} & a_{34} \\ a_{42} & a_{44} \end{vmatrix}$$

A special algebraic complement is that of a single element, called the *cofactor* of that element. The cofactor of a_{ij} is

$$(-1)^{i+j}[\text{comp of } a_{ij}]$$

where the complement of a_{ij} is the determinant of order $(n - 1)$ obtained by striking out the ith row and jth column from D. In the remainder of this work, the cofactor of a_{ij} will be called A_{ij}.

1.4 Laplace's Development of a Determinant

Although the value of a determinant may be obtained from the definition as a sum of products, for any but one of very low order, say less than four or five, it is an impossibly tedious task and other methods must be obtained. One of the most useful is that of Laplace and it will be given here without proof.* In what follows, wherever the term *rows* appears the term *columns* may be substituted as well.

Choose any m rows from an nth-order determinant and form from these m rows all possible mth-order determinants by striking out $(n - m)$ columns. There will be

$$\frac{n!}{(n-m)!\,m!}$$

mth-order determinants in these m rows. If one forms the sum of products

*M. Bôcher, *Introduction to Higher Algebra*, New York: The Macmillan Company, 1938, pp. 24–26.

of all these mth-order determinants with their algebraic complements, then this sum is the value of the determinant. As an example, consider the expansion of a fourth-order determinant by minors of the first and third columns.

$$\begin{vmatrix} 2 & 0 & 1 & 5 \\ 1 & -2 & 0 & 1 \\ 3 & 1 & 1 & 2 \\ 1 & -1 & 0 & -1 \end{vmatrix} = \begin{vmatrix} 2 & 1 \\ 1 & 0 \end{vmatrix} (-1)^{1+3+1+2} \begin{vmatrix} 1 & 2 \\ -1 & -1 \end{vmatrix} + \begin{vmatrix} 2 & 1 \\ 3 & 1 \end{vmatrix} (-1)^{1+3+1+3} \begin{vmatrix} -2 & 1 \\ -1 & -1 \end{vmatrix}$$

$$+ \begin{vmatrix} 2 & 1 \\ 1 & 0 \end{vmatrix} (-1)^{1+3+1+4} \begin{vmatrix} -2 & 1 \\ 1 & 2 \end{vmatrix} + \begin{vmatrix} 1 & 0 \\ 3 & 1 \end{vmatrix} (-1)^{1+3+2+3} \begin{vmatrix} 0 & 5 \\ -1 & -1 \end{vmatrix}$$

$$+ \begin{vmatrix} 1 & 0 \\ 1 & 0 \end{vmatrix} (-1)^{1+3+2+4} \begin{vmatrix} 0 & 5 \\ 1 & 2 \end{vmatrix} + \begin{vmatrix} 3 & 1 \\ 1 & 0 \end{vmatrix} (-1)^{1+3+3+4} \begin{vmatrix} 0 & 5 \\ -2 & 1 \end{vmatrix}$$

$$= (-1)(-1)(1) + (-1)(1)(3) + (-1)(-1)(-5)$$
$$+ (1)(-1)(5) + (0)(1)(-5) + (-1)(-1)(10)$$
$$= -2$$

One recognizes that this determinant may be expanded by other schemes and the familiar expansion by elements of a single row may be used, the latter being a special case of the Laplace development.

Suppose that an nth-order determinant

$$D = \begin{vmatrix} a_{11} & a_{12} & \dots & a_{1n} \\ a_{21} & a_{22} & \dots & a_{2n} \\ \cdot & & & \\ \cdot & & & \\ \cdot & & & \\ a_{n1} & a_{n2} & \dots & a_{nn} \end{vmatrix}$$

is expanded by elements of the ith row in a Laplace development, then

$$D = \sum_{j=1}^{n} a_{ij} A_{ij}, \quad i = 1, 2, 3, \dots, n$$

Consider also the quantity

$$E_{ik} = \sum_{j=1}^{n} a_{ij} A_{kj}, \quad i = 1, 2, 3, \dots, n$$
$$k = 1, 2, 3, \dots, n$$

where, for any $i = k$, $E_{ii} = D$. Suppose, however, $i \neq k$ then what is the value of E_{ik}? Note that E_{ik} is the sum of products of elements of the ith row by the corresponding cofactors of elements of the kth row. It is clear that this is the expansion of a determinant in which the ith row and kth row are identical since A_{kj} contains the elements of the ith row of D. Such a determinant, with two identical rows, is zero, so that $E_{ik} = 0$, $i \neq k$. Hence,

$$\sum_{j=1}^{n} a_{ij} A_{kj} = \begin{matrix} D & i = k \\ 0 & i \neq k \end{matrix}$$

For columns

$$\sum_{i=1}^{n} a_{ij} A_{ik} = \begin{matrix} D & j = k \\ 0 & j \neq k \end{matrix}$$

We say that A_{kj} is an *alien cofactor* of a_{ij} if $k \neq i$, and similarly for A_{ik} if $k \neq j$, and the expansion represented by E_{ik} is the expansion of D by alien cofactors.

1.5 Solution of Linear Simultaneous Algebraic Equations

In the sequel, it will be found that a large number of problems in applied mathematics reduce to problems which require the solution of linear simultaneous algebraic equations; hence throughout this work, we shall frequently return to this problem. Although it is never used for large systems for computation, since Cramer's rule is a useful tool in many theoretical considerations, it will be presented here. We consider n equations in n unknowns, denoted by $x_1, x_2, \ldots, x_n$,

$$\begin{aligned} a_{11}x_1 + a_{12}x_2 + \cdots + a_{1n}x_n &= b_1 \\ a_{21}x_1 + a_{22}x_2 + \cdots + a_{2n}x_n &= b_2 \\ &\vdots \\ a_{n1}x_1 + a_{n2}x_2 + \cdots + a_{nn}x_n &= b_n \end{aligned} \tag{1.5.1}$$

Now multiply the first equation by A_{1j}, the second by A_{2j}, etc., and the nth by A_{nj} and add all the equations, obtaining

$$x_1 \sum_{i=1}^{n} a_{i1} A_{ij} + x_2 \sum_{x=1}^{n} a_{i2} A_{ij} + \cdots + x_j \sum_{i=1}^{n} a_{ij} A_{ij} + \cdots + x_n \sum_{i=1}^{n} a_{in} A_{ij} = \sum_{i=1}^{n} b_i A_{ij}$$

The coefficient of x_j is $\sum_{i=1}^{n} a_{ij} A_{ij}$ and is the expansion of D, the determinant formed from the coefficients in the equations. The other coefficients, however, are expansions of D by alien cofactors, and are zero. Therefore,

$$x_j = \frac{1}{D} \sum_{i=1}^{n} b_i A_{ij}, \quad D \neq 0$$

If one examines the term

$$D_j = \sum_{i=1}^{n} b_i A_{ij} \tag{1.5.2}$$

it is apparent that D_j is the value of a determinant which is identical with D except in place of the elements in the jth column, a_{ij}, b_i has been substituted where i is the row number. Thus

$$x_j = \frac{D_j}{D} \tag{1.5.3}$$

where

$$D_j = \begin{vmatrix} a_{11} & a_{12} & \cdots & a_{1,\,j-1} & b_1 & a_{1,\,j+1} & \cdots & a_{1n} \\ a_{21} & a_{22} & \cdots & a_{2,\,j-1} & b_2 & a_{2,\,j+1} & \cdots & a_{2n} \\ \vdots & & & & & & & \\ a_{n1} & a_{n2} & \cdots & a_{n,\,j-1} & b_n & a_{n,\,j+1} & \cdots & a_{nn} \end{vmatrix} \tag{1.5.4}$$

and this is Cramer's rule. Note that in the formula for x_j, D_j/D is defined only when D is not zero. Let us show now that x_j defined by Eq. (1.5.3) really is a solution to the set, Eq. (1.5.1). If one substitutes x_j into the ith equation of (1.5.1), obtaining

$$a_{i1} D_1 + a_{i2} D_2 + \cdots + a_{ij} D_j + \cdots + a_{in} D_n = b_i D \tag{1.5.5}$$

then it must be shown that this is an identity. However each D_j, as exhibited in Eq. (1.5.4), may be expanded by elements of the jth column, giving

$$D_j = b_1 A_{1j} + b_2 A_{2j} + \cdots + b_n A_{nj}$$

and this may be substituted into Equation (1.5.5) with the result

$$\begin{aligned} a_{i1}(b_1 A_{11} &+ b_2 A_{21} + \cdots + b_n A_{n1}) \\ &+ a_{i2}(b_1 A_{12} + b_2 A_{22} + \cdots + b_n A_{n2}) \\ &\vdots \\ &+ a_{ij}(b_1 A_{1j} + b_2 A_{2j} + \cdots + b_n A_{nj}) \\ &\vdots \\ &+ a_{in}(b_1 A_{1n} + b_2 A_{2n} + \cdots + b_n A_{nn}) = b_i D \end{aligned}$$

This may be rearranged to give

$$\begin{aligned} b_1(a_{i1} A_{11} &+ a_{i2} A_{12} + \cdots + a_{ij} A_{1j} + \cdots + a_{in} A_{1n}) \\ &+ b_2(a_{i1} A_{21} + a_{i2} A_{22} + \cdots + a_{ij} A_{2j} + \cdots + a_{in} A_{2n}) \\ &\vdots \\ &+ b_i(a_{i1} A_{i1} + a_{i2} A_{i2} + \cdots + a_{ij} A_{ij} + \cdots + a_{in} A_{in}) \\ &\vdots \\ &+ b_n(a_{i1} A_{n1} + a_{i2} A_{n2} + \cdots + a_{ij} A_{nj} + \cdots + a_{in} A_{nn}) = b_i D \end{aligned}$$

Now each of the extended terms in parentheses is zero, since it is an alien cofactor expansion of the determinant D, except the ith, which is the expansion of D itself and therefore the identity is established. Cramer's rule leads to a useful theoretical result in that it gives information on when a set of equations will have solutions. That is, for the set of equations in which the b_i are not all zero there can be no solution if $D = 0$. The existence of solutions will be discussed in detail later.

1.6 Multiplication of Determinants

Since a determinant is a number or at most a function there is no question whether multiplication is possible, the only question to be answered is whether there is a convenient algorithm for the product, which can be expressed as a determinant. We shall consider the multiplication of two determinants of the same order and for convenience will consider third-order determinants. The following expression is an identity, as can easily be verified by expanding the sixth-order determinant by Laplace's development of third-order minors of the first three rows, there being only one third-order determinant in the first three rows which is not zero.

$$\begin{vmatrix} a_{11} & a_{12} & a_{13} \\ a_{21} & a_{22} & a_{23} \\ a_{31} & a_{32} & a_{33} \end{vmatrix} \begin{vmatrix} b_{11} & b_{12} & b_{13} \\ b_{21} & b_{22} & b_{23} \\ b_{31} & b_{32} & b_{33} \end{vmatrix} = \begin{vmatrix} a_{11} & a_{12} & a_{13} & 0 & 0 & 0 \\ a_{21} & a_{22} & a_{23} & 0 & 0 & 0 \\ a_{31} & a_{32} & a_{33} & 0 & 0 & 0 \\ -1 & 0 & 0 & b_{11} & b_{12} & b_{13} \\ 0 & -1 & 0 & b_{21} & b_{22} & b_{23} \\ 0 & 0 & -1 & b_{31} & b_{32} & b_{33} \end{vmatrix} \tag{1.6.1}$$

Now note that in this sixth-order determinant if the first column is multiplied by b_{11} and added to the fourth column, the first column is left unchanged but the element b_{11} will change to zero. Similarly, if the first column is multiplied by b_{12} and added to the fifth column the element b_{12} will change to zero. It is evident, therefore, if the first three columns are multiplied by appropriate b_{ij}, the elements b_{ij} in the third-order minor in the lower right-hand corner may be reduced to zero. These operations leave the value of the sixth-order determinant unchanged but change its form to

$$\begin{vmatrix} a_{11} & a_{12} & a_{13} & (a_{11}b_{11}+a_{12}b_{21}+a_{13}b_{31}) & (a_{11}b_{12}+a_{12}b_{22}+a_{13}b_{32}) & (a_{11}b_{13}+a_{12}b_{23}+a_{13}b_{33}) \\ a_{21} & a_{22} & a_{23} & (a_{21}b_{11}+a_{22}b_{21}+a_{23}b_{31}) & (a_{21}b_{12}+a_{22}b_{22}+a_{23}b_{32}) & (a_{21}b_{13}+a_{22}b_{23}+a_{23}b_{33}) \\ a_{31} & a_{32} & a_{33} & (a_{31}b_{11}+a_{32}b_{21}+a_{33}b_{31}) & (a_{31}b_{12}+a_{32}b_{22}+a_{33}b_{32}) & (a_{31}b_{13}+a_{32}b_{23}+a_{33}b_{33}) \\ -1 & 0 & 0 & 0 & 0 & 0 \\ 0 & -1 & 0 & 0 & 0 & 0 \\ 0 & 0 & -1 & 0 & 0 & 0 \end{vmatrix}$$

This determinant may be expanded by third-order minors of the last three columns to give

$$\begin{vmatrix} (a_{11}b_{11}+a_{12}b_{21}+a_{13}b_{31}) & (a_{11}b_{12}+a_{12}b_{22}+a_{13}b_{32}) & (a_{11}b_{13}+a_{12}b_{23}+a_{13}b_{33}) \\ (a_{21}b_{11}+a_{22}b_{21}+a_{23}b_{31}) & (a_{21}b_{12}+a_{22}b_{22}+a_{23}b_{32}) & (a_{21}b_{13}+a_{22}b_{23}+a_{23}b_{33}) \\ (a_{31}b_{11}+a_{32}b_{21}+a_{33}b_{31}) & (a_{31}b_{12}+a_{32}b_{22}+a_{33}b_{32}) & (a_{31}b_{13}+a_{32}b_{23}+a_{33}b_{33}) \end{vmatrix} \tag{1.6.2}$$

and this is the product expressed in Eq. (1.6.1). Analysis of this product shows that the element in the ith row and jth column of the product determinant is obtained by multiplying the elements of the ith row of the first determinant by the corresponding elements of the jth column of the second and summing. Therefore

$$\begin{gathered} |a_{ij}|\,|b_{ij}| = |c_{ij}| \\ c_{ij} = \sum_{k=1}^{n} a_{ik}b_{kj} \quad i, j = 1 \text{ to } n \end{gathered} \tag{1.6.3}$$

for two nth-order determinants. Also

$$|b_{ij}|\,|a_{ij}| = |c_{ij}|$$

as may be easily verified by interchanging rows and columns in each determinant of the product.

1.7 Differentiation of a Determinant

Suppose the determinant D has elements a_{ij} which are functions of a variable t, as $a_{ij} = a_{ij}(t)$, then, in order to compute the derivative D with respect to t, we can write D from the definition

$$D = \sum(-1)^h(a_{1l_1}a_{2l_2}\ldots a_{nl_n})$$

and

$$\begin{aligned} \frac{dD}{dt} &= \sum(-1)^h\left(\frac{da_{1l_1}}{dt}a_{2l_2}\ldots a_{nl_n}\right) \\ &\quad + \sum(-1)^h\left(a_{1l_1}\frac{da_{2l_2}}{dt}\cdots a_{nl_n}\right) \\ &\quad\ \vdots \\ &\quad + \sum(-1)^h\left(a_{1l_1}a_{2l_2}\cdots\frac{da_{nl_n}}{dt}\right) \end{aligned}$$

so that the derivative of a determinant of order n is the sum of n determinants of the form,

$$\frac{dD}{dt} = D_1' + D_2' + \cdots + D_n'$$

in which

$$D'_j = \begin{vmatrix} a_{11} & a_{12} & \dots & a_{1j} & \dots & a_{1n} \\ a_{21} & a_{22} & \dots & a_{2j} & \dots & a_{2n} \\ \cdot & & & & & \\ \cdot & & & & & \\ \cdot & & & & & \\ \dfrac{da_{j1}}{dt} & \dfrac{da_{j2}}{dt} & \dots & \dfrac{da_{jj}}{dt} & \dots & \dfrac{da_{jn}}{dt} \\ \cdot & & & & & \\ \cdot & & & & & \\ \cdot & & & & & \\ a_{n1} & a_{n2} & \dots & a_{nj} & \dots & a_{nn} \end{vmatrix}$$

where the jth row has been differentiated.

EXERCISES

1. Solve the following systems of equations by Cramer's rule

(a)
$$\begin{aligned} x - 2y + 3z &= 2 \\ 2x - 3z &= 3 \\ x + y + z &= 6 \end{aligned}$$

(b)
$$\begin{aligned} x - y + z - w &= 1 \\ 3x + 2y + 4z + 2w &= 1 \\ 2x + 4y - 5z &= -1 \\ 3y - 2z + 3w &= 5 \end{aligned}$$

2. Reason that the equation of a plane in three dimensions is

$$\begin{vmatrix} x & y & z & 1 \\ x_1 & y_1 & z_1 & 1 \\ x_2 & y_2 & z_2 & 1 \\ x_3 & y_3 & z_3 & 1 \end{vmatrix} = 0$$

where (x_i, y_i, z_i), $i = 1, 2, 3$, is a point.

3. In the x,y plane show that the area of a triangle formed from the three points (x_i, y_i), $i = 1, 2, 3$, is given by

$$A = \frac{1}{2}\begin{vmatrix} x_1 & y_1 & 1 \\ x_2 & y_2 & 1 \\ x_3 & y_3 & 1 \end{vmatrix}$$

4. Evaluate the determinant

$$\begin{vmatrix} 3 & 5 & 2 & 4 \\ 1 & 1 & -1 & 6 \\ 2 & 3 & 5 & 1 \\ 2 & 1 & 4 & 8 \end{vmatrix}$$

(a) By Laplace's development using two rows, by Laplace's development using two columns, and by elements of a single row or column.

(b) By using the elementary operations to produce a number of zeros in rows and columns and then expanding.

5. Show that

$$\begin{vmatrix} 1+c_1 & 1 & 1 & 1 \\ 1 & 1+c_2 & 1 & 1 \\ 1 & 1 & 1+c_3 & 1 \\ 1 & 1 & 1 & 1+c_4 \end{vmatrix} = c_1c_2c_3c_4\left(1+\frac{1}{c_1}+\frac{1}{c_2}+\frac{1}{c_3}+\frac{1}{c_4}\right)$$

6. Show that

$$\begin{vmatrix} x+a_1 & a_2 & a_3 & a_4 \\ a_1 & x+a_2 & a_3 & a_4 \\ a_1 & a_2 & x+a_3 & a_4 \\ a_1 & a_2 & a_3 & x+a_4 \end{vmatrix} = x^3(x+a_1+a_2+a_3+a_4)$$

without expanding the determinant.

7. Show that

$$\begin{vmatrix} a_1-x & a_2 & a_3 & a_4 \\ a_2 & a_3-x & a_4 & a_1 \\ a_3 & a_4 & a_1-x & a_2 \\ a_4 & a_1 & a_2 & a_3-x \end{vmatrix} =$$

$$[(x^2-(a_1-a_3)^2-(a_2-a_4)^2]\,(x-a_1-a_2-a_3-a_4)\,(x-a_1+a_2-a_3+a_4)$$

without expanding the determinant.

8. Prove that

$$D = \begin{vmatrix} -1 & 0 & 0 & a \\ 0 & -1 & 0 & b \\ 0 & 0 & -1 & c \\ x & y & z & -1 \end{vmatrix} = 1-ax-by-cz$$

hence $D = 0$ is the equation of a plane.

Linear Algebraic Equations and Elementary Properties of Matrices

2

2.1 Definition

A *matrix* is an array of elements arranged in rows and columns as

$$\mathbf{A} = \begin{bmatrix} a_{11} & a_{12} & \dots & a_{1n} \\ a_{21} & a_{22} & \dots & a_{2n} \\ \cdot & \cdot & & \cdot \\ \cdot & \cdot & & \cdot \\ \cdot & \cdot & & \cdot \\ a_{m1} & a_{m2} & \dots & a_{mn} \end{bmatrix} = [a_{ij}]$$

where the square brackets are meant to differ from the straight vertical lines of a determinant. The elements a_{ij} are functions of some variable or variables, or real or complex numbers, or more abstrusely, are members of a field.* For our purposes, the elements a_{ij} will be real or complex numbers for the moment; later in this book, they will be functions. The foregoing matrix **A** is said to have *m* rows and *n* columns and is said to be an *m* by *n* matrix. Matrices will always be denoted by a boldface letter. If the matrix is square, that is, with *n* rows and *n* columns, it is referred to as a matrix of *n*th order. A special matrix containing only a single column is called a *vector* and is denoted in general by a small boldface letter. Capital letters will be reserved for matrices with more than one column (or row).

$$\mathbf{x} = \begin{bmatrix} x_1 \\ x_2 \\ \cdot \\ \cdot \\ \cdot \\ x_n \end{bmatrix}$$

In order to make matrices useful, operations on them must be defined; although in some cases these operations may appear to be defined in an arbitrary manner, it will be found that these operations will make matrices a considerable aid in multicomponent and multivariable problems.

2.2 Elementary Operations and Properties

Two matrices are said to be *equal* if they contain the same number of rows and the same number of columns and if corresponding elements are equal. If a matrix is multiplied by a scalar *k*, then each element is multiplied by *k* as

$$k[a_{ij}] = [ka_{ij}] = \begin{bmatrix} ka_{11} & ka_{12} & \dots & ka_{1n} \\ ka_{21} & ka_{22} & \dots & ka_{2n} \\ \cdot & & & \\ \cdot & & & \\ \cdot & & & \\ ka_{m1} & ka_{m2} & \dots & ka_{mn} \end{bmatrix}$$

[Note the difference in the definition of scalar multiplication for matrices and determinants.] Addition of two matrices may be performed if the two matrices have the same number of columns and the same number of rows and the *sum* is defined as the matrix whose elements are the sum of the corresponding elements of the individual matrices as

*A field satisfies the axioms that if r and s are in the field, then $r + s$, $r - s$, $r \cdot s$ are in the field, and if s is not zero $r \div s$ is also in the field.

$$[a_{ij}] + [b_{ij}] = [c_{ij}] = [a_{ij} + b_{ij}] = [b_{ij} + a_{ij}] = [b_{ij}] + [a_{ij}]$$
$$[a_{ij}] + [b_{ij}] + [c_{ij}] = [a_{ij} + b_{ij}] + [c_{ij}] = [a_{ij} + b_{ij} + c_{ij}]$$
$$= [a_{ij}] + [b_{ij} + c_{ij}]$$

The commutative and associative laws for addition are valid as shown since the numbers in each element commute and associate. *Subtraction* may be defined in an obvious way. A matrix each of whose elements is zero is called the *zero* matrix and is denoted by **0**. A zero vector will be denoted by **o**.

Many of the important uses of matrices depend upon the definition of multiplication. Two matrices **A** and **B** are said to be *conformable* in the order **AB** if **A** has the same number of columns as **B** has rows. Multiplication is defined only for conformable matrices. Given a matrix **A** with m rows and n columns and a matrix **B** with n rows and p columns, the product matrix **C** is a matrix of m rows and p columns in which the element in the ith row and jth column of **C** is obtained as the sum of products of the corresponding elements in the ith row of **A** and jth column of **B**,

$$[a_{ij}][b_{ij}] = [c_{ij}] = \left[\sum_{k=1}^{n} a_{ik}b_{kj}\right]$$

This is the same algorithm as that for obtaining the elements in the product of two determinants. Consider the following illustrations,

(a) $$\begin{bmatrix} a & b & c \\ e & f & g \end{bmatrix} \begin{bmatrix} x \\ y \\ z \end{bmatrix} = \begin{bmatrix} ax + by + cz \\ ex + fy + gz \end{bmatrix}$$

(b) $$[x \quad y \quad z] \begin{bmatrix} a & e \\ b & f \\ c & g \end{bmatrix} = [ax + by + cz \quad ex + fy + gz]$$

(c) $$\begin{bmatrix} x \\ y \\ z \end{bmatrix} [a \quad b \quad c] = \begin{bmatrix} ax & bx & cx \\ ay & by & cy \\ az & bz & cz \end{bmatrix}$$

(d) $$[x \quad y \quad z] \begin{bmatrix} a \\ b \\ c \end{bmatrix} = [ax + by + cz]$$

Example (a) shows that a 2 by 3 matrix multiplied by a 3 by 1 matrix gives a 2 by 1 matrix. The product taken in the reverse order

$$\begin{bmatrix} x \\ y \\ z \end{bmatrix} \begin{bmatrix} a & b & c \\ e & f & g \end{bmatrix}$$

is not defined because the matrices in this order are not conformable. Hence

$\mathbf{AB} \neq \mathbf{BA}$ because of lack of conformability. Even if $\mathbf{B}$ and $\mathbf{A}$ are conformable, the commutative law for multiplication may not be valid, for consider the following matrices:

$$\begin{bmatrix} 2 & 1 \\ 0 & -1 \end{bmatrix}\begin{bmatrix} 1 & -1 \\ 2 & 1 \end{bmatrix} = \begin{bmatrix} 4 & -1 \\ -2 & -1 \end{bmatrix}$$

$$\begin{bmatrix} 1 & -1 \\ 2 & 1 \end{bmatrix}\begin{bmatrix} 2 & 1 \\ 0 & -1 \end{bmatrix} = \begin{bmatrix} 2 & 2 \\ 4 & 1 \end{bmatrix}$$

This is a very important departure from conventionality and one that must always be kept in mind whenever it is necessary to take the product of matrices. The distributive law for multiplication is valid since*

$$\begin{aligned} \mathbf{A}[\mathbf{B} + \mathbf{C}] = [a_{ij}][b_{ij} + c_{ij}] &= \left[\sum_k a_{ik}(b_{kj} + c_{kj})\right] \\ &= \left[\sum_k a_{ik}b_{kj} + \sum_k a_{ik}c_{kj}\right] \\ &= \mathbf{AB} + \mathbf{AC} \end{aligned}$$

The associative law for multiplication may also be shown by

$$\begin{aligned} \mathbf{A}(\mathbf{BC}) = [a_{ij}]\left[\sum_k b_{ik}c_{kj}\right] &= \left[\sum_l a_{il} \sum_k b_{lk}c_{kj}\right] \\ &= \left[\sum_k \left(\sum_l a_{il}b_{lk}\right)c_{kj}\right] = (\mathbf{AB})\mathbf{C} \end{aligned}$$

A square matrix with non-zero elements on the main diagonal and zeroes elsewhere is called a *diagonal* matrix, and if, in particular, the diagonal elements are all unity, it is called the *idem matrix* or *unit* matrix:

$$\mathbf{I} = \begin{bmatrix} 1 & 0 & 0 & & \cdots & 0 \\ 0 & 1 & 0 & & & 0 \\ 0 & 0 & 1 & 0 & \cdots & 0 \\ \cdot & & & & & \\ \cdot & & & & & \\ \cdot & & & & & \\ 0 & 0 & & \cdots & \cdots & 1 \end{bmatrix}$$

2.3 The Adjoint and the Inverse Matrices

In the applications of matrices, certain other matrices related to a given matrix are found to be useful. The *transpose* of a matrix $\mathbf{A}$ is a matrix with rows and columns interchanged and is denoted by $\mathbf{A}^T$:

*We use the notation $\sum_k$ to indicate summation on the k index.

$$\mathbf{A}^T = \begin{bmatrix} a_{11} & a_{21} & \dots & a_{m1} \\ a_{12} & a_{22} & \dots & a_{m2} \\ \cdot & & & \\ \cdot & & & \\ \cdot & & & \\ a_{1n} & a_{2n} & \dots & a_{mn} \end{bmatrix} = [a_{ji}]$$

in the case of a vector, $\mathbf{x}^T$ is

$$\mathbf{x}^T = [x_1 \quad x_2 \quad \dots \quad x_n]$$

and is referred to as a *row vector*, whereas $\mathbf{x}$ itself is a column vector. Row vectors will consistently carry the superscript T in this work. Remember, notation tends to vary somewhat from book to book. Some books do not use the T superscript for a row vector; they leave the reader to determine from the context what is intended. For a first treatment, however, it seems less confusing to use the notation suggested here.

A second matrix of importance is the adjoint matrix and is defined for square matrices only. In a square matrix $\mathbf{A}$ with elements a_{ij} let A_{ij} be the cofactor of a_{ij} in the determinant of $\mathbf{A}$; then

$$adj\ \mathbf{A} = \begin{bmatrix} A_{11} & A_{21} & \dots & A_{n1} \\ A_{12} & A_{22} & \dots & A_{n2} \\ \cdot & & & \\ \cdot & & & \\ \cdot & & & \\ A_{1n} & A_{2n} & \dots & A_{nn} \end{bmatrix}$$

In the adjoint of $\mathbf{A}$, A_{ij} is in the transpose position of a_{ij}. The motivation for the definition of such a matrix becomes evident when one considers the product

$$\mathbf{A}\ adj\ \mathbf{A} = \begin{bmatrix} a_{11} & a_{12} & \dots & a_{1n} \\ a_{21} & a_{22} & \dots & a_{2n} \\ \cdot & & & \\ \cdot & & & \\ \cdot & & & \\ a_{n1} & a_{n2} & \dots & a_{nn} \end{bmatrix} \begin{bmatrix} A_{11} & A_{21} & \dots & A_{n1} \\ A_{12} & A_{22} & \dots & A_{n2} \\ & & & \\ & & & \\ & & & \\ A_{1n} & A_{2n} & \dots & A_{nn} \end{bmatrix}$$

The element in the ith row and jth column of the product is

$$\sum_k a_{ik} A_{jk}$$

In Section 1.4 it was shown that for $i \neq j$ this is an alien cofactor expansion of the determinant D, which is the determinant formed from the matrix $\mathbf{A}$. For $i = j$, it is D itself. Hence the product matrix has D's on the main diagonal and zeroes elsewhere,

$$\mathbf{A}\,adj\,\mathbf{A} = \begin{bmatrix} D & 0 & 0 & \dots & 0 \\ 0 & D & 0 & \dots & 0 \\ \cdot & & & & \cdot \\ \cdot & & & & \cdot \\ \cdot & & & & \cdot \\ 0 & 0 & & \dots & D \end{bmatrix} = D\mathbf{I}$$

A similar analysis will show that

$$(adj\,\mathbf{A})\mathbf{A} = D\mathbf{I}$$

and therefore **A** and adj **A** commute. Thus one sees here the reason for the definition of **I** as a unity matrix, for, if one sets

$$\frac{1}{D}\,adj\,\mathbf{A} = \mathbf{A}^{-1}$$

then $\mathbf{A}^{-1}$ plays the role of an inverse matrix provided that $D \neq 0$, since

$$\mathbf{A}\mathbf{A}^{-1} = \mathbf{I} = \mathbf{A}^{-1}\mathbf{A}$$

As an example of the inverse of a matrix consider

$$\mathbf{A} = \begin{vmatrix} 1 & 2 & -1 \\ 0 & 3 & 2 \\ 1 & -1 & 1 \end{vmatrix}$$

then

$$adj\,\mathbf{A} = \begin{vmatrix} 5 & -1 & 7 \\ 2 & 2 & -2 \\ -3 & 3 & 3 \end{vmatrix}; \; |\mathbf{A}| = 12$$

$$\mathbf{A}\,adj\,\mathbf{A} = \begin{vmatrix} 12 & 0 & 0 \\ 0 & 12 & 0 \\ 0 & 0 & 12 \end{vmatrix} = 12\mathbf{I}$$

A notation frequently used for the determinant of a square matrix **A** is $|\mathbf{A}|$ and this will be adopted henceforth. The notation $|\mathbf{A}| = \det \mathbf{A}$ is also used. A matrix **A** is said to be *nonsingular* if $|\mathbf{A}| \neq 0$ and *singular* if $|\mathbf{A}| = 0$. If **A** is singular,

$$\mathbf{A}\,adj\,\mathbf{A} = \mathbf{0}$$

A singular matrix has an adjoint but no inverse.

We are now in a position to define other matrices, for example, the product **AA** will be denoted by $\mathbf{A}^2$ and

$$\mathbf{A}^n = \underbrace{\mathbf{A}\mathbf{A} \quad \dots \quad \mathbf{A}}_{n \text{ factors}}$$

Also

$$\mathbf{A}^n\mathbf{A}^m = \underbrace{\mathbf{AA} \quad \dots \quad \mathbf{A}}_{n \text{ factors}}\underbrace{\mathbf{AA} \quad \dots \quad \mathbf{A}}_{m \text{ factors}} = \mathbf{A}^{n+m}$$

so that the law of exponents for matrices holds for integral exponents. Similarly, for nonsingular matrices,

$$\mathbf{A}^{-n} = (\mathbf{A}^{-1})^n$$

A direct calculation will show that **I** is the unity matrix also in the sense that

$$\mathbf{IA} = \mathbf{AI} = \mathbf{A}$$

so that

$$\mathbf{A}^0 = \mathbf{I}$$

Also

$$\mathbf{A}^n\mathbf{A}^{-m} = \underbrace{\mathbf{AA} \quad \dots \quad \mathbf{A}}_{n \text{ factors}}\underbrace{\mathbf{A}^{-1}\mathbf{A}^{-1} \quad \dots \quad \mathbf{A}^{-1}}_{m \text{ factors}}$$
$$= \mathbf{A}^{n-m}$$

so that the law of exponents holds for positive and negative exponents for nonsingular matrices **A**. Since multiplication by a scalar and addition have been defined, the matrix polynomial

$$c_0\mathbf{I} + c_1\mathbf{A} + c_2\mathbf{A}^2 + \dots + c_n\mathbf{A}^n$$

has a meaning and is a matrix having the same order as **A**. For nonsingular matrices **A** the general matrix polynomial

$$\sum_{j=0}^{m} c_j\mathbf{A}^{-j} + \sum_{k=1}^{n} d_k\mathbf{A}^k$$

has a definite meaning and can be computed. In this connection it should be stressed that **A** is always a square matrix.

2.4 The Reversal Rule for Transposes and Inverses

Given two conformable matrices **A** and **B** in the order **AB** it is necessary frequently to compute the inverse and transpose of the product. Now

$$(\mathbf{AB})^{-1}(\mathbf{AB}) = \mathbf{I}$$

then postmultiplying (multiplying on the right by $\mathbf{B}^{-1}$)

$$(\mathbf{AB})^{-1}(\mathbf{ABB}^{-1}) = \mathbf{B}^{-1}$$

or

$$(\mathbf{AB})^{-1}\mathbf{A} = \mathbf{B}^{-1}$$

Also postmultiplying by $\mathbf{A}^{-1}$

$$(\mathbf{AB})^{-1} = \mathbf{B}^{-1}\mathbf{A}^{-1}$$

It follows also that

$$(\mathbf{ABC})^{-1} = \mathbf{C}^{-1}(\mathbf{AB})^{-1} = \mathbf{C}^{-1}\mathbf{B}^{-1}\mathbf{A}^{-1}$$

and the inverse of a product is the product of the inverses in the reverse order. To find the transpose of a product is not quite as straightforward, since the definition of multiplication must be used. Here

$$(\mathbf{AB})^T = [[a_{ij}][b_{ij}]]^T = \left[\sum_k a_{ik}b_{kj}\right]^T$$
$$= \left[\sum_k a_{jk}b_{ki}\right] = \left[\sum_k b_{ki}a_{jk}\right]$$
$$= \mathbf{B}^T\mathbf{A}^T$$

and the transpose of a product is the product of the transposes. Also

$$(\mathbf{ABC})^T = \mathbf{C}^T\mathbf{B}^T\mathbf{A}^T.$$

2.5 Rank of a Matrix

From a given matrix many determinants may be formed by striking out whole rows and whole columns leaving a square array from which a determinant may be formed. Given, for example,

$$\begin{bmatrix} a_{11} & a_{12} & a_{13} & a_{14} & a_{15} \\ a_{21} & a_{22} & a_{23} & a_{24} & a_{25} \\ a_{31} & a_{32} & a_{33} & a_{34} & a_{35} \\ a_{41} & a_{42} & a_{43} & a_{44} & a_{45} \end{bmatrix}$$

if the a_{ij} are constants or functions, then striking out a column leaves a fourth-order determinant and there are five such. Striking out two columns and one row leaves a third-order determinant and there are forty different such. Striking out three columns and two rows leaves a second-order determinant and there are sixty of these and there are twenty first-order determinants. Many of these determinants may be zero. Suppose now in a general m by n matrix that all determinants, formed by striking out whole rows and whole columns, of order greater than r are zero but there is one determinant of order r which is not zero, then the matrix is said to have rank r. For example, in the matrix

$$\begin{bmatrix} 2 & 1 & 3 & 4 \\ -1 & 1 & -2 & -1 \\ 0 & 3 & -1 & 2 \end{bmatrix}$$

all third-order determinants are zero but there is a second-order determinant (in fact, several) which is not zero and therefore this matrix has rank two.

The concept of the rank of a matrix occurs repeatedly in theoretical considerations and it is worth our efforts to study it in some detail. For example, what can be said about the rank of the product of two matrices

AB if the rank of each is known? Also, if the rank of a matrix is known, what other conclusions may be drawn about that matrix? In the following we shall attempt to answer these questions.

Given the two matrices in the product

$$\mathbf{AB} = \begin{bmatrix} a_{11} & a_{12} & \dots & a_{1n} \\ a_{21} & a_{22} & \dots & a_{2n} \\ \cdot & & & \\ \cdot & & & \\ \cdot & & & \\ a_{m1} & a_{m2} & \dots & a_{mn} \end{bmatrix} \begin{bmatrix} b_{11} & b_{12} & \dots & b_{1p} \\ b_{21} & b_{22} & \dots & b_{2p} \\ \cdot & & & \\ \cdot & & & \\ \cdot & & & \\ b_{n1} & b_{n2} & \dots & b_{np} \end{bmatrix}$$

$$= \begin{bmatrix} c_{11} & c_{12} & \dots & c_{1p} \\ c_{21} & c_{22} & \dots & c_{2p} \\ \cdot & & & \\ \cdot & & & \\ \cdot & & & \\ c_{m1} & c_{m2} & \dots & c_{mp} \end{bmatrix} = \mathbf{C}$$

the rank of **C** is determined by examining the determinants formed by striking out rows and columns. Any determinant of order k in **C** is a linear combination of determinants of order k taken from **B**. It is also a linear combination of determinants of order k taken from **A**. For definiteness consider the following case:

$$\begin{bmatrix} a_{11} & a_{12} & a_{13} \\ a_{21} & a_{22} & a_{23} \\ a_{31} & a_{32} & a_{33} \end{bmatrix} \begin{bmatrix} b_{11} & b_{12} & b_{13} & b_{14} \\ b_{21} & b_{22} & b_{23} & b_{24} \\ b_{31} & b_{32} & b_{33} & b_{34} \end{bmatrix} = \begin{bmatrix} c_{11} & c_{12} & c_{13} & c_{14} \\ c_{21} & c_{22} & c_{23} & c_{24} \\ c_{31} & c_{32} & c_{33} & c_{34} \end{bmatrix}$$

and consider the second-order determinant,

$$\begin{vmatrix} c_{11} & c_{13} \\ c_{31} & c_{33} \end{vmatrix} = \begin{vmatrix} a_{11}b_{11} + a_{12}b_{21} + a_{13}b_{31} & a_{11}b_{13} + a_{12}b_{23} + a_{13}b_{33} \\ a_{31}b_{11} + a_{32}b_{21} + a_{33}b_{31} & a_{31}b_{13} + a_{32}b_{23} + a_{33}b_{33} \end{vmatrix}$$

By the rules for addition and scalar multiplication for determinants,

$$\begin{vmatrix} c_{11} & c_{13} \\ c_{31} & c_{33} \end{vmatrix} = b_{11}\begin{vmatrix} a_{11} & c_{13} \\ a_{31} & c_{33} \end{vmatrix} + b_{21}\begin{vmatrix} a_{12} & c_{13} \\ a_{32} & c_{33} \end{vmatrix} + b_{31}\begin{vmatrix} a_{13} & c_{13} \\ a_{33} & c_{33} \end{vmatrix}$$

The first determinant on the right may be decomposed in the same way

$$\begin{vmatrix} a_{11} & c_{13} \\ a_{31} & c_{33} \end{vmatrix} = b_{13}\begin{vmatrix} a_{11} & a_{11} \\ a_{31} & a_{31} \end{vmatrix} + b_{23}\begin{vmatrix} a_{11} & a_{12} \\ a_{31} & a_{32} \end{vmatrix} + b_{33}\begin{vmatrix} a_{11} & a_{13} \\ a_{31} & a_{33} \end{vmatrix}$$

Each of the other two determinants may be decomposed in a similar fashion and it is evident then that each second-order determinant of **C** is a linear combination of second-order determinants of **A** taken from the same two rows of **A** as occurred in **C**. If one interchanges rows and columns, the second-order determinant of **C** may be written

$$\begin{vmatrix} c_{11} & c_{31} \\ c_{13} & c_{33} \end{vmatrix} = \begin{vmatrix} a_{11}b_{11} + a_{12}b_{21} + a_{13}b_{31} & a_{31}b_{11} + a_{32}b_{21} + a_{33}b_{31} \\ a_{11}b_{13} + a_{12}b_{23} + a_{13}b_{33} & a_{31}b_{13} + a_{32}b_{23} + a_{33}b_{33} \end{vmatrix}$$

The same type of decomposition may be performed here with the result that this determinant may be expressed as a linear combination of the second-order determinants of **B** taken from the same columns of **B** as occurred in **C** (the first and third). Therefore if all second-order determinants of **A** or **B** are zero, then all second-order determinants in **C** will be zero. A similar result could also be shown for third-order determinants; that is, every third-order determinant of **C** is a linear combination of the third-order determinants of **A** (only one) and of **B**. Obviously, every first-order determinant of **C** (an element) is a linear combination of the elements of the rows of **A** and the columns of **B**. In the general case, any determinant of order k in C can be decomposed into a linear combination of k-rowed determinants of **A** taken from the same rows of **A** as **C** or a linear combination of k-columned determinants of **B** taken from the same columns of **B** as **C**. Since all the determinants which occur in **C** occur in **A** and **B**, it follows that the rank of **C** is less than or equal to the smaller of the rank of **A** and the rank of **B**; that is,

$$r_{\mathbf{C}} \leq \begin{Bmatrix} r_{\mathbf{A}} \\ r_{\mathbf{B}} \end{Bmatrix}$$

where $r_{\mathbf{A}}$, $r_{\mathbf{B}}$, $r_{\mathbf{C}}$ stand for the ranks of the respective matrices.

A special case of this result is of some interest. Suppose that **A** is a nonsingular square matrix and **B** is such that **A** and **B** are conformable in that order. Since A is nonsingular, its rank is n. If R is the rank of **AB** and $r_{\mathbf{B}}$ is the rank of **B**, then

$$R \leq \mathbf{r_B}$$

Because **A** is nonsingular, however, one may write

$$\mathbf{A}^{-1}\mathbf{AB} = \mathbf{B}$$

Now the rank of **B** must be less than or equal to the rank of $\mathbf{A}^{-1}$, which is n, or the rank of **AB** which is R, or

$$r_{\mathbf{B}} \leq R$$

Therefore

$$r_{\mathbf{B}} = R$$

[Since $\mathbf{AA}^{-1} = \mathbf{I}$, the rank of **I**, which is n, is less than or equal to either the rank of **A** or the rank of $\mathbf{A}^{-1}$. But the rank of **A** for **A** nonsingular is n, so therefore the rank of $\mathbf{A}^{-1}$ must also be n, indicating that $\mathbf{A}^{-1}$ is nonsingular. This was obvious, however, since $\mathbf{A}^{-1}$ has an inverse and is therefore nonsingular.] We have derived here the important result that the rank of a matrix is invariant under premultiplication by a nonsingular square matrix. A similar result may be shown for postmultiplication.

2.6 Elementary Matrix Operations on Rows and Columns

We shall now show that certain elementary operations may be performed on matrices by matrix operators, and since these operators are nonsingular, the rank of the matrix will be unaffected. We will be interested primarily in premultiplicative matrices which will perform operations on particular rows of a matrix. Similar matrices may also be defined which, when used as postmultipliers, will have the same effect upon columns.

The matrix **I** has no effect when used as a premultiplier; when, however, two rows of **I** are interchanged, its rank remains the same but its effect as a premultiplier is to interchange the same two rows in the multiplicand. For example,

$$\begin{bmatrix} 1 & 0 & 0 & 0 \\ 0 & 0 & 1 & 0 \\ 0 & 1 & 0 & 0 \\ 0 & 0 & 0 & 1 \end{bmatrix} \begin{bmatrix} a_{11} & a_{12} & a_{13} & a_{14} & a_{15} \\ a_{21} & a_{22} & a_{23} & a_{24} & a_{25} \\ a_{31} & a_{32} & a_{33} & a_{34} & a_{35} \\ a_{41} & a_{42} & a_{43} & a_{44} & a_{45} \end{bmatrix} = \begin{bmatrix} a_{11} & a_{12} & a_{13} & a_{14} & a_{15} \\ a_{31} & a_{32} & a_{33} & a_{34} & a_{35} \\ a_{21} & a_{22} & a_{23} & a_{24} & a_{25} \\ a_{41} & a_{42} & a_{43} & a_{44} & a_{45} \end{bmatrix}$$

A convenient notation for such a premultiplier is $\mathbf{I}_{23}$, indicating that the second and third rows of **I** have been interchanged. In general, $\mathbf{I}_{ij}$ will be an idem matrix with rows i and j interchanged and its effect as a premultiplier is to interchange the ith and jth rows of the multiplicand. Note that $|\mathbf{I}_{ij}| = -1$. If the matrix $\mathbf{I}_{ij}$ is used as a postmultiplier (not on the preceding one), it has the same effect on columns, that is, interchanges the same columns as have been interchanged in $\mathbf{I}_{ij}$.

Another elementary matrix of interest is the idem matrix modified by the insertion of a scalar k in position (i, j), that is, at the intersection of the ith row and jth column. If $i = j$, that is, if the one on the main diagonal is changed to a k, then obviously, when such a matrix is used as a premultiplier, the multiplicand will remain unchanged except that the elements of the ith row will be multiplied by k. In the case

$$\begin{bmatrix} 1 & 0 & 0 & 0 \\ 0 & 1 & k & 0 \\ 0 & 0 & 1 & 0 \\ 0 & 0 & 0 & 1 \end{bmatrix} \begin{bmatrix} a_{11} & a_{12} & a_{13} & a_{14} & a_{15} \\ a_{21} & a_{22} & a_{23} & a_{24} & a_{25} \\ a_{31} & a_{32} & a_{33} & a_{34} & a_{35} \\ a_{41} & a_{42} & a_{43} & a_{44} & a_{45} \end{bmatrix} = \begin{bmatrix} a_{11} & a_{12} & a_{13} & a_{14} & a_{15} \\ a_{21}+ka_{31} & a_{22}+ka_{32} & a_{23}+ka_{33} & a_{24}+ka_{34} & a_{25}+ka_{35} \\ a_{31} & a_{32} & a_{33} & a_{34} & a_{35} \\ a_{41} & a_{42} & a_{43} & a_{44} & a_{45} \end{bmatrix}$$

A convenient notation for this matrix is $\mathbf{J}_{23}$ and it has the effect of multiplying the third row by k and adding it to the second row. J_{ij} multiplies the jth row by k and adds it to the ith row. Clearly $|\mathbf{J}_{ij}| = 1$, except when $k = 0$ and $i = j$. $\mathbf{J}_{ij}$ used as a postmultiplier will multiply the ith column by k and add it to the jth column.

As an application of the matrices $\mathbf{J}_{ij}$ consider their effect on the matrix

$$\mathbf{A} = \begin{bmatrix} a_{11} & a_{12} & a_{13} & a_{14} & a_{15} \\ a_{21} & a_{22} & a_{23} & a_{24} & a_{25} \\ a_{31} & v_{32} & a_{33} & a_{34} & a_{35} \\ a_{41} & a_{42} & a_{43} & a_{44} & a_{45} \end{bmatrix}$$

when appropriately used. Suppose $a_{11} \neq 0$, then $\mathbf{J}_{21}$ with $k = -a_{21}/a_{11}$ will change $\mathbf{A}$ to

$$\mathbf{A}_1 = \begin{bmatrix} a_{11} & a_{12} & a_{13} & a_{14} & a_{15} \\ 0 & b_{22} & b_{23} & b_{24} & b_{25} \\ a_{31} & a_{32} & a_{33} & a_{34} & a_{35} \\ a_{41} & a_{42} & a_{43} & a_{44} & a_{45} \end{bmatrix}$$

where the b_{ij} are the result of the operation performed on that row. If a_{11} had been zero then there is a matrix $\mathbf{I}_{12}$ which would have produced a similar result.

Similarly, there are matrices $\mathbf{J}_{31}$, $\mathbf{J}_{41}$ which will give when applied to A_1

$$\mathbf{A}_3 = \begin{bmatrix} a_{11} & a_{12} & a_{13} & a_{14} & a_{15} \\ 0 & b_{22} & b_{23} & b_{24} & b_{25} \\ 0 & b_{32} & b_{33} & b_{34} & b_{35} \\ 0 & b_{42} & b_{43} & b_{44} & b_{45} \end{bmatrix}$$

Now suppose that $b_{22} \neq 0$. If it is zero, then there is a matrix $\mathbf{I}_{24}$ which will interchange the second and fourth rows. There are matrices $\mathbf{J}_{32}$, $\mathbf{J}_{42}$, $\mathbf{J}_{12}$, with $k = -(b_{32}/b_{22})$, $k = -(b_{42}/b_{22})$, and $k = -(a_{12}/b_{22})$, respectively, to give

$$\mathbf{A}_6 = \begin{bmatrix} a_{11} & 0 & c_{13} & c_{14} & c_{15} \\ 0 & b_{22} & b_{23} & b_{24} & b_{25} \\ 0 & 0 & c_{33} & c_{34} & c_{35} \\ 0 & 0 & c_{43} & c_{44} & c_{45} \end{bmatrix}$$

If it should happen that $b_{22} = 0$, $b_{32} = 0$, $b_{42} = 0$ then there are postmultiplicative matrices which will interchange the second column with the third, fourth, or fifth columns to produce an element in the (2, 2) position which is not zero. Hence a non-zero element can be produced at position (2, 2) or else all the elements in the lower right-hand three by four matrix are zero.

We now examine the third row. If c_{33} is zero, we proceed as just described. There will be a sequence of premultiplicative matrices $\mathbf{J}_{43}$, $\mathbf{J}_{23}$, $\mathbf{J}_{13}$ perhaps combined with a sequence of postmultiplicative matrices which will give

$$\mathbf{A}_9 = \begin{bmatrix} a_{11} & 0 & 0 & c_{14} & c_{15} \\ 0 & b_{22} & 0 & c_{24} & c_{25} \\ 0 & 0 & c_{33} & c_{34} & c_{35} \\ 0 & 0 & 0 & d_{44} & d_{45} \end{bmatrix}$$

and if $d_{44} \neq 0$, there will be a sequence of matrices $\mathbf{J}_{14}$, $\mathbf{J}_{24}$, $\mathbf{J}_{34}$ which when applied to $\mathbf{A}_9$ will give

$$\mathbf{A}_{12} = \begin{bmatrix} a_{11} & 0 & 0 & 0 & d_{15} \\ 0 & b_{22} & 0 & 0 & d_{25} \\ 0 & 0 & c_{33} & 0 & d_{35} \\ 0 & 0 & 0 & d_{44} & d_{45} \end{bmatrix}$$

Thus there will be a premultiplicative matrix and a postmultiplicative matrix, both rank preserving, which will reduce the matrix $\mathbf{A}$ to diagonal form in the first four columns and rows. It is evident that in the sequence of matrix transformations a complete row or rows of zeros might be produced at any given time. In this case that row would be interchanged with the fourth row and omitted from further manipulation.

Our main conclusion is that there exist rank-preserving matrices, pre- and postmultiplicative, made up of a product of $\mathbf{J}_{ij}$'s and $\mathbf{I}_{ij}$'s which will transform an arbitrary m by n matrix to the following diagonal form.

$$\begin{bmatrix} a_{11} & 0 & 0 & \dots & 0 & a_{1,\,m+1} & \dots & a_{1n} \\ 0 & a_{22} & 0 & \dots & 0 & a_{2,\,m+1} & \dots & a_{2n} \\ 0 & 0 & a_{33} & & & & & \\ \vdots & & & \ddots & & \vdots & & \\ 0 & 0 & \dots & 0 & a_{mm} & a_{m,\,m+1} & \dots & a_{mn} \end{bmatrix}$$

where there may be p, $p < m$, whole rows of zeroes in the last p rows. During these row operations, it is possible that a whole column of zeroes might appear, for example, in the third column and hence $a_{33} = 0$. Further columns of zeroes might appear so that on the diagonal some of the elements might be zero. If postmultiplicative matrices are used, the columns of zeroes may be moved to the right with the result that the preceding form will be obtained with $a_{ii} \neq 0$, $i = 1$ to p.

2.7 Sylvester's Law of Nullity

Consider now two square matrices $\mathbf{A}$ and $\mathbf{B}$ with ranks $r_{\mathbf{A}}$ and $r_{\mathbf{B}}$, respectively, and let the rank of the product $\mathbf{AB}$ be $r_{\mathbf{AB}}$. The square matrix $\mathbf{A}$ may be premultiplied by a nonsingular matrix which will produce $p_{\mathbf{A}} = n - r_{\mathbf{A}}$ complete rows of zeroes. Similarly there is a postmultiplicative matrix which when applied to $\mathbf{B}$ will reduce it to diagonal form and there will be $p_{\mathbf{B}}$ columns, all elements of which are zero. Let these two transformed matrices be denoted by $\mathbf{A}'$ and $\mathbf{B}'$. The product $\mathbf{A}'\mathbf{B}'$ will be a matrix which will have a number of columns of zeroes. The number of columns of zeroes in $\mathbf{A}'\mathbf{B}'$ cannot be greater than $p_{\mathbf{A}} + p_{\mathbf{B}}$ or

$$p_{A'B'} \leq p_{A'} + p_{B'}$$

(If **A** and **B** are nonsingular, the equality can hold.)
Now $p_{A'} = n - r_A$, $p_{B'} = n - r_B$, $p_{A'B'} = n - r_{AB}$,

so $$n - r_{AB} \leq n - r_A + n - r_B$$

and therefore $$r_{AB} \geq r_A + r_B - n \tag{2.7.1}$$

which, when combined with,

$$r_{AB} \leq \begin{Bmatrix} r_A \\ r_B \end{Bmatrix}$$

is Sylvester's law of nullity.

An interesting application of Sylvester's law of nullity is to the product

$$\mathbf{A} \text{ adj } \mathbf{A}$$

in the case in which **A** is singular but of rank $(n - 1)$. If **A** is nonsingular, then adj **A** is also nonsingular. If **A** is singular and of rank $n - 1$, however, from the formula

$$\mathbf{A} \text{ adj } \mathbf{A} = |\mathbf{A}|\mathbf{I}$$

it follows that since $|\mathbf{A}| = 0$

$$\mathbf{A} \text{ adj } \mathbf{A} = \mathbf{0}$$

and the zero matrix has rank zero. The question is "What is the rank of adj **A**?" in this case. Equation (2.7.1) reduces to

$$0 \geq r_A + r_{\text{adj}A} - n$$

or $\quad 0 \geq (n - 1) + r_{\text{adj}A} - n \quad$ and $\quad r_{\text{adj}A} \leq 1$

But it cannot be less than one, for since the rank of **A** was $(n - 1)$, there is at least one determinant A_{ij}, a cofactor of a_{ij} in $|\mathbf{A}|$, which is not zero. Consequently

$$r_{\text{adj}A} = 1$$

if the rank of A is $(n - 1)$.

2.8 Linear Simultaneous Algebraic Equations: The General Theory

The set of linear simultaneous algebraic equations in which the number of equations is equal to the number of unknowns is of the form

$$\begin{aligned} a_{11}x_1 + a_{12}x_2 + \cdots + a_{1n}x_n &= b_1 \\ a_{21}x_1 + a_{22}x_2 + \cdots + a_{2n}x_n &= b_2 \\ &\vdots \\ a_{n1}x_1 + a_{n2}x_2 + \cdots + a_{nn}x_n &= b_n \end{aligned}$$

From Cramer's rule we know that there is a solution if the determinant of the matrix of the coefficients is not zero. Using the definition of matrix multiplication and equality, the foregoing set of equations may be written

$$\mathbf{Ax} = \mathbf{b}$$

and premultiplying by $\mathbf{A}^{-1}$ we obtain

$$\mathbf{x} = \mathbf{A}^{-1}\mathbf{b}$$

as a compact way of writing the solution for $|\mathbf{A}| \neq 0$. In the sequel, we shall discuss methods of obtaining the inverse of a matrix. Many problems arise in applied mathematics in which the number of equations is not equal to the number of unknowns, hence it is desirable to have available general theorems on when there will or will not be solutions of such equations.

We now consider the general set of linear simultaneous algebraic equations in which the number of equations may not be the same as the number of unknowns

$$\begin{aligned} a_{11}x_1 + a_{12}x_2 + \cdots + a_{1n}x_n &= b_1 \\ a_{21}x_1 + a_{22}x_2 + \cdots + a_{2n}x_n &= b_2 \\ &\vdots \\ a_{m1}x_1 + a_{m2}x_2 + \cdots + a_{mn}x_n &= b_m \end{aligned} \tag{2.8.1}$$

These may be written

$$\mathbf{Ax} = \mathbf{b} \tag{2.8.2}$$

Our purpose here is not only to show how to find solutions but also to indicate under what conditions there will be solutions. It seems obvious that if there are more unknowns than equations there may be many solutions. On the other hand, if there are more equations than unknowns, the possibility must exist that the set of equations will be self-contradictory or incompatible. *By a solution of Eq.* (*2.8.1*) *or* (*2.8.2*), *we will mean a vector* $\mathbf{x}$ *which when substituted into these equations will produce an identity.*

We proceed by the Gauss-Jordan method of elimination. In Eq. (2.8.1) we shall assume that $a_{11} \neq 0$. If it is zero, then renumber the unknowns so that the first coefficient is not zero. The first equation can then be written

$$x_1 + \frac{a_{12}}{a_{11}}x_2 + \frac{a_{13}}{a_{11}}x_3 + \cdots + \frac{a_{1n}}{a_{11}}x_n = \frac{b_{11}}{a_{11}}$$

and if we multiply this equation by a_{21} and subtract it from the second equation, we obtain a new equation with the variable x_1 eliminated as

$$b_{22}x_2 + b_{23}x_3 + \cdots + b_{2n}x_n = c_2$$

This same procedure may be repeated in the third, fourth, . . . , nth equations to give a modified system of the form

$$\begin{aligned} x_1 + b_{12}x_2 + b_{13}x_3 + \cdots + b_{1n}x_n &= c_1 \\ b_{22}x_2 + b_{23}x_3 + \cdots + b_{2n}x_n &= c_2 \\ b_{32}x_2 + b_{33}x_3 + \cdots + b_{3n}x_n &= c_3 \\ &\vdots \\ b_{m2}x_2 + b_{m3}x_3 + \cdots + b_{mn}x_n &= c_m \end{aligned} \tag{2.8.3}$$

Suppose now that $b_{22} \neq 0$ and divide the second of these equations by b_{22} and multiply it by b_{12} and subtract it from the first equation. This will eliminate x_2 from the first equation and the same procedure may be used to eliminate x_2 from each of the other equations. A new system of equations then results which has the form

$$\begin{aligned} x_1 + \quad + c_{13}x_3 + c_{14}x_4 + \cdots + c_{1n}x_n &= d_1 \\ x_2 + c_{23}x_3 + c_{24}x_4 + \cdots + c_{2n}x_n &= d_2 \\ + c_{33}x_3 + x_{34}x_4 + \cdots + c_{3n}x_n &= d_3 \\ &\vdots \\ c_{m3}x_3 + c_{m4}x_4 + \cdots + c_{mn}x_n &= d_m \end{aligned}$$

Continuing in this fashion, we shall finally reach a point when the set of equations will have the following general form:

$$\begin{aligned} x_1 - \gamma_{1,\,r+1}x_{r+1} - \gamma_{1,\,r+2}x_{r+2} \quad \cdots \quad - \gamma_{1n}x_n &= \alpha_1 \\ x_2 - \gamma_{2,\,r+1}x_{r+1} - \gamma_{2,\,r+2}x_{r+2} \quad \cdots \quad - \gamma_{2n}x_n &= \alpha_2 \\ &\vdots \\ x_r - \gamma_{r,\,r+1}x_{r+1} - \gamma_{r,\,r+2}x_{r+2} \quad \cdots \quad - \gamma_{rn}x_n &= \alpha_r \\ 0 &= \alpha_{r+1} \\ 0 &= \alpha_{r+2} \\ &\vdots \\ 0 &= \alpha_m \end{aligned} \tag{2.8.4}$$

Let us consider now the set of Eq. (2.8.4). If $r < m$, the only way there can be solutions to this set of equations is for $\alpha_{r+1} = 0$, $\alpha_{r+2} = 0, \ldots, \alpha_n = 0$, otherwise there would be an inconsistency. The solution can then be written in the form

$$\begin{aligned}
x_1 &= \gamma_{1,\, r+1}x_{r+1} + \gamma_{1,\, r+2}x_{r+2} + \cdots + \gamma_{1n}x_n + \alpha_1 \\
x_2 &= \gamma_{2,\, r+1}x_{r+1} + \gamma_{2,\, r+1}x_{r+2} + \cdots + \gamma_{2n}x_n + \alpha_2 \\
&\cdot \\
&\cdot \\
&\cdot \\
x_r &= \gamma_{r,\, r+1}x_{r+1} + \gamma_{r,\, r+1}x_{r+2} + \cdots + \gamma_{rn}x_n + \alpha_r
\end{aligned} \tag{2.8.5}$$

where, if one sets arbitrary values for $x_{r+1}, x_{r+2}, \ldots, x_n$, the values of $x_1, x_2, \ldots, x_r$ are uniquely determined. If $r = m$, then the values of $x_{m+1}, x_{m+2}, \ldots, x_n$ may be arbitrarily set and $x_1, x_2, \ldots, x_m$ are uniquely determined. If $m > n$, then $r \leq n < m$ certainly. If in this case $r = n$, in order to have solutions $\alpha_{n+1} = 0$, $\alpha_{n+2} = 0, \ldots, \alpha_m = 0$ necessarily and a unique solution will exist for then $x_j = \alpha_j$, $j = 1$ to n. On the other hand, if $r < n < m$, then in order to have consistent solutions $\alpha_{r+1} = 0$, $\alpha_{r+2} = 0, \ldots, \alpha_m$ and the last $m - r$ equations provide no useful information as far as the solution is concerned.

This information may be summarized very neatly in terms of the ranks of two matrices. Write the matrix of the coefficients as

$$\mathbf{A} = \begin{bmatrix} a_{11} & a_{12} & \cdots & a_{1n} \\ a_{21} & a_{22} & \cdots & a_{2n} \\ \cdot \\ \cdot \\ \cdot \\ a_{m1} & a_{m2} & \cdots & a_{mn} \end{bmatrix}$$

and write another matrix, called the *augmented* matrix of **A** as

$$\text{aug}\,\mathbf{A} = \begin{bmatrix} a_{11} & a_{12} & \cdots & a_{1n} & b_1 \\ a_{21} & a_{22} & \cdots & a_{2n} & b_2 \\ \cdot \\ \cdot \\ \cdot \\ a_{m1} & a_{m2} & \cdots & a_{mn} & b_m \end{bmatrix}$$

Exactly the same operations performed on the set of Eq. (2.8.1) may be performed by rank-preserving pre-and/or post-multiplicative matrices on **A** and aug **A**, with the result that **A** and aug **A** become matrices

$$\mathbf{B} = \begin{bmatrix} 1 & 0 & 0 & \cdots & 0 & -\gamma_{1,\,r+1} & -\gamma_{1,\,r+2} & \cdots & -\gamma_{1n} \\ 0 & 1 & 0 & \cdots & 0 & -\gamma_{2,\,r+1} & -\gamma_{2,\,r+2} & \cdots & -\gamma_{2n} \\ 0 & 0 & 1 & & & -\gamma_{3,\,r+1} & -\gamma_{3,\,r+2} & \cdots & -\gamma_{3n} \\ \vdots & & & \ddots & \vdots & & & & \\ 0 & 0 & \cdots & & 1 & -\gamma_{r,\,r+1} & -\gamma_{r,\,r+2} & \cdots & -\gamma_{rn} \\ 0 & 0 & \cdots & & 0 & 0 & 0 & & 0 \\ \vdots & & & & & 0 & 0 & & 0 \\ & & & & & & & & \vdots \\ 0 & 0 & \cdots & & 0 & \cdots & \cdots & \cdots & 0 \end{bmatrix}$$

and

$$\text{aug}\,\mathbf{B} = \begin{bmatrix} 1 & 0 & 0 & \cdots & 0 & -\gamma_{1,\,r+1} & -\gamma_{1,\,r+2} & \cdots & -\gamma_{1n} & \alpha_1 \\ 0 & 1 & 0 & \cdots & 0 & -\gamma_{2,\,r+1} & -\gamma_{2,\,r+2} & \cdots & -\gamma_{2n} & \alpha_2 \\ 0 & 0 & 1 & \cdots & 0 & -\gamma_{3,\,r+1} & -\gamma_{3,\,r+2} & \cdots & -\gamma_{3n} & \alpha_3 \\ \vdots & & & & & & & & & \vdots \\ 0 & 0 & 0 & \cdots & 1 & -\gamma_{r,\,r+1} & -\gamma_{r,\,r+2} & \cdots & -\gamma_{rn} & \alpha_r \\ 0 & 0 & 0 & \cdots & 0 & 0 & 0 & \cdots & 0 & \alpha_{r+1} \\ 0 & 0 & & \cdots & 0 & 0 & 0 & \cdots & 0 & \alpha_{r+2} \\ & & & & & & & & & \vdots \\ 0 & 0 & & \cdots & 0 & 0 & 0 & \cdots & 0 & \alpha_m \end{bmatrix}$$

If $\alpha_{r+1} = 0$, $\alpha_{r+2} = 0, \ldots, \alpha_m = 0$, the set of Eq. (2.8.4) will have a solution and it follows then that rank **B** equals rank aug **B** and rank **A** equals rank aug **A**, necessarily. Let us show now that the condition is also sufficient; that is, if rank **A** equals rank aug **A** the set of equations will have a solution. Now the rank of **B** is certainly r. Since by hypothesis the rank of aug **B** is also r, every determinant of order $r + 1$ must be zero. Determinants of order $r + 1$ which can be extracted are

$$\begin{vmatrix} 1 & 0 & 0 & \dots & 0 & \alpha_1 \\ 0 & 1 & 0 & \dots & 0 & \alpha_2 \\ 0 & 0 & 1 & \dots & 0 & \alpha_3 \\ \cdot & & & & & \\ \cdot & & & & & \\ \cdot & & & & & \\ 0 & 0 & 0 & \dots & 1 & \alpha_r \\ 0 & 0 & 0 & \dots & 0 & \alpha_j \end{vmatrix} = \alpha_j, \quad j = r+1, r+2, \dots m$$

on expanding by elements of the last row, and hence α_j must be zero. But if $\alpha_j = 0, j = r + 1$ to m, then the set of equations has solutions since then there is no inconsistency. Thus we arrive at the most important theorem on linear simultaneous algebraic equations: *The necessary and sufficient condition that a set of linear simultaneous equations have solutions is that the rank of the matrix of the coefficients be the same as the rank of the augmented matrix of the coefficients.*

A second important theorem which may be deduced from the foregoing discussion is the following: *If the rank of the matrix of the coefficients in a set of linear simultaneous algebraic equations is r and is the same as the rank of the augmented matrix, and if n is the number of unknowns, the values of $(n - r)$ of the unknowns may be arbitrarily assigned and the remaining r unknowns are uniquely determined, provided that the matrix of the coefficients of the remaining r unknowns has rank r.*

A set of equations as in Eq. (2.8.1) is said to be *homogeneous* if all the b_i, $i = 1$ to m, are zero. In this case the rank of **A** is always the same as the rank of aug **A** and therefore such a system always has solutions for, to be sure, if all the x_i are zero, the equations are satisfied. Such a solution is called the *trivial solution* and is not very interesting. A more interesting question is determining under what conditions a set of homogeneous linear simultaneous algebraic equations will have a nontrivial solution, that is, a set of x_i which satisfy the equations, not all x_i being zero. If n is the number of unknowns and if rank $\mathbf{A} = r < n$, then the set of Eq. (2.8.1) may be written in the form of Eq. (2.8.5), with each $\alpha_j = 0$, that is,

$$\begin{aligned} x_1 &= \gamma_{1,\,r+1}x_{r+1} + \gamma_{1,\,r+2}x_{r+2} + \cdots + \gamma_{1n}x_n \\ x_2 &= \gamma_{2,\,r+1}x_{r+1} + \gamma_{2,\,r+2}x_{r+2} + \cdots + \gamma_{2n}x_n \\ &\cdot \\ &\cdot \\ &\cdot \\ x_r &= \gamma_{r,\,r+1}x_{r+1} + \gamma_{r,\,r+2}x_{r+2} + \cdots + \gamma_{rn}x_n \end{aligned} \tag{2.8.6}$$

and $x_{r+1}, x_{r+2}, \dots, x_n$ may be assigned at pleasure and $x_1, x_2, \dots, x_r$ determined uniquely. On the other hand, if the set of equations has a nontrivial solution and $m < n$ then $r < n$. If $m > n$ and there is a nontrivial solution, then it must be possible to reduce the system to n or fewer equations and from Cramer's rule, the determinant of the coefficients must be zero and therefore the rank must be less than n. Thus: *the necessary and sufficient condition that a set of homogeneous linear simultaneous equations in n variables have other than the trivial solution is that the rank of the matrix of coefficients be less than n.*

2.9 Further Implications of Rank

Given a determinant and if it is known that one of its rows is a linear combination of other rows, then by decomposition of the determinant, using the theorem on addition of determinants, the original determinant must be zero since each of the determinants in the decomposition must contain two identical rows. Suppose now one considers a square matrix of order $r + 1$ whose determinant is zero and suppose also that in this matrix of order $r + 1$ there is a determinant of order r which is not zero. Is it possible to conclude that the omitted row (and column) is a linear combination of the other r rows (and columns)? If it is, then this is the converse of the foregoing notion.

For the sake of convenience, we assume that the determinant of order r which is not zero is in the upper left-hand corner of the matrix $\mathbf{Q}$,

$$\mathbf{Q} = \begin{bmatrix} a_{11} & a_{12} & \cdots & a_{1r} & a_{1,\, r+1} \\ a_{21} & a_{22} & \cdots & a_{2r} & a_{2,\, r+1} \\ \cdot & & & & \\ \cdot & & & & \\ \cdot & & & & \\ a_{r1} & a_{r2} & \cdots & a_{rr} & a_{r,\, r+1} \\ a_{r+1,\, 1} & a_{r+1,\, 2} & \cdots & a_{r+1,\, r} & a_{r+1,\, r+1} \end{bmatrix} \tag{2.9.1}$$

We ask the question: Is it possible to determine a set of constants β_i such that

$$\beta_1 a_{i1} + \beta_2 a_{i2} + \beta_3 a_{i3} + \cdots + \beta_r a_{ir} = a_{i,\, r+1} \qquad i = 1, 2, 3, \ldots, r \tag{2.9.2}$$

This is a set of linear algebraic equations whose determinant is not zero and so a set of β_i's can be obtained, therefore each $a_{i,\, r+1}$, $i = 1$ to r is a linear combination of each of the r elements in the same row and more, it is the same linear combination. Now does it follow that

$$\beta_1 a_{r+1,\, 1} + \beta_2 a_{r+1,\, 2} + \cdots + \beta_r a_{r+1,\, r} = a_{r+1,\, r+1} \tag{2.9.3}$$

That is, is $a_{r+1,\, r+1}$ the same linear combination of the elements in its row? Now in this case we know that if

$$\mathbf{A} = \begin{bmatrix} a_{11} & a_{12} & \cdots & a_{1r} \\ a_{21} & a_{22} & \cdots & a_{2r} \\ \cdot & & & \\ \cdot & & & \\ \cdot & & & \\ a_{r1} & a_{r2} & \cdots & a_{rr} \end{bmatrix}$$

and if A_j is used as the determinant corresponding to $\mathbf{A}$ but with the jth column replaced by the $r + 1$st column of $\mathbf{Q}$, Eq. (2.9.1), then

$$\beta_j = \frac{A_j}{|\mathbf{A}|}$$

by Cramer's rule. If β_j is substituted into Eq. (2.9.3)

$$A_1 a_{r+1,\,1} + A_2 a_{r+1,\,2} + \cdots + A_r a_{r+1,\,r} = a_{r+1,\,r+1}|\mathbf{A}| \tag{2.9.4}$$

and we must show that this is an identity. Consider now the expansion of $|\mathbf{Q}|$ by elements of the last row. Since this determinant is zero, it may be written as

$$\begin{aligned}(-1)^{r+2}a_{r+1,\,1}|A_{r+1,\,1}| + (-1)^{r+3}a_{r+1,\,2}|A_{r+1,\,2}| + \cdots \\ + (-1)^{2r+2}a_{r+1,\,r+1}|A_{r+1,\,r+1}| = 0\end{aligned} \tag{2.9.5}$$

where $|A_{r+1,\,j}|$ is the *complement* of $a_{r+1,\,j}$ (not the cofactor) in $|\mathbf{Q}|$. However, the quantities A_j and $|A_{r+1,\,j}|$ are related and a little examination will show that $|A_{r+1,j}|$ and A_j are the same determinants but with certain columns interchanged. In fact, $|A_{r+1,\,r+1}|$ is the same as $|\mathbf{A}|$ and $|A_{r+1,\,r}|$ is the same as A_r. $|A_{r+1,\,r-1}|$ would be the same as A_{r-1} if the $(r - 1)$st and rth columns were interchanged, and in fact one can show that

$$\begin{aligned}
|A_{r+1,\,r+1}| &= |\mathbf{A}| \\
|A_{r+1,\,r}| &= A_r \\
|A_{r+1,\,r-1}| &= (-1)A_{r-1} \\
|A_{r+1,\,r-2}| &= (-1)^2 A_{r-2} \\
&\cdot \\
&\cdot \\
&\cdot \\
|A_{r+1,\,2}| &= (-1)^{r-2} A_2 \\
|A_{r+1,\,1}| &= (-1)^{r-1} A_1
\end{aligned}$$

so that Eq. (2.9.5) reduces to

$$-a_{r+1,\,1}A_1 - a_{r+1,\,2}A_2 - a_{r+1,\,3}A_3 \cdots - a_{r+1,\,r}A_r + a_{r+1,\,r+1}|\mathbf{A}| = 0$$

and this is the same as Eq. (2.9.4) and the conjectured identity has been established. What has been done here can obviously also be done for columns. Thus, if a determinant is equal to zero, at least one of its columns must be a linear combination of the other columns in the determinant.

Suppose now that one has a general m by n matrix $\mathbf{A}$ and suppose its rank is r, then every determinant of order $r + 1$ and greater which may be formed from rows and columns must be zero. It follows then from the preceding analysis that every row and column other than those represented in the determinant of order r which is not zero can be represented as a linear combination of those r rows and columns. The importance of this fact cannot be overemphasized, for much of what follows will depend upon it. Consider such a general matrix

$$\begin{bmatrix} a_{11} & a_{12} & a_{13} & \cdots & a_{1r} & \cdots & a_{1n} \\ a_{21} & a_{22} & a_{23} & \cdots & a_{2r} & \cdots & a_{2n} \\ \cdot \\ \cdot \\ \cdot \\ a_{r1} & a_{r2} & a_{r3} & \cdots & a_{rr} & \cdots & a_{rn} \\ a_{m1} & a_{m2} & a_{m3} & \cdots & a_{mr} & \cdots & a_{mn} \end{bmatrix}$$

and suppose it has rank r. If the non-zero determinant is not in the upper left-hand corner, the matrix may be operated on by rank-preserving matrices to interchange rows and columns so that it can be shifted to the upper left-hand corner. Then any row in the last $m - r$ may be expressed as a linear combination of the first r rows; that is, there exist r constants α_1, $\alpha_2 \ldots$, α_r such that

$$a_{pj} = \alpha_1 a_{1j} + \alpha_2 a_{2j} + \alpha_3 a_{3j} + \cdots + \alpha_r a_{rj} \quad \text{for} \quad j = 1, 2, 3, \ldots, n$$

where p is one of the numbers $r + 1$, $r + 2, \ldots, m$ and where the set of numbers $\alpha_1, \alpha_2, \ldots, \alpha_r$ will be, in general, different for each p.

It is a simple exercise to show that if a determinant is not zero then no one of its rows or columns can be a linear combination of the remaining rows and columns.

2.10 The Solution of Homogeneous Linear Equations

In many applications it will be necessary to find solutions to a set of linear equations

$$\sum_{j=1}^{n} a_{ij}x_j = 0, \quad \mathbf{Ax} = \mathbf{o} \tag{2.10.1}$$

There is always the trivial solution, $\mathbf{x} = \mathbf{o}$, and we know from the theorems of Section 2.8 that if the equations are to have nontrivial solutions, $\mathbf{x} \neq \mathbf{o}$, the rank of the matrix of the coefficients must be less than n. Assume first that the rank of $\mathbf{A}$ is $n - 1$. In this case it is easy to find a set of x_j's which will satisfy the equations if one refers to Section 1.4. The expansion of $|\mathbf{A}|$ is

$$\sum_{j=1}^{n} a_{ij}A_{ij} = 0 \quad \text{for each } i$$

Also
$$\sum_{j=1}^{n} a_{ij}A_{kj} = 0 \quad \text{for each } i \text{ and each } k$$
since this represents an alien cofactor expansion. Therefore a solution is
$$x_j = A_{kj} \quad j = 1, 2, \ldots, n$$
for each k. Note that the vector,
$$\mathbf{x} = \begin{bmatrix} x_1 \\ x_2 \\ \cdot \\ \cdot \\ \cdot \\ x_n \end{bmatrix} = \begin{bmatrix} A_{k1} \\ A_{k2} \\ \cdot \\ \cdot \\ \cdot \\ A_{kn} \end{bmatrix}$$
which is a solution, is the kth column of adj $\mathbf{A}$, and at first glance there are n solution vectors $\mathbf{x}$. In Section (2.7), however, it was shown that if $\mathbf{A}$ has rank $(n - 1)$, adj $\mathbf{A}$ has rank one and therefore each column of adj $\mathbf{A}$ can only be a constant multiple of each of the other columns and the solution of Eq. (2.10.1) is
$$\mathbf{x}_k = c\mathbf{A}_k$$
where c is any non-zero constant and $\mathbf{A}_k$ is any non-zero column of adj $\mathbf{A}$.

Suppose now the rank of $\mathbf{A}$ is r, then we know that in reality there are only r equations, from Section 2.8, in the set of Eq. (2.10.1). These may be written
$$\begin{aligned} a_{11}x_1 + a_{12}x_2 + \cdots + a_{1r}x_r &= -a_{1,\,r+1}x_{r+1} \cdots - a_{1n}x_n \\ a_{21}x_1 + a_{22}x_2 + \cdots + a_{22}x_r &= -a_{2,\,r+1}x_{r+1} \cdots - a_{2n}x_n \\ &\;\;\vdots \\ a_{r1}x_1 + a_{r2}x_2 + \cdots + a_{rr}x_r &= -a_{r,\,r+1}x_{r+1} \cdots - a_{rn}x_n \end{aligned} \tag{2.10.2}$$
where the matrix of the coefficients of $x_1, x_2, \ldots, x_r$ has rank r. We know that we can choose the values of $x_{r+1}, x_{r+2}, \ldots, x_n$ arbitrarily, and then the values of the unknowns $x_1, x_2, \ldots, x_r$ are uniquely determined. The question then arises as to how many such solutions there are since x_{r+1}, $x_{r+2}, \ldots, x_n$ can take on any values at all. The set of Eq. (2.10.2) can be solved by Cramer's rule, and if one examines the determinants in detail and makes the appropriate decomposition of the numerator determinants using the addition rule for determinants, it follows that $x_1, x_2, \ldots, x_n$ are linear combinations of $x_{r+1}, x_{r+2}, \ldots, x_n$

or
$$\begin{aligned} x_1 &= \gamma_{1,\,r+1}x_{r+1} + \gamma_{1,\,r+2}x_{r+2} + \cdots + \gamma_{1n}x_n \\ x_2 &= \gamma_{2,\,r+1}x_{r+1} + \gamma_{2,\,r+2}x_{r+2} + \cdots + \gamma_{2n}x_n \\ &\;\;\vdots \\ x_r &= \gamma_{r,\,r+1}x_{r+1} + \gamma_{r,\,r+2}x_{r+2} + \cdots + \gamma_{rn}x_n \end{aligned}$$

This was also the result obtained in Section 2.8, Eq (2.8.6). Suppose that we make $n - \mathrm{r} + 1$ choices for the variables $x_{r+1}, x_{r+2}, \ldots, x_n$, then these solutions may be denoted by

$$\begin{matrix} x_1^{(0)} & x_2^{(0)} & \cdots & x_r^{(0)} & x_{r+1}^{(0)} & x_{r+2}^{(0)} & \cdots & x_n^{(0)} \\ x_1^{(1)} & x_2^{(1)} & \cdots & x_r^{(1)} & x_{r+1}^{(1)} & x_{r+2}^{(1)} & \cdots & x_n^{(1)} \\ x_1^{(2)} & x_2^{(2)} & \cdots & x_r^{(2)} & x_{r+1}^{(2)} & x_{r+2}^{(2)} & \cdots & x_n^{(2)} \\ \cdot & & & & & & & \\ \cdot & & & & & & & \\ \cdot & & & & & & & \\ x_1^{(n-r)} & x_2^{(n-r)} & \cdots & x_r^{(n-r)} & x_{r+1}^{(n-r)} & x_{r+2}^{(n-r)} & \cdots & x_n^{(n-r)} \end{matrix}$$

where each row is one of the $n - r + 1$ solutions.

Considered as a matrix, this array of terms, each row of which is a solution, has rank $(n - r)$, for each of the first r columns are linear combinations of the last $(n - r)$ columns and therefore one of the rows must be a linear combination of the other $(n - r)$ rows. Hence we can say that if $(n - r)$ sets of quantities $x_{r+1}^{(j)}, x_{r+2}^{(j)}, \ldots, x_n^{(j)}$, $j = 1$ to $n - r$ are chosen so that their determinant is not zero, then any other choice of the same $(n - r)$ quantities will be a linear combination of the $(n - r)$ sets originally chosen. The set of $n - r$ solutions, $x_1^{(j)}, x_2^{(j)}, \ldots, x_n^{(j)}$, $j = 1$ to $n - r$, is then called a *fundamental set* of solutions for the equations. It is clear that there are infinitely many fundamental sets since the elements of the set may be chosen arbitrarily but in a given set there cannot be more than $(n - r)$ independent ones.

It is interesting now to consider the nonhomogeneous set of equations

$$\mathbf{A}\mathbf{x} = \mathbf{b} \tag{2.10.3}$$

where $\mathbf{A}$ is an m by n matrix, $\mathbf{x}$ is a vector with n elements, and b a vector with m elements. Suppose the ranks of $\mathbf{A}$ and aug $\mathbf{A}$ are the same, say r. Then it is a simple matter to generate a solution to this equation which contains $(n - r)$ arbitrary constants. Let $\mathbf{x}_p$ be the solution obtained, by setting $x_{r+1}, x_{r+2}, \ldots, x_n$ each equal to zero, for the system

$$\begin{aligned} a_{11}x_1 + a_{12}x_2 + \cdots + a_{1r}x_r &= b_1 \\ a_{21}x_1 + a_{22}x_2 + \cdots + a_{2r}x_r &= b_2 \\ &\cdot \\ &\cdot \\ &\cdot \\ a_{r1}x_1 + a_{r2}x_2 + \cdots + a_{rr}x_r &= b_r \end{aligned}$$

then

$$\mathbf{x} = \mathbf{x}_p + \alpha_1\mathbf{x}_1 + \alpha_2\mathbf{x}_2 + \cdots + \alpha_{n-r}\mathbf{x}_{n-r} \tag{2.10.4}$$

where the $\mathbf{x}_j$, $j = 1$ to $n - r$ is a fundamental set of solutions of

$$\mathbf{A}\mathbf{x} = \mathbf{o}$$

Direct substitution into Eq. (2.10.3) gives

$$\mathbf{A}\left[\mathbf{x}_p + \sum_1^{n-r} \alpha_j \mathbf{x}_j\right] = \mathbf{A}\mathbf{x}_p + \sum_1^{n-r} \alpha_j \mathbf{A}\mathbf{x}_j$$
$$= \mathbf{b} + \mathbf{o}$$

The solution given in Eq. (2.10.4) is called the *general solution.*

2.11 Comparison of Solution of Linear Equations by Determinants and Elimination

In Section 2.8 in connection with the proof of the fundamental theorems on linear simultaneous algebraic equations, the Gauss-Jordan method of elimination was described. Although there are many methods of solution for such equations and although many more will probably be devised in the future, it seems likely that this one will prevail. In later sections of this work other techniques will be described. In order to gain some insight into the expenditure of effort required to solve a given system of equations we shall compare the number of multiplications required to solve n equations in n unknowns by the method of elimination with Cramer's rule. Let us consider the number of multiplications required to pass from the system of Eq. (2.8.1) to the Eq. (2.8.3) with $m = n$, where a division is also counted as a multiplication. To get the first equation in Eq. (2.8.3), n multiplications are required. To get the succeeding $(n - 1)$ equations, the first equation must be multiplied successively by $a_{21}, a_{31}, \ldots, a_{n1}$ and subtracted from each of the others. This requires $n\,(n - 1)$ multiplications (not counting multiplication by unity). Therefore to get Eq. (2.8.3)

$$n + n(n - 1)$$

multiplications are needed. Now we must repeat this same procedure on the second equation and the number of multiplications required is

$$(n - 1) + (n - 1)(n - 1)$$

The number of multiplications required in the next step is

$$(n - 2) + (n - 1)(n - 2)$$

The total number of multiplications required to obtain each x_j explicitly is

$$\sum_1^n n + (n - 1)\sum_1^n n = n\sum_1^n n = \tfrac{1}{2}n^2(n + 1)$$

or for very large n to a very good approximation $n^3/2$. Therefore there are about $n^3/2$ multiplications required to diagonalize a system of n equations and n unknowns so that each x_j is given explicitly without further manipulation.

Consider now Cramer's rule. Each determinant has $n!$ terms in each of which there are n terms in a product requiring $(n - 1)$ multiplications.

There are $(n + 1)$ determinants to be evaluated and then n divisions making a total of

$$(n - 1)(n + 1)(n!) + n$$

equivalent multiplications, or for large n, very accurately, $n^2n!$ multiplications. In order to get some idea about what this means in terms of computing time on a large digital computer, suppose that the time required for a single multiplication is 0.0001 sec and suppose we neglect addition and bookkeeping time in the computer as being essentially the same by the two methods. For $n = 100$, $n^3/2$, is 500,000 or the computing time is 50 sec, a not unreasonable figure. Now for Cramer's rule, $n^2n!$ may be approximated for large n by Stirling's approximation for the factorial which is

$$\log_e n! = (n + \tfrac{1}{2}) \log_e n - n$$

If $n = 100$,

$$100^2(100!) \cong 10^{162}$$

so that 10^{158} sec are required or approximately $3(10)^{150}$ years and thus the statement made earlier that Cramer's rule is never used for large systems is brought into proper perspective.

The method which is used on most computers is the method of elimination as described in Section 2.8 but with two slight modifications. After the elimination of x_1 from each of the last $(n - 1)$ equations, a system like Eq. (2.8.3) is obtained. The last $(n - 1)$ equations are then treated in the same way as the n were without eliminating x_2 from the first equation. The last $(n - 2)$ equations are then treated the same way and eventually a system of the form

$$\begin{aligned}
x_1 + \alpha_{12}x_2 + \alpha_{13}x_3 + \cdots + \alpha_{1n}x_n &= C_1 \\
x_2 + \alpha_{23}x_3 + \cdots + \alpha_{2n}x_n &= C_2 \\
&\cdot \\
&\cdot \\
&\cdot \\
x_{n-1} + \alpha_{n-1,n}x_n &= C_{n-1} \\
x_n &= C_n
\end{aligned}$$

is obtained, giving x_n explicitly. $x_{n-1}, x_{n-2}, \ldots, x_1$ are then obtained by back substitution in the equations. The second modification which is found desirable in numerical work is to scan the matrix $\mathbf{A}$ and pick out the largest a_{ij}, in absolute value; the corresponding variable is x_j. This variable is then eliminated first from the other $(n - 1)$ equations. Once this is done the matrix of the remaining $(n - 1)$ equations is scanned for largest coefficient and the variable corresponding is then eliminated from the other $(n - 2)$ equations. This method has the effect of reducing round-off error and is called the method of "pivotal condensation" and the largest a_{ij} at each stage is called the *pivot*.

2.12. Computation of the Inverse Matrix by the Method of Elimination

The method of elimination has another advantage and that is if a number of systems of equations of the form

$$\mathbf{Ax} = \mathbf{b}$$

must be solved in which $\mathbf{b}$ may change from system to system but $\mathbf{A}$ remains the same, then effectively all the systems may be solved at once, since the operations performed on different $\mathbf{b}$ will be the same and the results for different $\mathbf{b}$ may be stored appropriately in the computer memory. This leads directly to a method for the inversion of matrices, for suppose the systems of equations are

$$\mathbf{Ax} = \mathbf{e}_j$$

where $\mathbf{e}_j$ are the unit vectors

$$\mathbf{e}_1 = \begin{bmatrix} 1 \\ 0 \\ 0 \\ \cdot \\ \cdot \\ \cdot \\ 0 \end{bmatrix}, \mathbf{e}_2 = \begin{bmatrix} 0 \\ 1 \\ 0 \\ \cdot \\ \cdot \\ \cdot \\ 0 \end{bmatrix}, \ldots, \mathbf{e}_n = \begin{bmatrix} 0 \\ 0 \\ \cdot \\ \cdot \\ \cdot \\ 0 \\ 1 \end{bmatrix}$$

and $\mathbf{e}_j$ has a one in the jth position but zeroes elsewhere. Let the solutions to these n systems be $\mathbf{x}_1, \mathbf{x}_2, \ldots, \mathbf{x}_n$ corresponding to $\mathbf{e}_1, \mathbf{e}_2, \ldots, \mathbf{e}_n$, respectively. The solutions may be written, where a row vector whose elements are column vectors is a matrix, as

$$\mathbf{A}[\mathbf{x}_1\mathbf{x}_2 \ldots \mathbf{x}_n] = [\mathbf{e}_1\mathbf{e}_2 \ldots \mathbf{e}_n]$$

as may be verified by expanding this equation, and

$$\begin{bmatrix} a_{11} & a_{12} & \ldots & a_{1n} \\ a_{21} & a_{22} & \ldots & a_{2n} \\ \cdot & & & \\ \cdot & & & \\ \cdot & & & \\ a_{n1} & a_{n2} & \ldots & a_{nn} \end{bmatrix} \begin{bmatrix} x_{11} & x_{12} & \ldots & x_{1n} \\ x_{21} & x_{22} & \ldots & x_{2n} \\ \cdot & & & \\ \cdot & & & \\ \cdot & & & \\ x_{n1} & x_{n2} & \ldots & x_{nn} \end{bmatrix} = \begin{bmatrix} 1 & 0 & 0 & \ldots & 0 \\ 0 & 1 & 0 & \ldots & 0 \\ \cdot & & & & \\ \cdot & & & & \\ \cdot & & & & \\ 0 & 0 & 0 & \ldots & 1 \end{bmatrix}$$

and hence the matrix of solutions is the inverse of $\mathbf{A}$,

$$\mathbf{A}^{-1} = \begin{bmatrix} x_{11} & x_{12} & \ldots & x_{1n} \\ x_{21} & x_{22} & \ldots & x_{2n} \\ \cdot & & & \\ \cdot & & & \\ \cdot & & & \\ x_{n1} & x_{n2} & \ldots & x_{nn} \end{bmatrix}$$

This procedure is very interesting because of the circumlocution which results in the terminology of linear equation solving. It is generally assumed that linear algebraic equations are solved by inverting the matrix of the coefficients with resultant multiplication of the inverse by the right-hand side column vector. As a matter of fact most matrix inversion routines for computers involve the method of elimination for systems just described.

2.13 Derivation of Linear Equations by Method of Least Squares

In this section, we consider the estimation of the coefficients in an equation of the form

$$y = a_0 + a_1x_1 + a_2x_2 + \cdots + a_mx_m$$

when one is given n sets of data of the form

$$\{y_i, x_{1i}\, x_{2i}, \ldots, x_{mi}\}, \quad i = 1, 2, 3, \ldots, n$$

and where we consider the case in which $n > m + 1$. If $n = m + 1$ then obviously the $\{a_j\}$ may be determined more or less exactly. The method proposed by Gauss is as follows: Let the estimated values* of $\{a_j\}$ be $\{b_j\}$ and form the residuals R_i,

$$R_i = y_i - b_0 - b_1x_{1i} - b_2x_{2i} \cdots - b_mx_{mi}$$

for each i from 1 to n and form the sum of squares,

$$Q = \sum_{i=1}^{n} R_i^2 = \sum_{i=1}^{n} (y_i - b_0 - b_1x_{1i} - b_2x_{2i} \quad \ldots \quad b_mx_{mi})^2$$

Now Q is a function of $m + 1$ parameters b_j and the sum of squares will have its smallest value if

$$\frac{\partial Q}{\partial b_j} = 0 \quad \text{for all} \quad j$$

or (where all summations are on i from one to n)

$$\frac{\partial Q}{\partial b_j} = -2 \sum_i [y_i - b_0 - b_1x_{1i} - b_2x_{2i} \cdots -b_mx_{mi}]x_{ji} = 0;$$

$$j = 1, 2, 3, \ldots, m$$

$$\frac{\partial Q}{\partial b_0} = 2 \sum_i [y_i - b_0 - b_1x_{1i} - b_2x_{2i} \cdots -b_mx_{mi}] = 0$$

so

$$\sum_i x_{ji}y_i - b_0 \sum_i x_{ji} - b_1 \sum_i x_{1i}x_{ji} - b_2 \sum_i x_{2i}x_{ji} \cdots -b_m \sum_i x_{mi}x_{ji} = 0$$

$$\sum_i y_i - b_0n - b_1 \sum_i x_{1i} - b_2 \sum_i x_{2i} \cdots -b_m \sum_i x_{mi} = 0 \tag{2.13.1}$$

Now the last equation may be written

*In this book whenever the notation $\{a_j\}$ occurs, it will refer to a set of values of a_j where the number of a_j in the set is assumed to be obvious.

$$b_0 = \frac{1}{n}\left[\sum_i y_i - b_1 \sum_i x_{1i} - b_2 \sum_i x_{2i} \cdots - b_m \sum_i x_{mi}\right] \qquad (2.13.2)$$
$$= \tilde{y} - b_1\tilde{x}_1 - b_2\tilde{x}_2 \cdots - b_m\tilde{x}_m$$

where the tildes indicate the means of the various quantities.

Now Eq. (2.13.1) may be written

$$(\tilde{y} - b_1\tilde{x}_1 - b_2\tilde{x}_2 \cdots - b_m\tilde{x}_m)\sum_i x_{ji} + b_1 \sum_i x_{ji}x_{1i} + b_2 \sum_i x_{ji}x_{2i} \cdots + b_m \sum_i x_{mi}x_{ji} = \sum_i x_{ji}y_i$$

and consider the term, for example,

$$-b_2\tilde{x}_2 \sum_i x_{ji} + b_2 \sum_i x_{ji}x_{2i} = b_2 \sum_i x_{ji}(x_{2i} - \tilde{x}_2)$$
$$= b_2 \sum_i (x_{ji} - \tilde{x}_j)(x_{2i} - \tilde{x}_2)$$

since

$$\sum_i \tilde{x}_j(x_{2i} - \tilde{x}_2) = \sum_i \tilde{x}_j x_{2i} - \sum_i \tilde{x}_j\tilde{x}_2 = 0$$

Now call

$$S_{kj} = \sum_i (x_{ji} - \tilde{x}_j)(x_{ki} - \tilde{x}_k)$$

then

$$\begin{aligned} b_1S_{11} + b_2S_{12} + b_3S_{13} + \cdots + b_mS_{1m} &= S_{1y} \\ b_1S_{21} + b_2S_{22} + b_3S_{23} + \cdots + b_mS_{2m} &= S_{2y} \\ &\vdots \\ b_1S_{m1} + b_2S_{m2} + b_3S_{m3} + \cdots + b_mS_{mm} &= S_{my} \end{aligned}$$

or

$$\mathbf{Sb} = \mathbf{S}_y$$

Taking the inverse of **S**

$$\mathbf{b} = \mathbf{S}^{-1}\mathbf{S}_y$$

and let

$$\mathbf{S}^{-1} = \begin{bmatrix} c_{11} & c_{12} & \cdots & c_{1m} \\ c_{21} & c_{22} & \cdots & c_{2m} \\ \vdots & & & \\ c_{m1} & c_{m2} & \cdots & c_{mm} \end{bmatrix}$$

and

$$b_j = c_{ji}S_{1y} + c_{j2}S_{2y} + \ldots + c_{jm}S_{my}, \quad j = 1, 2, \ldots, m$$

therefore the $\{b_j\}$ are determined uniquely since now b_0 may also be determined from Eq. (2.13.2)

Under suitable hypotheses about the normal distribution of the $\{y_i\}$ one can then compute the variances and covariances of the b_j and so obtain informaton about their significance. Our only point here is that these quantities are related to the elements c_{jk} of the inverse matrix as

$$\text{var}\,(b_j) = \sigma^2 c_{jj}$$
$$\text{cov}\,(b_j b_k) = \sigma^2 c_{jk}$$

where σ^2 is the variance of the observations on y_i and var and cov stand for *variance* and *covariance*, respectively. The correlation coefficients between the b_j and b_k are computed from

$$\rho_{jk} = \frac{c_{jk}}{\sqrt{c_{jj}c_{kk}}}$$

In these formulae note that $\mathbf{S}$ and $\mathbf{S}^{-1}$ are symmetric matrices, $S_{kj} = S_{jk}$, $c_{kj} = c_{jk}$.

2.14 Related Matrices

A matrix is said to be *symmetric* if $a_{ij} = a_{ji}$ and therefore $\mathbf{A}^T = \mathbf{A}$; a matrix may be symmetric only if it is square. If the matrix $\mathbf{A}$ has complex elements, then the matrix where a_{ij} is replaced by its complex conjugate is A^*. If $\mathbf{A}^{*T} = \mathbf{A}$, then the matrix is said to be Hermitian. An *orthogonal matrix* is one for which $\mathbf{A}^T = \mathbf{A}^{-1}$. A unitary matrix is one which satisfies the property $\mathbf{A}^{*T} = \mathbf{A}^{-1}$. In a skew symmetric matrix $\mathbf{A}^T = -\mathbf{A}$.

2.15 Partitioned Matrices

In some applications, and particularly in linear programming to be treated in Chapter 4, it is convenient to use partitioned matrices. For example, consider the fourth-order matrix partitioned as shown

$$\mathbf{M} = \left[\begin{array}{cc|cc} a_{11} & a_{12} & a_{13} & a_{14} \\ a_{21} & a_{22} & a_{23} & a_{24} \\ a_{31} & a_{32} & a_{33} & a_{34} \\ \hline a_{41} & a_{42} & a_{43} & a_{44} \end{array}\right] = \begin{bmatrix} \mathbf{A} & \mathbf{B} \\ \mathbf{C} & \mathbf{D} \end{bmatrix}$$

where

$$\mathbf{A} = \begin{bmatrix} a_{11} & a_{12} \\ a_{21} & a_{22} \\ a_{31} & a_{32} \end{bmatrix} \qquad \mathbf{B} = \begin{bmatrix} a_{13} & a_{14} \\ a_{23} & a_{24} \\ a_{33} & a_{34} \end{bmatrix}$$

$$\mathbf{C} = [\,a_{41} \quad a_{42}\,] \qquad \mathbf{D} = [\,a_{43} \quad a_{44}\,]$$

Thus the matrix $\mathbf{M}$ now has as elements submatrices. A little reflection will show that **M** may be multiplied by another matrix **N** also in partitioned form if each of the pairs of submatrices is conformable in the proper order as shown in

$$\mathbf{MN} = \begin{bmatrix} \mathbf{A} & \mathbf{B} \\ \mathbf{C} & \mathbf{D} \end{bmatrix} \begin{bmatrix} \mathbf{E} & \mathbf{F} \\ \mathbf{G} & \mathbf{H} \end{bmatrix} = \begin{bmatrix} \mathbf{AE} + \mathbf{BG} & \mathbf{AF} + \mathbf{BH} \\ \mathbf{CE} + \mathbf{DG} & \mathbf{CF} + \mathbf{DH} \end{bmatrix}$$

Further, two partitional matrices **M** and **N** may be added if they are partitioned in the same way so that addition of corresponding submatrices is defined.

2.16 Differentiation and Integration of Matrices

If one considers the elements of a matrix **A** as functions of a variable t, $a_{ij} = a_{ij}(t)$ then the derivative of **A** with respect to t is defined as the matrix each of whose elements has been differentiated with respect to t; that is,

$$\frac{d\mathbf{A}}{dt} = \left[\frac{da_{ij}}{dt}\right]$$

Similarly, integration of a matrix is defined as

$$\int_a^b \mathbf{A}\, dt = \left[\int_a^b a_{ij}(t)\, dt\right]$$

EXERCISES

1. Given the matrix

$$\mathbf{A} = \begin{bmatrix} 1 & -2 & 3 \\ 2 & 0 & -3 \\ 1 & 1 & 1 \end{bmatrix}$$

(a) Compute $\mathbf{A}^2$, $\mathbf{A}^4$.

(b) Compute adj **A** and $\mathbf{A}^{-1}$.

2. Carry out the following triple product in two ways and note the difference in numbers of operations

$$\begin{bmatrix} 3 & 1 & -1 & 6 \\ 4 & -6 & 9 & 7 \\ -4 & 7 & 3 & 1 \end{bmatrix} \begin{bmatrix} 2 & 3 \\ 6 & 8 \\ -4 & 3 \\ 2 & 5 \end{bmatrix} \begin{bmatrix} 3 & 2 \\ -1 & 2 \end{bmatrix}$$

3. Solve the following system of equations:

$$\begin{aligned} -x_1 + 2x_2 + 3x_3 &= 1 \\ -2x_1 + 4x_2 - x_3 &= -3 \\ -x_1 + 2x_2 - 4x_3 &= -4 \\ -5x_1 + 10x_2 - 6x_3 &= -10 \end{aligned}$$

4. Solve the following system of equations:

$$\begin{aligned} x_1 + 4x_2 - x_3 &= -5 \\ 5x_1 + 2x_2 - 3x_3 &= -1 \\ -2x_1 + x_2 + x_3 &= -2 \\ -x_1 + 5x_2 &= -2 \end{aligned}$$

5. Solve the following system of equations:

$$\begin{aligned} -x_1 + 4x_2 - 3x_3 + 2x_4 &= 3 \\ 2x_1 - x_2 + 2x_3 + x_4 &= 1 \\ -3x_1 + 2x_2 - x_3 + 3x_4 &= 4 \\ -7x_1 + x_2 - x_3 + 3x_4 &= 4 \end{aligned}$$

6. Solve the following system of equations;

$$\begin{aligned} 2x_1 - x_2 + x_3 + 2x_4 &= 5 \\ -x_1 + 2x_2 + 3x_3 + x_4 &= -3 \\ x_1 + 4x_2 + 11x_3 + x_4 &= 1 \end{aligned}$$

7. Solve the following system of equations:

$$\begin{aligned} 8x_1 + 11x_2 + 3x_3 - 2x_4 &= 0 \\ 3x_1 + 2x_2 + 2x_3 - x_4 &= 0 \\ -x_1 + 7x_2 - 3x_3 + x_4 &= 0 \\ -11x_1 + 4x_2 - 12x_3 + 5x_4 &= 0 \end{aligned}$$

How many linearly independent solutions are there in a fundamental set?

8. Solve the following system of equations:

$$\begin{aligned} 3x_1 - 3x_2 + 5x_3 + 2x_4 &= 0 \\ x_1 - x_2 + x_3 + 4x_4 &= 0 \\ 4x_1 - 2x_2 + 3x_3 + 3x_4 &= 0 \end{aligned}$$

How many linearly independent solutions are there in a fundamental set?

9. Solve the following system of equations:

$$\begin{aligned} x_1 + x_2 + 3x_3 &= 0 \\ 2x_1 + x_2 + 4x_3 &= 0 \\ x_1 - x_2 - x_3 &= 0 \end{aligned}$$

10. Reduce the following matrix to "diagonal form" by producing zeroes as shown in the text by elementary transformations on rows and columns.

$$\begin{bmatrix} -2 & 1 & 1 & 3 & -1 \\ -4 & -3 & 7 & -1 & 2 \\ 2 & -4 & 2 & 3 & 1 \\ 5 & 2 & -7 & -2 & -1 \end{bmatrix}$$

11. Determine for what values of λ the following set of equations will have nontrival solutions:

$$\begin{aligned} -\lambda x_1 + x_2 + 3x_3 &= 0 \\ -3x_1 - 2x_2 + 4x_3 &= 0 \\ \lambda x_1 + 4x_2 + 2\lambda x_3 &= 0 \end{aligned}$$

12. Three planes in three-dimensional space may intersect in a point or line, may be parallel, etc. Determine how the various configurations are related to the rank of the matrix of the coefficients and the rank of the augmented matrix.

13. It is known that the rank of the following matrix is three. Show that the fourth column, for example, is a linear combination of the first three columns.

$$\begin{bmatrix} 1 & 3 & 0 & 5 \\ -1 & 1 & 1 & 0 \\ 2 & -1 & 2 & 5 \\ -1 & 2 & -1 & -1 \end{bmatrix}$$

and that the fourth row is a linear combination of the first three rows.

Vectors and Matrices 3

3.1 Introduction

In Chapter 2 the analysis has been more or less classical in approach; in the present and succeeding chapters, a more modern view will prevail. We shall consider special matrices of one column and m rows, called *vectors*. For present purposes a *vector* is defined as an ordered m-tuple of elements, where the elements are real or complex numbers or functions. We can consider a vector as a point in m-dimensional space or as a directed line segment from the origin to that point. The elements are then referred to as *coordinates* or *components* and both terms will be used.

Consider, for example, the relation

$$\mathbf{y} = \mathbf{A}\mathbf{x} \tag{3.1.1}$$

where

$$\mathbf{y} = \begin{bmatrix} y_1 \\ y_2 \\ \cdot \\ \cdot \\ \cdot \\ y_m \end{bmatrix}, \quad \mathbf{x} = \begin{bmatrix} x_1 \\ x_1 \\ \cdot \\ \cdot \\ \cdot \\ x_n \end{bmatrix}, \quad \mathbf{A} = \begin{bmatrix} a_{11} & a_{12} & \cdots & a_{1n} \\ a_{21} & a_{22} & \cdots & a_{2n} \\ \cdot & & & \\ \cdot & & & \\ \cdot & & & \\ a_{m1} & a_{m2} & \cdots & a_{mn} \end{bmatrix}$$

Equation (3.1.1) is a transformation of the points in an n-dimensional space to the points in an m-dimensional space; that is, to each vector in one space there corresponds a vector $\mathbf{y}$ in the second space. If $m \neq n$ then the inverse of the transformation presented by Eq. (3.1.1) is not defined and hence one cannot find the vector $\mathbf{x}$ which corresponds to a given $\mathbf{y}$. The transformation from $\mathbf{x}$ to $\mathbf{y}$ is also referred to as a *mapping* of one space onto the other. When the matrix $\mathbf{A}$ is square and nonsingular, the inverse transformation

$$\mathbf{x} = \mathbf{A}^{-1}\mathbf{y}$$

also exists and the mapping is one point to one point in each direction. Such a mapping is called an *isomorphism*. $\mathbf{A}$ is called the *matrix of the transformation*. If $\mathbf{x}$ is related to another transformation by the relation

$$\mathbf{x} = \mathbf{B}\mathbf{z}$$

Then $\mathbf{z}$ and $\mathbf{y}$ are related by

$$\mathbf{y} = \mathbf{A}\mathbf{B}\mathbf{z}$$

and the matrix $\mathbf{AB}$ is the matrix of the transformation from $\mathbf{z}$ to $\mathbf{y}$.

3.2 Sets of Vectors

For the moment we will consider a collection of vectors which contains a finite number. These vectors will be denoted by $\mathbf{a}_i$ where $\{\mathbf{a}_i\}$ will refer to the whole set or collection of vectors

$$\mathbf{a}_i = \begin{bmatrix} a_{1i} \\ a_{2i} \\ \cdot \\ \cdot \\ \cdot \\ a_{mi} \end{bmatrix}$$

The second subscript refers to the vector; the first subscript designates the component. We consider the vector collection

$$\begin{bmatrix} a_{11} \\ a_{21} \\ a_{31} \\ \cdot \\ \cdot \\ \cdot \\ a_{m1} \end{bmatrix} \begin{bmatrix} a_{12} \\ a_{22} \\ a_{32} \\ \cdot \\ \cdot \\ \cdot \\ a_{m2} \end{bmatrix} \begin{bmatrix} a_{13} \\ a_{23} \\ a_{33} \\ \cdot \\ \cdot \\ \cdot \\ a_{m3} \end{bmatrix} \quad \cdots \quad \begin{bmatrix} a_{1n} \\ a_{2n} \\ a_{3n} \\ \cdot \\ \cdot \\ \cdot \\ a_{mn} \end{bmatrix}$$

and we will call the collection a *vector space*. The term vector space is therefore a general one and refers to a particular collection of vectors and will change from one situation to another depending upon the collection of vectors under consideration.

A set of n vectors $\{\mathbf{a}_i\}$ is said to be *linearly dependent* if there exist n constants c_j, $j = 1$ to n, not all zero such that

$$c_1\mathbf{a}_1 + c_2\mathbf{a}_2 + c_3\mathbf{a}_3 + \cdots + c_n\mathbf{a}_n = \mathbf{0} \tag{3.2.1}$$

If no such set of constants exists, the vectors are said to be *linearly independent*. Equation (3.2.1) can be written

$$\begin{aligned} c_1 a_{11} + c_2 a_{12} + \cdots + c_n a_{1n} &= 0 \\ c_1 a_{21} + c_2 a_{22} + \cdots + c_n a_{2n} &= 0 \\ &\vdots \\ c_1 a_{m1} + c_2 a_{m2} + \cdots + c_n a_{mn} &= 0 \end{aligned} \tag{3.2.2}$$

The question of linear dependence of n vectors then reduces to the question whether this set of equations has nontrivial solutions in the c_j's. The answer to this question follows directly from the last theorem of Section 2.8: The necessary and sufficient condition that a set of linear homogeneous simultaneous algebraic equations in n variables has other than the trivial solution is that the rank of the matrix of the coefficients be less than n. Hence, if $m < n$ the set of vectors is always linearly dependent. If $m \geqslant n$, the matrix must be examined in detail to determine its rank. Suppose now that the rank of the matrix is r, then we shall adopt the terminology that the rank of the set of vectors is r and we shall also speak of the matrix of the vector set. Let the determinant of order r which is not zero have its columns made up from elements of the first r vectors. If this is not the case the vectors may be renumbered. Now if a_s is any one of remaining $(n - r)$ vectors there is a set of constants α_j, $j = 1$ to r, and an α_s such that

$$\alpha_1 \mathbf{a}_1 + \alpha_2 \mathbf{a}_2 + \cdots + \alpha_r \mathbf{a}_r + \alpha_s \mathbf{a}_s = \mathbf{0}$$

Since α_s cannot be zero

$$\mathbf{a}_s = -\frac{\alpha_1}{\alpha_s}\mathbf{a}_1 - \frac{\alpha_2}{\alpha_s}\mathbf{a}_2 - \cdots - \frac{\alpha_r}{\alpha_s}\mathbf{a}_r$$

or

$$\mathbf{a}_s = \alpha_1'\mathbf{a}_1 + \alpha_2'\mathbf{a}_2 + \cdots + \alpha_r'\mathbf{a}_r$$

and any one of the remaining $(n - r)$ vectors may be expressed as a linear combination of the first r linearly independent vectors. The set of linearly independent vectors is said to form a *basis* in the space or to *span* the space, and the vector space under consideration is said to be of dimension r. The term *dimension* as used here is the number of linearly independent vectors in the space and not the number of components in each vector in the space. It must be stressed that the basis is taken with respect to the particular space in question, and if other vectors are introduced into the space of vectors the basis may not be complete; that is, the basis may have to be enlarged to accommodate the new vector. We adopt the terminology that a given basis is complete with respect to a certain space. With the kind of spaces to be discussed in this work the question of completeness of the basis seems trivial. In more advanced works, however, the idea of a finite dimensional vector space is generalized to an infinite dimensional function space and the idea of a basis and its completeness is a subtle one.

Consider a vector space containing four vectors

$$\begin{bmatrix} 1 \\ 0 \\ -1 \end{bmatrix}, \begin{bmatrix} 2 \\ -1 \\ 1 \end{bmatrix}, \begin{bmatrix} 4 \\ -1 \\ -1 \end{bmatrix}, \begin{bmatrix} -1 \\ 1 \\ -2 \end{bmatrix}$$

The rank of the corresponding matrix is not three but two and *any* two vectors in the space may be taken as a basis since every pair of vectors contains a non-zero second-order determinant. If $\mathbf{a}_1$ and $\mathbf{a}_2$ are taken as the vectors in the basis, a simple calculation will show that

$$\mathbf{a}_3 = 2\mathbf{a}_1 + \mathbf{a}_2$$
$$\mathbf{a}_4 = \mathbf{a}_1 - \mathbf{a}_2$$

If $\mathbf{a}_1$ and $\mathbf{a}_3$ are taken as a basis,

$$\mathbf{a}_2 = \mathbf{a}_3 - 2\mathbf{a}_1$$
$$\mathbf{a}_4 = 3\mathbf{a}_1 - \mathbf{a}_3$$

Suppose now that a fifth vector is introduced into the set of vectors composing the space, say

$$\begin{bmatrix} 1 \\ -1 \\ 0 \end{bmatrix}$$

The matrix of the set of vectors is

$$\begin{bmatrix} 1 & 2 & 4 & -1 & 1 \\ 0 & -1 & -1 & 1 & -1 \\ -1 & 1 & -1 & -2 & 0 \end{bmatrix}$$

Any third-order determinant not including the fifth column is zero. Any third-order determinant including the fifth column is not zero. Thus any basis for this space must include $\mathbf{a}_5$ but any two of the others will suffice. It is clear then that the basis for a space may not be unique so that a given vector space may have many bases but the representation of a vector in that space in terms of the basis once it has been chosen is unique.

3.3 Application to Chemical Equilibrium

The problem in chemical thermodynamics of determining the number of independent components in a system at equilibrium if the system contains a large number of chemically reacting species may be involved and tedious and as frequently presented appears to be one which does not have a systematic method of solution. The number of independent components in an equilibrium mixture is the minimum number of substances the amounts of which are necessary to specify the composition of a phase. It is also stated as the minimum number of substances required to prepare the phase. Suppose that a phase at equilibrium contains n_j moles of a substance M_j. Each chemical substance M_j is made up of atoms A_i and let the number of atoms of A_i in M_j be a_{ij}. A table may be made up listing M_j along the top and A_i vertically at the left so that the element at the intersection of the ith row and jth column a_{ij} is a small integer or zero representing the number of atoms of A_i in M_j.

$$\begin{array}{cccccc} & M_1 & M_2 & M_3 & \cdots & M_n \\ A_1 & a_{11} & a_{12} & a_{13} & \cdots & a_{1n} \\ A_2 & a_{21} & a_{22} & a_{23} & \cdots & a_{2n} \\ \cdot & & & & & \\ \cdot & & & & & \\ \cdot & & & & & \\ A_m & a_{m1} & a_{m2} & a_{m3} & \cdots & a_{mn} \end{array} \tag{3.3.1}$$

We have assumed that there are n chemical species in the mixture and m atomic species included in all of the M_j. The chemical substance would then be written

$$M_j = (A_1)_{a_{1j}}(A_2)_{a_{2j}} \cdots (A_m)_{a_{mj}} \tag{3.3.2}$$

There is a redundancy here since a_{ij} contains all the information that $(A_i)_{a_{ij}}$ does. Hence the chemical formula for M_j is

$$a_{1j}a_{2j}a_{3j} \cdots a_{mj} \tag{3.3.3}$$

For chemical manipulations and study of reactions and structure, this would not be a very useful representation but for our purposes it is almost as good as the conventional formula, save for its lack of a history. Now Eq. (3.3.3) is a row vector, $\mathbf{M}_j^T$, and therefore the chemical substances M_j in Eq. (3.3.1) are vectors in a vector space which we might call the *equilibrium space* and the matrix of the vectors the atomic matrix.

Suppose the rank of the atomic matrix is r and again for convenience suppose the determinant of order r which is not zero is in the first r columns. Then the number of independent vectors in the equilibrium space is r and each of the vectors in the space may be represented as a linear combination of vectors in the basis; that is,

$$\begin{aligned} \mathbf{M}_{r+1} &= \alpha_{1,\,r+1}\mathbf{M}_1 + \alpha_{2,\,r+1}\mathbf{M}_2 + \cdots + \alpha_{r,\,r+1}\mathbf{M}_r \\ \mathbf{M}_{r+2} &= \alpha_{1,\,r+2}\mathbf{M}_1 + \alpha_{2,\,r+2}\mathbf{M}_2 + \cdots + \alpha_{r,\,r+2}\mathbf{M}_r \\ &\ \ \vdots \\ \mathbf{M}_n &= \alpha_{1n}\mathbf{M}_1 + \alpha_{2n}\mathbf{M}_2 + \cdots + \alpha_{rn}\mathbf{M}_r \end{aligned} \tag{3.3.4}$$

These vector relations are, however, nothing more than relations among chemical substances, that is, chemical reactions, and they account for all the chemical substances in the equilibrium mixture; furthermore $n - r$ is the *minimum* number of reactions which can account for all of the substances since each reaction contains a substance which the other one does not. The α_{ij} in Eq. (3.3.4) are the stoichiometric coefficients in the reactions. The minimal set of reactions is, in general, not unique but we know from thermodynamics that this is not necessary since the thermodynamic functions, free energy, enthalpy, entropy, equilibrium constant, etc., are state functions and their calculation is not dependent upon the path which resulted in the formation of the substances themselves.

If in the reaction mixture there is an infinitesimal shift in the equilibrium with a resulting change in the composition, let dn_j be the change in the number of moles of M_j, then from conservation of atomic species,

$$\begin{aligned}
a_{11}\,dn_1 + a_{12}\,dn_2 + a_{13}\,dn_3 + \cdots + a_{1n}\,dn_n &= 0\\
a_{21}\,dn_1 + a_{22}\,dn_2 + a_{23}\,dn_3 + \cdots + a_{2n}\,dn_n &= 0\\
&\;\vdots\\
a_{m1}\,dn_1 + a_{m2}\,dn_2 + a_{m3}\,dn_3 + \cdots + a_{mn}\,dn_n &= 0
\end{aligned}$$

and, of course, these can be integrated from some initial state n_{j0} to give

$$\begin{aligned}
a_{11}n_1 + a_{12}n_2 + \cdots + a_{1n}n_n &= a_{11}n_{10} + a_{12}n_{20} + \cdots + a_{1n}n_{n0}\\
a_{21}n_1 + a_{22}n_2 + \cdots + a_{2n}n_n &= a_{21}n_{10} + a_{22}n_{20} + \cdots + a_{2n}n_{n0}\\
&\;\vdots\\
a_{m1}n_1 + a_{m2}n_2 + \cdots + a_{mn}n_n &= a_{m1}n_{10} + a_{m2}n_{20} + \cdots + a_{mn}n_{n0}
\end{aligned} \tag{3.3.5}$$

or

$$\mathbf{An} = \mathbf{An}_0$$

and because the rank of the atomic matrix $\mathbf{A}$ is r there are only r equations in this set rather than m. Since in such a system there are $n - r$ reactions there are $n - r$ equilibrium relations which may be written symbolically

$$\begin{aligned}
f_1(n_1, n_2, \ldots, n_n) &= 0\\
f_2(n_1, n_2, \ldots, n_n) &= 0\\
&\;\vdots\\
f_{n-r}(n_1, n_2, \ldots, n_n) &= 0
\end{aligned}$$

These $n - r$ equations must be solved simultaneously with the r-independent atomic balances in order to obtain the composition of the equilibrium mixture resulting from some initial composition. It is obvious that the $n - r$

equilibrium relations will serve to define $(n - r)$ values of the n_j in terms of r values of the remaining n_j so that the composition of any equilibrium mixture is completely determined if these r n_j's are known.

Equation (3.3.5) can be put into more convenient form by defining some new variables. In the reaction involving M_{r+1}, let $x_{r+1}^{(r+1)}$ be the moles of M_{r+1} produced when $x_1^{(r+1)}, x_2^{(r+1)}, \ldots, x_r^{(r+1)}$ moles of $M_1, M_2, \ldots, M_r$ respectively, are converted. From Eq. (3.3.4)

$$\frac{x_i^{(r+1)}}{\alpha_{i,\, r+1}} = \frac{x_j^{(r+1)}}{\alpha_{j,\, r+1}} = \frac{x_{r+1}^{(r+1)}}{1}$$

Define a new reaction variable

$$x^{(r+1)} = \frac{x_i^{(r+1)}}{\alpha_{i,\, r+1}}$$

which is an intrinsic conversion variable and does not refer to a particular species. For the other reactions, one may do likewise

$$x^{(r+2)} = \frac{x_i^{(r+2)}}{\alpha_{i, r+2}}$$

$$\vdots$$

$$x^{(n)} = \frac{x_i^{(n)}}{\alpha_{i,\, n}}$$

and we obtain

$$\begin{aligned}
n_1 &= n_{10} - \alpha_{1,\, r+1}\, x^{(r+1)} - \alpha_{1,\, r+2}\, x^{(r+2)} \ldots - \alpha_{1n} x^{(n)} \\
n_2 &= n_{20} - \alpha_{2,\, r+1}\, x^{(r+1)} - \alpha_{2,\, r+2}\, x^{(r+2)} \ldots - \alpha_{2n} x^{(n)} \\
&\vdots \\
n_r &= n_{r0} - \alpha_{r,\, r+1}\, x^{(r+1)} - \alpha_{r,\, r+2}\, x^{(r+2)} \ldots - \alpha_{rn} x^{(n)}
\end{aligned}$$

$$\begin{aligned}
n_{r+1} &= n_{r+1,\, 0} + x^{(r+1)} \\
n_{r+2} &= n_{r+2,\, 0} + x^{(r+2)} \\
&\vdots \\
n_n &= n_{n0} + x^{(n)}
\end{aligned}$$

and this expresses the composition of the reaction mixture in terms of $(n - r)$ quantities $x^{(j)}$, $j = r + 1, r + 2, \ldots, n$. When n_j is substituted into the equilibrium relations there will be $n - r$ relations among $(n - r)$ unknowns $x^{(j)}$.

As an example of the calculation of the number of independent reactions consider a reaction mixture consisting of O_2, H_2, CO, CO_2, H_2CO, CH_3OH, C_2H_5OH, $(CH_3)_2CO$, CH_3CHO, CH_4, H_2O:

	O_2	H_2	CO	CO_2	H_2CO	CH_3OH	C_2H_5OH	$(CH_3)_2CO$	CH_4	CH_3CHO	H_2O
O	2	0	1	2	1	1	1	1	0	1	1
H	0	2	0	0	2	4	6	6	4	4	2
C	0	0	1	1	1	1	2	3	1	2	0

The rank of this system is 3, there are 11 chemical species, the minimum number of reactions possible is 8, the number of independent components is 3, so that the composition of the mixture may be described by specifying the three mole numbers of the independent components.

3.4 Application to Dimensional Analysis

In a well-posed physical problem described by mathematical equations, the equations, and hence the solution, may be written in terms of dimensionless groups, the number of dimensionless groups being considerably fewer than the number of variables and parameters in the system. This has a great advantage since it means that, although information may be needed on one particular system, the relation among the dimensionless groups may be found experimentally by studying the same problem but with another and perhaps more convenient system. Hence an important problem is determining the minimum number of independent dimensionless groups from a specified number of variables and parameters. For example, one is aware that in a problem of heat conduction in a rod in the transient state the length, initial temperature, heat capacity per unit of volume, thermal conductivity, time, end temperatures, and position in the rod are related, and we assume that the variables and parameters must appear in the final formula in dimensionless groups. We desire to determine the minimum number of such groups.

Let the physical quantities which have relevance in the problem be denoted by $P_1, P_2, \ldots, P_n$ where these may be entities like viscosity, thermal conductivity, surface tension, diameter, heat capacity, etc., and suppose the fundamental quantities are $m_1, m_2, \ldots, m_n$, where these are mass, length, time, temperature, etc. The expression $[P_j]$ will be the dimension of the physical quantity P_j so that

$$[P_j] = m_1^{\alpha_{1j}} m_2^{\alpha_{2j}} \ldots m_m^{\alpha_{mj}}$$

where α_{1j} might be the number of length units in P_j. The α_{ij} are small positive or negative integers or zero. The dimensional matrix may be written

$$\begin{array}{cccccc} & [P_1] & [P_2] & [P_3] & \ldots & [P_n] \\ m_1 & \alpha_{11} & \alpha_{12} & \alpha_{13} & \ldots & \alpha_{1n} \\ m_2 & \alpha_{21} & \alpha_{22} & \alpha_{23} & \ldots & \alpha_{2n} \\ \cdot & & & & & \\ \cdot & & & & & \\ \cdot & & & & & \\ m_m & \alpha_{m1} & \alpha_{m2} & \alpha_{m3} & \ldots & \alpha_{mn} \end{array}$$

where α_{ij} is the number of fundamental quantities m_i involved in P_j (for example, if P_j were velocity and m_i were length, α_{ij} would be one). Considered as a vector space, the number of linear independent vectors is r and each of the remaining $n - r$ vectors can be expressed as a linear combination of the r independent ones, thus

$$[P_j] = \sum_{i=1}^{r} w_{ij}[P_i], \quad j = r + 1, r + 2, \ldots, n$$

where the w_{ij} are constants. This equation is a vector equation written in terms of dimensional vectors. The equation written in terms of the physical quantities themselves is

$$P_j = \prod_{i=1}^{r} P_i^{w_{ij}}$$

or in other words

$$P_j^{-1} \prod_{i=1}^{r} P_i^{w_{ij}}$$

is a dimensionless group. There cannot be more than $(n - r)$ independent ones, and, further, this is the minimum number of independent ones since each one contains a quantity which does not appear in the others. There are many such groups and, for correlation of experimental data to obtain a dimensionless empirical correlation one uses the most convenient.

As an example consider that posed at the beginning of this section, determining the number of dimensionless groups in a problem in transient heat conduction in a rod with initial temperature T_i and end temperature T_0. The pertinent variables with their dimensions are

L = length of rod,	$[L] = l$
x = position of rod,	$[x] = l$
τ = time,	$[\tau] = t$
$c\rho$ = heat capacity per unit of volume,	$[c\rho] = \dfrac{m}{lt^2\theta}$
$T_0 - T_i$ = end temperatures,	$[T_0 - T_i] = \theta$
$T - T_i$ = temperature at position x at time τ,	$[T - T_i] = \theta$
k = thermal conductivity,	$[k] = \dfrac{ml}{t^3\theta}$

where the fundamental units used are mass (m), length (l), time (t), and temperature (θ). The zero temperature point will be taken as the initial temperature. The dimensional matrix is

	L	x	τ	k	$c\rho$	$T - T_i$	$T_0 - T_i$
m	0	0	0	1	1	0	0
l	1	1	0	1	−1	0	0
t	0	0	1	−3	−2	0	0
θ	0	0	0	−1	−1	1	1

The rank of this matrix is four; hence there must be three dimensionless groups. In this case the dimensionless groups may be determined by inspection of the table. A suitable set of three dimensionless groups is

$$\frac{T - T_i}{T_0 - T_i}, \quad \frac{x}{L}, \quad \frac{k\tau}{L^2 c\rho}$$

3.5 Expansion of Vectors in a Basis

In the previous sections we have frequently said that a vector may be expanded in terms of a basis but we have not said how this may be accomplished. Obviously, if a basis is $\mathbf{a}_1, \mathbf{a}_2, \ldots, \mathbf{a}_r$ and if

$$\mathbf{b} = \alpha_1 \mathbf{a}_1 + \alpha_2 \mathbf{a}_2 + \cdots + \alpha_r \mathbf{a}_r$$

then the $\{\alpha_j\}$ may be determined by the solution of a set of equations,

$$\begin{aligned}
b_1 &= \alpha_1 a_{11} + \alpha_2 a_{12} + \cdots + \alpha_r a_{1r} \\
b_2 &= \alpha_1 a_{21} + \alpha_2 a_{22} + \cdots + \alpha_r a_{2r} \\
&\;\vdots \\
b_r &= \alpha_1 a_{r1} + \alpha_2 a_{r2} + \cdots + \alpha_r a_{rr} \\
&\;\vdots \\
b_m &= \alpha_1 a_{m1} + \alpha_2 a_{m2} + \cdots + \alpha_r a_{mr}
\end{aligned} \tag{3.5.1}$$

where it is assumed that the basis is complete with respect to $\mathbf{b}$. If, as is assumed the rank is r then the $\{\alpha_j\}$ may be determined by solution of the first r simultaneous equations if the matrix of their coefficients has a nonvanishing determinant. The $\{\alpha_j\}$ determined in this way will then also satisfy the last $(n - r)$ equations.

A method which has great utility is one in which a simpler basis is used and the basis is then changed one vector at a time until the expansion in terms of the desired basis is obtained. Consider the set of vectors with m components

$$\mathbf{e}_1 = \begin{bmatrix} 1 \\ 0 \\ 0 \\ \cdot \\ \cdot \\ \cdot \\ 0 \end{bmatrix} \quad \mathbf{e}_2 = \begin{bmatrix} 0 \\ 1 \\ 0 \\ \cdot \\ \cdot \\ \cdot \\ 0 \end{bmatrix}, \ldots, \quad \mathbf{e}_m = \begin{bmatrix} 0 \\ 0 \\ 0 \\ \cdot \\ \cdot \\ 0 \\ 1 \end{bmatrix} \tag{3.5.2}$$

where $\mathbf{e}_j$ is the vector with one in the jth position and zeroes elsewhere. Then any vector in the original vector space consisting of $\mathbf{a}_1, \mathbf{a}_2, \ldots, \mathbf{a}_n$ may be expanded in terms of the set $\{\mathbf{e}_j\}$ for

$$\mathbf{a}_j = a_{1j}\mathbf{e}_1 + a_{2j}\mathbf{e}_2 + \cdots + a_{mj}\mathbf{e}_m, \quad \text{for all } j \tag{3.5.3}$$

and

$$\mathbf{b} = b_1\mathbf{e}_1 + b_2\mathbf{e}_2 + \cdots + b_m\mathbf{e}_m \tag{3.5.4}$$

The set of vectors $\{\mathbf{e}_j\}$ is called a *set of unit vectors* of mth order and *any* vector with m components may be expressed or expanded in terms of this set. To Eq. (3.5.4), we add and subtract a constant multiple of one of the vectors of the set $\{\mathbf{a}_j\}$, say, $\mathbf{a}_r$

$$\mathbf{b} = \sum_i b_i\mathbf{e}_i - \alpha\mathbf{a}_r + \alpha\mathbf{a}_r$$

But

$$\mathbf{a}_r = a_{1r}\mathbf{e}_1 + a_{2r}\mathbf{e}_2 + \cdots + a_{rr}\mathbf{e}_r + \cdots + a_{mr}\mathbf{e}_m \tag{3.5.5}$$

so

$$\mathbf{b} = \sum_i (b_i - \alpha a_{ir})\mathbf{e}_i + \alpha\mathbf{a}_r$$

Thus one obtains an expansion of the vector $\mathbf{b}$ in terms of a basis which includes $\mathbf{a}_r$ and a vector is removed, the vector to be removed depending upon the choice of α. If $\mathbf{e}_s$ is to be removed, then one chooses

$$\alpha = \frac{b_s}{a_{sr}}$$

so

$$\mathbf{b} = \sum_i \left(b_i - \frac{b_s}{a_{sr}} a_{ir}\right)\mathbf{e}_i + \frac{b_s}{a_{sr}}\mathbf{a}_r \tag{3.5.6}$$

Now each of the other vectors, not in the new basis, will have a different expansion. In particular, from Eq. (3.5.5),

$$\mathbf{e}_s = \frac{1}{a_{sr}}[\mathbf{a}_r - a_{1r}\mathbf{e}_1 - a_{2r}\mathbf{e}_2 \ldots - a_{s-1,r}\mathbf{e}_{s-1} - a_{s+1,r}\mathbf{e}_{s+1} + \cdots - a_{mr}\mathbf{e}_m] \tag{3.5.7}$$

The expansion of any vector $\mathbf{a}_j$ can be written from Eq. (3.5.6) as

$$\mathbf{a}_j = \sum_i \left(a_{ij} - \frac{a_{sj}}{a_{sr}} a_{ir}\right)\mathbf{e}_i + \frac{a_{sj}}{a_{sr}}\mathbf{a}_r, \quad j \neq r \tag{3.5.8}$$

$$\mathbf{a}_r = 0\mathbf{e}_1 + 0\mathbf{e}_2 + 0\mathbf{e}_3 + \cdots + 0\mathbf{e}_m + \mathbf{a}_r$$

This procedure may be conveniently summarized in a table (Table 3.5.1). The vectors are listed along the top of the table and the columns are the components of these vectors in terms of the unit vectors. Along the left-hand

TABLE 3.5.1

	$\mathbf{a}_1$	...	$\mathbf{a}_j$	...	$\mathbf{a}_r$	$\mathbf{e}_1$	...	$\mathbf{e}_s$	...	$\mathbf{b}$
$\mathbf{e}_1$	a_{11}		a_{1j}		a_{1r}	1		0		$\mathbf{b}_1$
$\mathbf{e}_2$	a_{21}		a_{2j}		a_{2r}	0		0		$\mathbf{b}_2$
$\mathbf{e}_3$	a_{31}		a_{3j}		a_{3r}	0		0		$\mathbf{b}_3$
⋮										
$\mathbf{e}_s$	a_{s1}		a_{sj}		a_{sr}	0		1		$\mathbf{b}_s$
⋮										
$\mathbf{e}_m$	a_{ms}		a_{mj}		a_{mr}	0		0		$\mathbf{b}_m$

side are listed the unit vectors; note particularly, that a row does not represent the expansion of the unit vectors in terms of the vectors listed along the top except for the unit vectors themselves.

From Eq. (3.5.6), (3.5.7), and (3.5.8) we can now determine the expansion coefficients in terms of the new basis and a new table, Table 3.5.2, can be

TABLE 3.5.2

	$\mathbf{a}_1$	$\mathbf{a}_j$	$\mathbf{a}_r$	$\mathbf{e}_1$	...	$\mathbf{e}_s$	...	$\mathbf{b}$
$\mathbf{e}_1$	$a_{11} - \dfrac{a_{s1}}{a_{sr}} a_{1r}$	$a_{1j} - \dfrac{a_{sj}}{a_{sr}} a_{1r}$	0	1		$-\dfrac{a_{1r}}{a_{sr}}$		$b_1 - \dfrac{b_s}{a_{sr}} a_{1r}$
$\mathbf{e}_2$	$a_{21} - \dfrac{a_{s1}}{a_{sr}} a_{2r}$	$a_{2j} - \dfrac{a_{sj}}{a_{sr}} a_{2r}$	0	0		$-\dfrac{a_{2r}}{a_{sr}}$		$b_2 - \dfrac{b_s}{a_{sr}} a_{2r}$
⋮								
$\mathbf{a}_r$	$\dfrac{a_{s1}}{a_{sr}}$	$\dfrac{a_{sj}}{a_{sr}}$	1	0		$\dfrac{1}{a_{sr}}$		$\dfrac{b_s}{a_{sr}}$
⋮								
$\mathbf{e}_m$	$a_{m1} - \dfrac{a_{s1}}{a_{sr}} a_{mr}$	$a_{mj} - \dfrac{a_{sj}}{a_{sr}} a_{mr}$	0	0		$-\dfrac{a_{mr}}{a_{sr}}$		$b_m - \dfrac{b_s}{a_{sr}} a_{mr}$

constructed. In order to simplify the writing only a few vectors are included in this table. Table 3.5.2 gives the coefficients as determined from Eq. (3.5.6), (3.5.7), and (3.5.8). We will now show that these entries in the table can be produced by operating on the rows of the matrix with elementary matrix operators. The entries opposite $\mathbf{a}_r$ may be obtained by dividing the row opposite $\mathbf{e}_s$ in Table 3.5.1 by a_{sr}. Now note under $\mathbf{a}_r$ all but the element opposite $\mathbf{a}_r$ are zero. In order to produce a zero opposite $\mathbf{e}_2$, for example, and under $\mathbf{a}_r$ we divide the row opposite $\mathbf{e}_s$ in Table 3.5.1 by a_{sr}, multiply it by a_{2r}, and subtract it from the second row. In Table 3.5.2 note that this produces the elements in the second row. Therefore a simple numerical algorithm allows us to find expansions of vectors in terms of the basis. This same procedure can now be repeated, one vector after another, until the desired expansion is obtained. In order to illustrate the method, we consider a simple example in which there are five vectors each with four components.

	$\mathbf{a}_1$	$\mathbf{a}_2$	$\mathbf{a}_3$	$\mathbf{a}_4$	$\mathbf{a}_5$	$\mathbf{e}_1$	$\mathbf{e}_2$	$\mathbf{e}_3$	$\mathbf{e}_4$
$\mathbf{e}_1$	1	−1	2	0	1	1	0	0	0
$\mathbf{e}_2$	−1	0	1	−2	2	0	1	0	0
$\mathbf{e}_3$	0	−1	3	−2	3	0	0	1	0
$\mathbf{e}_4$	1	−2	5	−2	4	0	0	0	1

The rank of the set $\{\mathbf{a}_j\}$ is two. Every third- and fourth-order determinant is zero. All the second-order determinants containing the first two rows are non-zero. Consider the expansion of

$$\mathbf{a}_5 = \alpha \mathbf{a}_4 + \beta \mathbf{a}_3$$

We shall first expand $\mathbf{a}_5$ into the unit vectors, then remove $\mathbf{e}_2$ and $\mathbf{e}_1$ successively, replacing $\mathbf{e}_2$ with $\mathbf{a}_4$ and $\mathbf{e}_1$ with $\mathbf{a}_3$.

	$\mathbf{a}_1$	$\mathbf{a}_2$	$\mathbf{a}_3$	$\mathbf{a}_4$	$\mathbf{a}_5$	$\mathbf{e}_1$	$\mathbf{e}_2$	$\mathbf{e}_3$	$\mathbf{e}_4$
$\mathbf{e}_1$	1	−1	2	0	1	1	0	0	0
$\mathbf{a}_4$	$\frac{1}{2}$	0	$-\frac{1}{2}$	1	−1	0	$-\frac{1}{2}$	0	0
$\mathbf{e}_3$	1	−1	2	0	1	0	−1	1	0
$\mathbf{e}_4$	2	−2	4	0	2	0	−1	0	1

Where we have operated on the rows of the first table to produce a one at the intersection of the second row and fourth column and zeroes at all other positions of the fourth column. The second step is to replace $\mathbf{e}_1$ with $\mathbf{a}_3$, and we operate on rows to produce a one at the intersection of the first row and third column. There results

	$\mathbf{a}_1$	$\mathbf{a}_2$	$\mathbf{a}_3$	$\mathbf{a}_4$	$\mathbf{a}_5$	$\mathbf{e}_1$	$\mathbf{e}_2$	$\mathbf{e}_3$	$\mathbf{e}_4$
$\mathbf{a}_3$	$\frac{1}{2}$	$-\frac{1}{2}$	1	0	$\frac{1}{2}$	$\frac{1}{2}$	0	0	0
$\mathbf{a}_4$	$\frac{3}{4}$	$-\frac{1}{4}$	0	1	$-\frac{3}{4}$	$\frac{1}{4}$	$-\frac{1}{2}$	0	0
$\mathbf{e}_3$	0	0	0	0	0	−1	−1	1	0
$\mathbf{e}_4$	0	0	0	0	0	−2	−1	0	1

so that

$$\begin{aligned}
\mathbf{a}_5 &= \tfrac{1}{2}\mathbf{a}_3 - \tfrac{3}{4}\mathbf{a}_4 + 0\mathbf{e}_3 + 0\mathbf{e}_4 \\
\mathbf{a}_2 &= -\tfrac{1}{2}\mathbf{a}_3 - \tfrac{1}{4}\mathbf{a}_4 + 0\mathbf{e}_3 + 0\mathbf{e}_4 \\
\mathbf{a}_1 &= \tfrac{1}{2}\mathbf{a}_3 + \tfrac{3}{4}\mathbf{a}_4 + 0\mathbf{e}_3 + 0\mathbf{e}_4 \\
\mathbf{e}_1 &= \tfrac{1}{2}\mathbf{a}_3 + \tfrac{1}{4}\mathbf{a}_4 - \mathbf{e}_3 - 2\mathbf{e}_4 \\
\mathbf{e}_2 &= -\tfrac{1}{2}\mathbf{a}_4 - \mathbf{e}_3 - \mathbf{e}_4
\end{aligned}$$

Examination of these results shows that what is supposed to happen occurs. The vectors $\{\mathbf{a}_j\}$ may be expanded in terms of two vectors; the expansion of the vectors $\mathbf{e}_1$ and $\mathbf{e}_2$ will require enough vectors so that their additional effective components may be represented in the expansion.

3.6 Application of Vector Expansion to the Solution of Linear Equations and Matrix Inversion

We now consider n linear equations in n unknowns

$$\mathbf{A}\mathbf{x} = \mathbf{b}$$

and where the columns of $\mathbf{A}$ will be considered as vectors so that the preceding matrix equation may be written as a vector equation

$$\mathbf{a}_1 x_1 + \mathbf{a}_2 x_2 + \cdots + \mathbf{a}_n x_n = \mathbf{b}$$

and written in this way a set of linear equations takes on a new aspect. The $\{x_j\}$ may be considered to be the coefficients in the expansion of $\mathbf{b}$ in terms of $\{\mathbf{a}_j\}$. With this in mind, the technique of the previous section may be used to obtain the expansion. As an example consider the equations

$$x_1 - x_2 + 2x_3 = -3$$
$$2x_1 + x_2 - x_3 = 5$$
$$-x_1 + 3x_2 + 3x_3 = 2$$

Using the previous paragraph, the first table would be

	$\mathbf{a}_1$	$\mathbf{a}_2$	$\mathbf{a}_3$	$\mathbf{e}_1$	$\mathbf{e}_2$	$\mathbf{e}_3$	$\mathbf{b}$
$\mathbf{e}_1$	1	-1	2	1	0	0	-3
$\mathbf{e}_2$	2	1	-1	0	1	0	5
$\mathbf{e}_3$	-1	3	3	0	0	1	2

The final table, after changing vectors one at a time, is

	$\mathbf{a}_1$	$\mathbf{a}_2$	$\mathbf{a}_3$	$\mathbf{e}_1$	$\mathbf{e}_2$	$\mathbf{e}_3$	$\mathbf{b}$
$\mathbf{a}_1$	1	0	0	$\frac{6}{25}$	$\frac{9}{25}$	$-\frac{1}{25}$	1
$\mathbf{a}_3$	0	0	1	$\frac{7}{25}$	$-\frac{2}{25}$	$\frac{3}{25}$	-1
$\mathbf{a}_2$	0	1	0	$-\frac{1}{5}$	$\frac{1}{5}$	$\frac{1}{5}$	2

hence the solution is

$$x_1 = 1, \quad x_2 = 2, \quad x_3 = -1$$

Also

$$\mathbf{e}_1 = \tfrac{6}{25}\mathbf{a}_1 - \tfrac{1}{5}\mathbf{a}_2 + \tfrac{7}{25}\mathbf{a}_3$$
$$\mathbf{e}_2 = \tfrac{9}{25}\mathbf{a}_1 + \tfrac{1}{5}\mathbf{a}_2 - \tfrac{2}{25}\mathbf{a}_3$$
$$\mathbf{e}_3 = -\tfrac{1}{25}\mathbf{a}_2 + \tfrac{1}{5}\mathbf{a}_2 + \tfrac{3}{25}\mathbf{a}_3$$

Consider the expansion

$$\mathbf{a}_j = a_{1j}\mathbf{e}_1 + a_{2j}\mathbf{e}_2 + a_{3j}\mathbf{e}_j + \cdots + a_{mj}\mathbf{e}_m \tag{3.5.5}$$

and suppose it is possible to find the expansion

$$\mathbf{e}_j = b_{1j}\mathbf{a}_1 + b_{2j}\mathbf{a}_2 + b_{3j}\mathbf{a}_3 + \cdots + b_{mj}\mathbf{a}_m \tag{3.6.1}$$

where it is assumed that the $\{\mathbf{e}_j\}$ and $\{\mathbf{a}_j\}$ have the same number of components. Now if Eq (3.6.1) is substituted into Eq. (3.5.5), then

$$
\begin{aligned}
\mathbf{a}_j &= a_{1j}(b_{11}\mathbf{a}_1 + b_{21}\mathbf{a}_2 + \cdots + b_{m1}\mathbf{a}_m) \\
&\quad + a_{2j}(b_{12}\mathbf{a}_1 + b_{22}\mathbf{a}_2 + \cdots + b_{m2}\mathbf{a}_m) \\
&\quad \vdots \\
&\quad + a_{mj}(b_{1m}\mathbf{a}_1 + b_{2m}\mathbf{a}_2 + \cdots + b_{mm}\mathbf{a}_m)
\end{aligned}
$$

or

$$
\begin{aligned}
\mathbf{a}_j &= (a_{1j}b_{11} + a_{2j}b_{12} + \cdots + a_{mj}b_{1m})\,\mathbf{a}_1 \\
&\quad + (a_{1j}b_{21} + a_{2j}b_{22} + \cdots + a_{mj}b_{2m})\mathbf{a}_2 \\
&\quad \vdots \\
&\quad + (a_{1j}b_{m1} + a_{2j}b_{m2} + \cdots + a_{mj}b_{m2})\mathbf{a}_m
\end{aligned}
$$

It follows therefore that

$$\delta_{ij} = a_{1j}b_{i1} + a_{2j}b_{i2} + a_{3j}b_{i3} + \cdots + a_{mj}b_{im}$$

where $\delta_{ij} = 0$, $i \neq j$, and $\delta_{ii} = 1$. Hence,

$$
\begin{bmatrix}
b_{11} & b_{12} & b_{13} & \dots & b_{1m} \\
b_{21} & b_{22} & b_{23} & \dots & b_{2m} \\
\vdots & & & & \\
b_{m1} & b_{m2} & b_{m3} & \dots & b_{mm}
\end{bmatrix}
\begin{bmatrix}
a_{11} & a_{12} & a_{13} & \dots & a_{1m} \\
a_{21} & a_{22} & a_{23} & \dots & a_{2m} \\
\vdots & & & & \\
a_{m1} & a_{m2} & a_{m3} & \dots & a_{mm}
\end{bmatrix}
=
\begin{bmatrix}
1 & 0 & 0 & \dots & 0 \\
0 & 1 & 0 & \dots & 0 \\
0 & 0 & 1 & \dots & 0 \\
\vdots & & & & \\
0 & 0 & 0 & \dots & 1
\end{bmatrix}
$$

or

$$\mathbf{BA} = \mathbf{I}$$

so that the matrices **A** and **B** are inverses of each other, provided that they are nonsingular.

Therefore the inverse of the matrix of the coefficients in the preceding problem is

$$
\tfrac{1}{25}\begin{vmatrix}
6 & 9 & -1 \\
-5 & 5 & 5 \\
7 & -2 & 2
\end{vmatrix}
$$

This gives a scheme for matrix inversion and a method for the solution of simultaneous algebraic equations. The next chapter will treat the problem of linear programming and the technique of Section 3.5 will be exploited in the so-called simplex method.

3.7 Application to Chemical Reaction Systems

In complex chemical reaction systems, where the number of chemical species is very large and where the reactions are specified kinetically, it is desirable to have criteria for determining the minimum number of equations which will describe the reaction system. We shall consider the empty tubular reactor, the analysis of which is essentially the same as the batch reactor.

We shall suppose the fluid flows in plug flow and all diffusive and conductive effects may be neglected. The heat transfer may be assumed to be characterized by a heat transfer coefficient at the wall. Let the running variable down the axis of the tube be x. The reactions which take place in the body of the fluid are given by the equations

$$\sum_{i=1}^{n} a_{ij} A_i = 0, \quad j = 1, 2, \ldots, m$$

where the usual thermodynamic convention that a_{ij} is positive for a product and negative for a reactant is adopted. There are m reactions and n chemical species. The nomenclature for the problem is summarized as follows:

G = total mass flow rate
T = temperature of reaction mixture
t = ambient temperature
h = heat transfer coefficient at tube wall
a = cross-sectional area of reactor
p = pressure
P = perimeter of reactor
$\tilde{h}_i$ = partial molar enthalpy of ith species
ΔH_j = heat of reaction of jth reaction
A_i = ith chemical species
C_{pi} = molar heat capacity of ith species
a_{ij} = stoichiometric coefficient of ith species in jth reaction
f_{ij} = rate of formation of ith species in jth reaction in moles per unit volume per unit time
τ = shearing stress at wall
f = friction factor
g_i = moles of species i per unit mass of reaction mixture
u = linear velocity along the axis of the tube
w = mass velocity of reaction mixture
τ = shearing stress at the tube wall

In order to derive the equations of conservation consider the idealized reactor

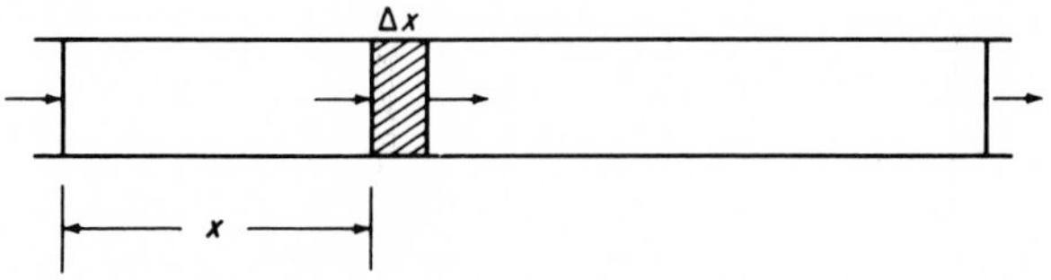

The convective inflow for species i is

$$(Gg_i)_x$$

where the notation means $(Gg_i)_x$ evaluated at position x. The ouflow by convection is

$$(Gg_i)_{x+\Delta x}$$

The rate at which species i is being formed by chemical reaction in the jth reaction is

$$(a\,\Delta x\, f_{ij})_{x'}, \quad x < x' < x + \Delta x$$

and in all the reactions

$$(a\,\Delta x \sum_j f_{ij})_{x'}, \quad x < x' < x + \Delta x$$

where the summation is over all reactions, $j = 1$ to m.

The conservation equation is

$$(Gg_i)_x - (Gg_i)_{x+\Delta x} + (a\,\Delta x \sum_j f_{ij})_{x'} = 0$$

and in the limit as $\Delta x \longrightarrow 0$

$$G\frac{dg_i}{dx} = a \sum_j f_{ij} \tag{3.7.1}$$

The reaction rate expressions f_{ij} for a particular reaction are not independent but are related by the expression

$$\frac{f_{ij}}{a_{ij}} = \frac{f_{kj}}{a_{kj}} = f_j$$

where f_j is the intrinsic rate and is related to f_{ij} by

$$f_{ij} = a_{ij} f_j$$

If this is substituted into Eq. (3.7.1)

$$G\frac{dg_i}{dx} = a \sum_j a_{ij} f_j$$

or

$$w\frac{dg_i}{dx} = \sum_j a_{ij} f_j, \quad i = 1 \text{ to } n \tag{3.7.2}$$

There are n such equations, one for each chemical species including inerts, but obviously this is too many, for we know that, for a single reaction, one such equation will suffice since there are relations connecting the mass concentrations of the species. The matrix of the stoichiometric coefficients is

$$\mathbf{A} = \begin{bmatrix} a_{11} & a_{12} & \cdots & a_{1m} \\ a_{21} & a_{22} & \cdots & a_{2m} \\ \cdot & & & \\ \cdot & & & \\ \cdot & & & \\ a_{n1} & a_{n2} & \cdots & a_{nm} \end{bmatrix}, \quad (n \text{ by } m \text{ matrix})$$

and suppose its rank is r. Suppose for convenience the determinant of order

r which is not zero is in the first r rows. Then there exist constants such that each of the row vectors

$$\mathbf{b}_k^T = [a_{k1} \quad a_{k2} \quad \dots \quad a_{km}], \quad k = r+1, r+2, \dots, n$$

is a linear combination of the first r row vectors $\mathbf{b}_k^T$, $k = 1$ to r. Let these constants be denoted by

$$\begin{aligned} &\beta_{1,r+1}, \quad \beta_{2,r+1}, \dots, \beta_{r,r+1}, \quad \text{for row } r+1 \\ &\beta_{1,r+2}, \quad \beta_{2,r+2}, \dots, \beta_{r,r+2}, \quad \text{for row } r+2 \\ &\quad\vdots \\ &\beta_{1,n}, \quad \beta_{2,n}, \dots, \beta_{r,n}, \quad \text{for row } n \end{aligned}$$

hence

$$\mathbf{b}_k^T = \sum_j \beta_{jk}\mathbf{b}_j^T, \quad k = r+1, r+2, \dots, n \tag{3.7.3}$$

The set of Eq. (3.7.2) may be written in matrix-vector form

$$w\frac{d\mathbf{g}}{dx} = \mathbf{Af} \tag{3.7.4}$$

where

$$\mathbf{g} = \begin{bmatrix} g_1 \\ g_2 \\ \cdot \\ \cdot \\ \cdot \\ g_n \end{bmatrix}, \quad \mathbf{f} = \begin{bmatrix} f_1 \\ f_2 \\ \cdot \\ \cdot \\ \cdot \\ f_m \end{bmatrix}$$

There is, however, a nonsingular square matrix $\mathbf{B}$ such that, when it operates on $\mathbf{A}$, will produce in the last $(n - r)$ rows

$$\mathbf{b}_k^T - \sum_j \beta_{jk}\mathbf{b}_j^T = \mathbf{0}^T$$

Then,

$$w\mathbf{B}\frac{d\mathbf{g}}{dx} = \mathbf{BAf}$$

and the last $(n - r)$ rows of the vector on the left-hand side will be

$$\frac{dg_k}{dx} - \sum_j \beta_{jk}\frac{dg_j}{dx} = 0, \quad k = r+1, r+2, \dots, n$$

and these are differential equations which may be integrated to give

$$g_k - g_{k0} = \sum_j \beta_{jk}(g_j - g_{j0}), \quad k = r+1, r+2, \dots, n \tag{3.7.5}$$

Hence $n - r$ of the mass concentrations may be expressed in terms of the other r. The system of mass conservation differential equations contains only r members and this is the rank of $\mathbf{A}$.

Referring again to the figure, we write an equation for conservation of energy which for our case will be an enthalpy balance. Here

$$(G \sum_i g_i \tilde{h}_i)_x$$

is the rate at which enthalpy flows into the element and

$$(G \sum_i g_i \tilde{h}_i)_{x+\Delta x}$$

is the rate of efflux of enthalpy. The rate at which heat flows into the element through the tube wall is

$$[P \, \Delta x h(t - T)]_{x'}, \quad x < x' < x + \Delta x$$

where t is the ambient temperature and h the over-all heat transfer coefficient. The conservation equation is

$$G(\sum_i g_i \tilde{h}_i)_x - G(\sum_i g_i \tilde{h}_i)_{x+\Delta x} + [Ph(t - T)]_{x'} \, \Delta x = 0, \quad x < x' < x + \Delta x$$

and in the limit as $\Delta x \longrightarrow 0$

$$G \frac{d}{dx} \sum_i g_i \tilde{h}_i - hP(t - T) = 0$$

and

$$G \sum_i \tilde{h}_i \frac{dg_i}{dx} + G \sum_i g_i \frac{d\tilde{h}_i}{dx} - hP(t - T) = 0$$

From Eq. (3.7.4) and since

$$\frac{d\tilde{h}_i}{dx} = C_{pi} \frac{dT}{dx}$$

where C_{pi} is the molar heat capacity, we have

$$a\mathbf{h}^T \mathbf{A}\mathbf{f} + G \frac{dT}{dx} \sum_i g_i C_{pi} - hP(t - T) = 0$$

where

$$\mathbf{h} = \begin{bmatrix} \tilde{h}_1 \\ \tilde{h}_2 \\ \cdot \\ \cdot \\ \cdot \\ \tilde{h}_n \end{bmatrix}$$

and is the species partial molar enthalpy vector. The quantity $\sum_i g_i C_{pi}$ is the molar heat capacity per unit mass of reaction mixture. Call it c_{pa}, then, since the first term is a scalar,

$$a\mathbf{f}^T \mathbf{A}^T \mathbf{h} + G c_{pa} \frac{dT}{dx} - hP(t - T) = 0$$

The quantity $\mathbf{A}^T\mathbf{h}$ is a vector whose components are the heats of reaction as may be directly verified, so

$$\mathbf{A}^T\mathbf{h} = \begin{bmatrix} \Delta H_1 \\ \Delta H_2 \\ \cdot \\ \cdot \\ \cdot \\ \Delta H_m \end{bmatrix} = \mathbf{\Delta H}$$

Thus,

$$a\mathbf{f}^T\,\mathbf{\Delta H} + Gc_{pa}\frac{dT}{dx} - hP(t - T) = 0 \tag{3.7.6}$$

and this is the conservation of energy equation, neglecting kinetic energy, etc.

The momentum balance may be written easily, for in the lumped constant system, one assumes that all the resistance to flow is concentrated at the wall. The momentum balance is

$$(pa)_x - (pa)_{x+\Delta x} + (Gu)_x - (Gu)_{x+\Delta x} - \tau_{x'}P\,\Delta x = 0, \quad x < x' < x + \Delta x$$

and in the limit

$$\frac{dp}{dx} + \rho u\frac{du}{dx} + \tau\frac{P}{a} = 0 \tag{3.7.8}$$

where ρ is the density and p is the pressure.

The friction factor f may be defined in terms of the shearing stress τ by

$$\tau = f\frac{\rho u^2}{8} = f\frac{w^2}{8\rho}$$

and f is a function of the Reynolds number $f = f(\text{Re})$.

We therefore have $r + 2$ differential equations

$$\frac{dT}{dx} = -\frac{h}{wc_{pa}}\frac{P}{a}(T - t) - \frac{1}{wc_{pa}}\mathbf{f}^T\,\mathbf{\Delta H} \tag{3.7.9}$$

$$\frac{dg_i}{dx} = \frac{1}{w}\sum_j a_{ij}f_j, \quad i = 1 \text{ to } r \tag{3.7.10}$$

$$\frac{dp}{dx} + w\frac{du}{dx} = -fP\frac{w^2}{8\rho a} \tag{3.7.11}$$

and the equations

$$w = \rho u \tag{3.7.12}$$

$$M = \left(\sum_i g_i\right)^{-1} \tag{3.7.13}$$

$$pM = \rho RT \tag{3.7.14}$$

where M is the molecular weight of the reaction mixture.

From Eq. (3.7.12), we can write

$$\rho\frac{du}{dx} + u\frac{d\rho}{dx} = 0$$

and from the logarithmic derivative in Eq. (3.7.14)

$$\frac{1}{\rho}\frac{d\rho}{dx} = \frac{1}{p}\frac{dp}{dx} + \frac{1}{M}\frac{dM}{dx} - \frac{1}{T}\frac{dT}{dx}$$

Combining these with Eq. (3.7.11), that equation may be written

$$\left(1 - \frac{w^2 RT}{p^2 M}\right)\frac{dp}{dx} + \frac{w^2 R}{pM}\frac{dT}{dx} = -fP\frac{w^2}{8\rho a} + \frac{wRT}{p}\sum_i \sum_j a_{ij} f_j$$

or the $(r + 2)$ differential equations may be written

$$\frac{dg_i}{dx} = \frac{1}{w}\sum_j a_{ij} f_j \tag{3.7.15}$$

$$\frac{dT}{dx} = \frac{h}{wc_{pa}}\frac{P}{a}(t - T) - \frac{1}{wc_{pa}}\mathbf{f}^T\,\Delta\mathbf{H} \tag{3.7.16}$$

$$\frac{dp}{dx} = \left(1 - \frac{w^2 RT}{p^2 M}\right)^{-1}\left[-fP\frac{w^2}{8\rho a} + \frac{wRT}{p}\sum_i \sum_j a_{ij} f_j - \frac{hwR}{wc_{pa}pM}\frac{P}{a}(T - t) + \frac{wR}{pMc_{pa}}\mathbf{f}^T\,\Delta\mathbf{H}\right] \tag{3.7.17}$$

These must be solved in conjunction with Eqs. (3.7.5), (3.7.13), and (3.7.14). A word must be said about $\mathbf{f}_j$. The functions $\mathbf{f}_j$ are the rate expressions and will be empirical functions of partial pressures, concentrations, mole fractions (or some other measure of the molecular concentration), and temperature and total pressure. If, for example,

$$f_j = f_j(p_1, p_2, \ldots, p_n, T, p)$$

then, since

$$p_i = \frac{g_i p}{\sum g_i}, \quad x_i = \frac{g_i}{\sum g_i}, \quad c_i = \frac{g_i MP}{RT}$$

where p_i, x_i, and c_i are partial pressure, mole fraction, and concentration, respectively, each p_i may be expressed as a function of g_i. Because of Eq. (3.7.5) each f_j may be expressed in terms of r of the set $\{g_i\}$ and all the $\{g_{io}\}$. The right-hand sides of Eqs. (3.7.15), (3.7.16), and (3.7.17) are therefore functions which, although complicated, are amenable to computation.

EXERCISES

1. Consider the set of algebraic equations,

$$\begin{aligned} -x_1 + x_2 - x_3 + x_4 &= -1 \\ 2x_1 + 4x_2 + 2x_3 + 3x_4 &= 1 \\ -5x_2 + 4x_3 + 2x_4 &= -1 \\ 3x_1 - 2x_2 + 3x_3 &= 5 \end{aligned}$$

Write these equations in terms of a vector representation as in Section 3.6 and solve by the method of that section.

2. Find the inverse of the matrix

$$\begin{bmatrix} 1 & -1 & 2 & 1 \\ 0 & 1 & 0 & -1 \\ 2 & -1 & 3 & 1 \\ -1 & -1 & 0 & 3 \end{bmatrix}$$

by the method of Section 3.6.

3. A mixture in chemical equilibrium contains four molecular species M_j each of which contains three different kinds of atoms A_i. The atomic matrix is

	M_1	M_2	M_3	M_4
A_1	1	0	1	0
A_2	0	2	1	1
A_3	2	2	3	1

Determine the number of thermodynamic components and the number of independent reactions.

4. Rayleigh assumed that the viscosity μ of a gas depended only upon the mass of a molecule m, the mean velocity V of a molecule and the coefficient of repulsion K between molecules, where the repulsive force between molecules is of the form $F = Kr^{-n}$ and r is the distance between molecules, or there is a relation of the form

$$f(\mu, K, m, V) = 0$$

If the temperature θ is assumed to be proportional to the kinetic energy $\frac{1}{2}mV^2$ show by dimensional considerations that

$$\mu = \text{c}m^{1/2}K^{2/(1-n)}\theta^s, \quad s = \frac{1}{2} + \frac{2}{n-1}$$

were α is a constant.

5. Assume the pressure p in a gas depends upon the mass m of a molecule, the average molecular velocity V, and the number of molecules n per unit of volume. Devise a formula for the pressure. If in addition one assumes that the absolute temperature θ is proportional to the mean kinetic energy, $\theta = K_1 mV^2$ show that the equation

$$p = \frac{\theta\rho R}{M}$$

results where ρ is the density and M is the molecular weight.

6. Consider the vibration of a drop of fluid. Assume the frequency of vibration depends upon surface tension, mass density, and diameter. Determine the dependence of the frequency on each of these quantities by dimensional arguments.

7. In Section 3.3 the number of independent reactions and thermodynamic components were related to the rank of a certain matrix. Suppose now there are q phases in physical equilibrium each of which contains the n chemical species. Devise a relationship among q, n, m, and r.

8. In a certain experiment in condensation of a vapor on a smooth vertical tube, it is assumed that the pertinent variables are the heat transfer coefficient, h; the temperature difference, $\Delta\theta$, between the pipe wall and the saturation temperature; the thermal conductivity of the condensate, k; the specific weight, ρg; the latent heat of vaporization, $\rho\lambda$, per unit volume of condensate; the viscosity, μ, and the length of the pipe, L. Show that there are *three* dimensionless groups. Show also that a suitable set of independent groups is

$$\frac{h\lambda}{kg}, \quad \frac{k\mu g^2\,\Delta\theta}{\rho^2\lambda^4}, \quad \frac{gL}{\lambda}$$

9. In the pyrolysis of a low molecular weight hydrocarbon, the following species are present: C_2H_6, H, C_2H_5, CH_3, CH_4, H_2, C_2H_4, C_3H_8, C_4H_{10}. Determine the number of independent reactions and number of thermodynamic components.

Linear Programming 4

4.1 Introduction

In many chemical engineering process operations, one discovers that a given product or products can be manufactured by several paths. On the one hand, as in the petroleum industry, a number of basic blending stocks are manufactured and a wide variety of products are produced by blending them. For example, with gasoline there will be a final specification on such things as octane number, volatility, sulfur content, gum content, etc. Some of these may be maximum specifications, such as octane number; whereas others may be minimum specifications, such as sulfur content. Each blending stock will have these properties in definite amounts and each will, in general, have a different cost or price associated with it. The problem then arises as to how much of each blending stock should be mixed into the final product so that all the specifications may be met. The key to the problem of linear programming is the word "met," for it is clear that, since specifications are presented as maximum or minimum restrictions, the final product can be produced in a variety of ways and more importantly at a variety of costs and profits. Linear programming seeks to solve the problem of prescribing an optimal way, of preparing the final product—in this case, for minimum cost or maximum profit.

A second example which we shall consider in some detail is one of Fenech and Acrivos* in which chemical products can be produced by a variety of

*E. J. Fenech and A. Acrivos, The Application of Linear Programming to Design Problems, *Chem. Eng. Sci.*, **58** (1956), 93–98.

processing steps and one seeks to find the path which produces the maximum profit. The problem is posed as follows: In a chemical plant *four* raw materials are available, A, B, C, and D, and five processes are available to produce four products E, F, G, H. The raw materials are in limited supply, the processing costs are fixed but vary from process to process, and the selling prices of the products are fixed. The data for the problem are as follows:

Raw Materials	Maximum Available Supply	Cost per Pound
	(pounds per day)	(dollars)
A	400	1.50
B	300	2.00
C	100	4.50
D	250	2.50

The data on the processes are as follows :

Process	Pounds Raw Material Used				Pounds of Product	Selling Price of Product (dollars per pound)
	A	B	C	D		
1	2	1	0	0	3(E)	3.00
2	2	1	0	0	3(F)	$2.33\frac{1}{3}$
3	3	1	0	2	6(G)	3.75
4	2	7	3	3	15(H)	5.00
5	2	7	3	3	15(H)	5.00

Processes 4 and 5 differ only in processing cost. Process 4 has a maximum capacity of 75 lb of H per day. If more than 75 lb of H are produced, it must be produced in process 5. The processes are

	Process	Reaction Cost (dollars per pound)
1		
	$A + B \longrightarrow E$	1.50(A)
2		
	$A + B \longrightarrow F$	0.50(A)
3		
	$A + B \longrightarrow E$	1.50(A)
	$E + A + D \longrightarrow G$	1.00(G)
4		
	$A + B \longrightarrow F$	0.50(A)
	$F + B + C + D \longrightarrow H$	2.00(H)
5		
	$A + B \longrightarrow F$	0.50(A)
	$F + B + C + D \longrightarrow H$	2.20(H)
(Excess H capacity)		

In order to set the problem we define x_j as the net profit from the jth process per day. Then the total profit per day is

$$P = x_1 + x_2 + x_3 + x_4 + x_5$$

and it may be shown that the restrictions on the processes are

$$\begin{aligned} 2x_1 + 2x_2 + \tfrac{3}{2}x_3 + \tfrac{1}{3}x_4 + \tfrac{2}{3}x_5 &\leq 400 \\ x_1 + x_2 + \tfrac{1}{2}x_3 + \tfrac{7}{6}x_4 + \tfrac{7}{3}x_5 &\leq 300 \\ \tfrac{1}{2}x_4 + x_5 &\leq 100 \\ x_3 + \tfrac{1}{2}x_4 + x_5 &\leq 250 \\ x_4 &\leq 30 \end{aligned}$$

with

$$x_j \geq 0, \quad j = 1, 2, 3, 4, 5$$

Our problem then is to find values of the x_j which will satisfy all the restrictions but make P as large as possible. In this problem, obviously, many x_j will satisfy the restrictions.

Before solving this problem or proceeding to the general case it is instructive to consider an even simpler problem since it will contain the main features of the more complicated cases. Suppose that

$$P = 2x_1 + 3x_2$$

is to be maximized subject to the restrictions,

$$\begin{aligned} x_1 + x_2 &\leq 8 \\ x_1 + 2x_2 &\leq 10 \\ x_1, x_2 &\geq 0 \end{aligned}$$

In two dimensions there is a geometric solution which it is instructive to carry out. In Fig. 4.1.1 we plot the lines

$$\begin{aligned} x_2 &= -x_1 + 8 \\ x_2 &= -\tfrac{1}{2}x_1 + 5 \end{aligned}$$

and it is obvious that we seek values of x_1 and x_2 in the shaded area. Consider also the family of lines

$$2x_1 + 3x_2 = P$$

which is a parallel family with slope $-\frac{2}{3}$, and which moves away from the origin for P increasing. We then seek the line with largest P which contains an x_1 and x_2 which belongs to the shaded region, and certainly this is the one which touches at point S. The solution is clear then, for it requires the simultaneous solution of the straight-line equations to give

$$x_2 = 2 \qquad x_1 = 6$$

and the corresponding value of P is 18.

Note that if the profit function P had the form

$$P = 2x_1 + x_2$$

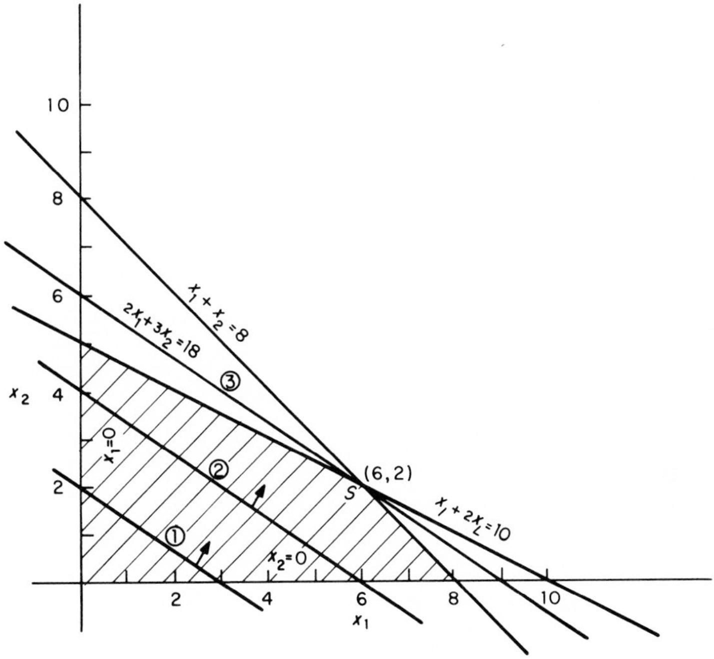

Figure 4.1.1 Two dimensional maximum problem in linear programming.

then the values of x_1 and x_2 which would satisfy the restrictions and maximize P would lie along a whole segment of a boundary of the shaded area rather than have a single point in common. Note two things in particular here: (1) the optimal solution lies *on* the boundary of the shaded region; (2) the shaded region is a *convex* region. This problem may be written in matrix-vector form

$$\max \mathbf{p}^T\mathbf{x}$$
$$\mathbf{A}\mathbf{x} \leq \mathbf{b}$$
$$\mathbf{x} \geq \mathbf{o}$$

where

$$\mathbf{p} = \begin{bmatrix} 2 \\ 3 \end{bmatrix}; \quad \mathbf{A} = \begin{bmatrix} 1 & 1 \\ 1 & 2 \end{bmatrix}; \quad \mathbf{b} = \begin{bmatrix} 8 \\ 10 \end{bmatrix}$$

An inequality in matrices or vectors has the obvious interpretation that corresponding elements satisfy the inequality. From this problem another problem, called the *dual*, may be defined as

$$\min(\mathbf{b}^T\mathbf{y})$$
$$\mathbf{A}^T\mathbf{y} \geq \mathbf{p}$$
$$\mathbf{y} \geq \mathbf{o}$$

or

$$\min(8y_1 + 10y_2)$$
$$y_1 + y_2 \geq 2$$
$$y_1 + 2y_2 \geq 3$$
$$y_1 \geq 0, \quad y_2 \geq 0$$

We can proceed in the same way by considering the two straight lines

$$y_2 = -y_1 + 2$$
$$y_2 = -\tfrac{1}{2}y_1 + \tfrac{3}{2}$$

These are plotted in Fig. 4.1.2, and those values of y_j which satisfy the restrictions must lie in the shaded area. The family of lines

$$8y_1 + 10y_2 = C$$

for C decreasing moves toward the region for positive C; hence the smallest C which contains a y_1 and y_2 in the shaded area touches at S or at the simultaneous solution of the two preceding equations,

$$y_1 = 1 \qquad y_2 = 1$$

and

$$C = 18$$

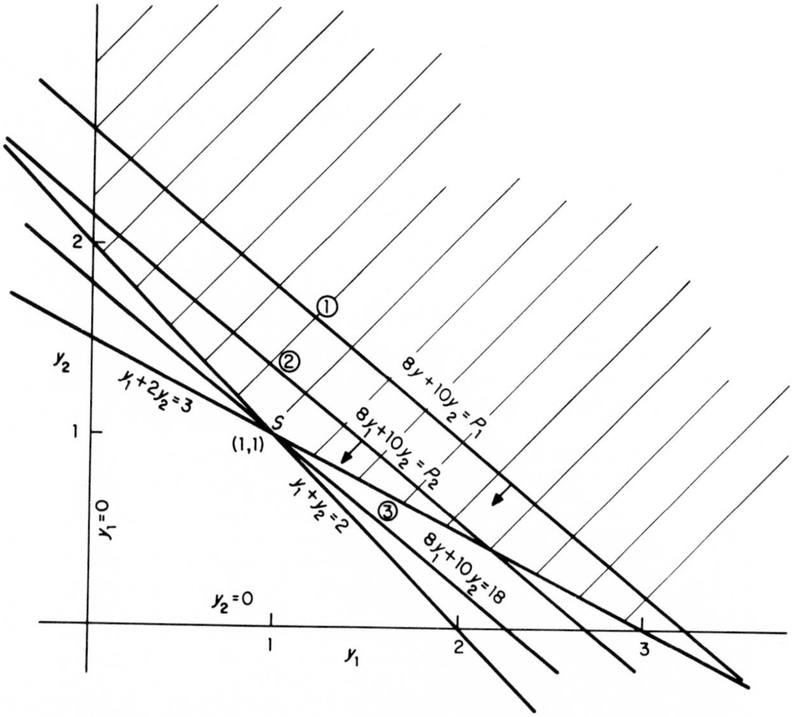

Figure 4.1.2 Two dimensional minimum problem in linear programming.

We note that

$$\max P = \min C$$

and that the solution of the minimum point lies on the boundary of a convex region of possible solutions.

Now we shall show that the solutions of more general problems have the same characteristics; that is, the possible solutions which satisfy the restrictions form a convex region, and the optimal solution always lies on the boundary of the convex region. Further, for each problem there is another problem, called the *dual*, for which the solution is the same in the sense that $\max P = \min C$.

4.2 The Problem and Its Dual

We shall consider first not the most general problem but rather one which has some simpler features. In a later section, we shall consider briefly a more general problem. We consider the linear function

$$P = p_1 x_1 + \cdots + p_n x_n$$

usually called the *profit function* or objective function, and m linear inequality restrictions or restraints

$$\begin{aligned} a_{11}x_1 + a_{12}x_2 + \cdots + a_{1n}x_n &\leq b_1 \\ a_{21}x_1 + a_{22}x_2 + \cdots + a_{2n}x_n &\leq b_2 \\ &\vdots \\ a_{m1}x_1 + a_{m2}x_2 + \cdots + a_{mn}x_n &\leq b_m \\ x_j \geq 0, \quad j = 1, 2, 3, \ldots, n & \end{aligned}$$

where each of the $b_j > 0$, and we ask for those $\{x_j\}$ which satisfy the inequalities and maximize P. This problem may be written in the short form

$$\begin{gathered} \max(\mathbf{p}^T\mathbf{x}) \\ \mathbf{A}\mathbf{x} \leq \mathbf{b}; \qquad \mathbf{b} > \mathbf{o} \\ \mathbf{x} \geq \mathbf{o} \end{gathered} \tag{LP}$$

where $\max(\mathbf{p}^T\mathbf{x})$ means that $\mathbf{p}^T\mathbf{x}$ is to be maximized over all feasible $\mathbf{x}$.

The dual of this problem is defined to be

$$\begin{gathered} \min(\mathbf{b}^T\mathbf{y}) \\ \mathbf{A}^T\mathbf{y} \geq \mathbf{p} \\ \mathbf{y} \geq \mathbf{o} \end{gathered} \tag{DLP}$$

where we ask for the minimum value of C

$$C = b_1 y_1 + b_2 y_2 + \cdots + b_m y_m$$

subject to the restrictions

$$\begin{aligned}
a_{11}y_1 + a_{21}y_2 + \cdots + a_{m1}y_m &\geq p_1 \\
a_{12}y_1 + a_{22}y_2 + \cdots + a_{m2}y_m &\geq p_2 \\
&\vdots \\
a_{1n}y_1 + a_{2n}y_2 + \cdots + a_{mn}y_m &\geq p_n
\end{aligned}$$

$$y_j \geq 0, \quad j = 1, 2, 3, \ldots, m$$

We refer to the first problem as LP (linear program) and the second as DLP (dual linear program). The word *linear* shows that the function P is linear in the variables and the inequalities are linear inequalities. The term *program* shows that the equations describe a "programming of activity." Either the LP or the DLP can occur in economic problems and occasionally both make economic sense in some context. If, however, one makes mathematical sense, in that it has a solution, then it may be shown that both make mathematical sense.

4.3 Transformation of the Problem

We shall consider the LP for the most part and turn to the DLP later. The restrictions may be written as equalities by adding quantities z_j as

$$\begin{aligned}
a_{11}x_1 + a_{12}x_2 + \cdots + a_{1n}x_n + z_1 &= b_1 \\
a_{21}x_1 + a_{22}x_2 + \cdots + a_{2n}x_n + z_2 &= b_2 \\
&\vdots \\
a_{m1}x_1 + a_{m2}x_2 + \cdots + a_{mn}x_n + z_m &= b_m
\end{aligned}$$

and if we let

$$x_j = \lambda_j, \quad j = 1 \text{ to } n$$

$$z_j = \lambda_{j+n}, \quad j = 1 \text{ to } m, \quad m + n = N$$

the inequality restrictions or restraints may be written

$$\begin{aligned}
a_{11}\lambda_1 + a_{12}\lambda_2 + \cdots + a_{1N}\lambda_N &= b_1 \\
a_{21}\lambda_1 + a_{22}\lambda_2 + \cdots + a_{2N}\lambda_N &= b_2 \\
&\vdots \\
a_{m1}\lambda_1 + a_{m2}\lambda_2 + \cdots + a_{mN}\lambda_N &= b_m
\end{aligned}$$

$$\lambda_j \geq 0, \qquad j = 1, 2, 3, \ldots, N$$

where $$a_{1,n+1} = a_{2,n+2} = a_{3,n+3} = \cdots = a_{mN} = 1$$

and all the other a_{ij} which have been introduced in this step are zero. The profit function P may now be written

$$P = p_1\lambda_1 + p_2\lambda_2 + \cdots + p_N\lambda_N = \mathbf{p}^T\boldsymbol{\lambda}$$

where $p_{n+1} = p_{n+2} = \ldots = p_N = 0$. If we now define*

$$\mathbf{P}_j = \begin{bmatrix} a_{1j} \\ a_{2j} \\ a_{3j} \\ \cdot \\ \cdot \\ \cdot \\ a_{mj} \end{bmatrix}; \qquad \mathbf{P}_o = \begin{bmatrix} b_1 \\ b_2 \\ b_3 \\ \cdot \\ \cdot \\ \cdot \\ b_m \end{bmatrix}$$

$$\boldsymbol{\lambda}^T = [\lambda_1\lambda_2\lambda_3 \ldots \lambda_N]$$

$$\mathbf{p}^T = [p_1p_2 \ldots p_N]$$

then the linear program may be written as

$$\mathbf{P}_0 = \sum_{j=1}^{N} \lambda_j\mathbf{P}_j \tag{4.3.1}$$

$$\max (\mathbf{p}^T\boldsymbol{\lambda}) \tag{4.3.2}$$

$$\boldsymbol{\lambda} \geq \mathbf{o} \tag{4.3.3}$$

and we ask what the coefficients must be in the expansion of the vector $\mathbf{P}_0$ in order that $\mathbf{p}^T\boldsymbol{\lambda}$ be a maximum with $\boldsymbol{\lambda} \geq \mathbf{o}$. Phrased in this way, since there are $m + n$ vectors in the expansion of a vector with m components, it is apparent that there can be many vectors $\boldsymbol{\lambda}$ which will satisfy the restraints since only m vectors are required in a basis. In fact it appears that there may be an infinity of expansions of $\mathbf{P}_0$.

Some terminology is needed here. Vectors $\boldsymbol{\lambda}$ which satisfy the restraints are referred to as *feasible solutions*. The feasible solution which maximizes $\mathbf{p}^T\boldsymbol{\lambda}$ is called the *optimal feasible solution.*

4.4 Convex Sets

Consider now a collection of points in a space of N dimensions, where N is arbitrary but fixed. We might think of the points in a sphere, rectangular parallelepiped, triangle, tetrahedron, or any other figure. Let the set of points under consideration be denoted by S. Suppose that $\mathbf{x}$ and $\mathbf{y}$ are two points which are members of S. If all the points on the line segment connecting $\mathbf{x}$ and $\mathbf{y}$ are also members of S, then S is said to be *convex*. A point on the segment connecting $\mathbf{x}$ and $\mathbf{y}$ may be written as

*Up to this point, vectors have been denoted by lower-case letters in bold face. In this chapter we depart slightly from this notation.

$$\mathbf{z} = \mu\mathbf{x} + (1 - \mu)\mathbf{y} \quad 1 \geq \mu \geq 0$$

where $\mu/(1 - \mu)$ is the division ratio of the line segment, as may be verified by a simple geometric construction. Thus the points in a sphere, tetrahedron, triangle, and in fact any nonreentrant body, form a convex set whereas a torus or star does not. A point which lies between two other points is called an *interior point*. The *extreme* points of a convex set are those points of the set which do not lie on a segment between any other two points of the set. The vertices of a triangle or tetrahedron, the surface points of a sphere are all examples of extreme points of convex sets. It seems apparent that, if one knows the extreme points of a convex set, then all the points of the set can be determined by a "filling in" process of drawing line segments between the extreme points and between all points lying on these line segments. For example, for a triangle, if the extreme points are $\mathbf{P}_1$, $\mathbf{P}_2$, $\mathbf{P}_3$, any other point $\mathbf{P}_0$ of the convex set is given by

$$\mathbf{P}_0 = \alpha_1\mathbf{P}_1 + \alpha_2\mathbf{P}_2 + \alpha_3\mathbf{P}_3$$

$$\alpha_1 \geq 0, \quad \alpha_2 \geq 0, \quad \alpha_3 \geq 0$$

$$\alpha_1 + \alpha_2 + \alpha_3 = 1$$

as may be shown by the following simple geometric construction: Suppose $\mathbf{P}_0$ is any point inside the triangle. Then a line from $\mathbf{P}_3$ through $\mathbf{P}_0$ will intersect the line connecting $\mathbf{P}_1$ and $\mathbf{P}_2$ at $\mathbf{P}'$. Clearly

$$\mathbf{P}' = \mu\mathbf{P}_1 + (1 - \mu)\mathbf{P}_2, \quad 1 \geq \mu \geq 0$$

$$\mathbf{P}_0 = \omega\mathbf{P}' + (1 - \omega)\mathbf{P}_3, \quad 1 \geq \omega \geq 0$$

and therefore $$\mathbf{P}_0 = \omega[\mu\mathbf{P}_1 + (1 - \mu)\mathbf{P}_2] + (1 - \omega)\mathbf{P}_3$$

with $$\omega\mu \geq 0, \quad \omega(1 - \mu) \geq 0, \quad 1 - \omega \geq 0$$

$$\omega\mu + \omega(1 - \mu) + 1 - \omega = 1$$

From this we take it as evident that, if the extreme points $\mathbf{P}_1, \mathbf{P}_2, \ldots, \mathbf{P}_k$ of a convex set are finite in number, any other point $\mathbf{P}_0$ may be expressed as

$$\mathbf{P}_0 = \sum_{j=1}^{k} \alpha_j\mathbf{P}_j$$

$$\alpha_j \geqslant 0, \quad \sum_{j=1}^{k} \alpha_j = 1$$

We are thinking here of sets whose boundary surfaces are planes or hyperplanes and this is certainly the case in linear programming problems. The restrictions written in equality form are hyperplanes and the feasible solutions will be those solutions which lie in a region bounded by these hyperplanes. Our first theorem in fact will be that the set of feasible solution vectors $\boldsymbol{\lambda}$ forms a convex set.

If $\boldsymbol{\lambda}_1$ and $\boldsymbol{\lambda}_2$ are two feasible solutions then, from Eq. (4.3.1),

$$\sum_{j=1}^{N} \lambda_{j1}\mathbf{P}_j = \mathbf{P}_0 \quad \text{and} \quad \sum_{j=1}^{N} \lambda_{j2}\mathbf{P}_j = \mathbf{P}_0$$

where λ_{j1} is the jth component of $\boldsymbol{\lambda}_1$, etc. Multiply the first by μ and the second by $1 - \mu$, where $1 \geq \mu \geq 0$, and add, giving

$$\mathbf{P}_0 = \sum_{j=1}^{N} [\mu\lambda_{j1} + (1 - \mu)\lambda_{j2}]\mathbf{P}_j$$

and since $\lambda_{j1} \geq 0$, $\lambda_{j2} \geq 0$, $\mu \geq 0$, $1 - \mu \geq 0$, then

$$\mu\lambda_{j1} + (1 - \mu)\lambda_{j2} \geq 0$$

so

$$\boldsymbol{\lambda} = \mu\boldsymbol{\lambda}_1 + (1 - \mu)\boldsymbol{\lambda}_2$$

is a feasible solution, and the set of feasible solutions is convex. (Note that all solutions and not only the non-negative ones form a convex set.)

4.5 Fundamental Theorems in Linear Programming

As pointed out earlier there may be an infinite number of feasible solutions. From Section 4.4, it is apparent that, in general, all the points of the convex set bounded by the hyperplanes defined by the restraints will be feasible solutions. In order to find the optimal feasible solutions we must find some way to reduce the number of candidates which may be optimal feasible solutions, and it is to this end that we now consider five theorems. These theorems will be proved since they are central to the method of solution known as the *Simplex method* of Dantzig.

THEOREM I: Given the extreme points* $\mathbf{P}_j$, $j = 1, 2, 3, \ldots, k$ of a convex polyhedron and a linear functional

$$f(\boldsymbol{\lambda}) = \sum_{j=1}^{N} \alpha_j\lambda_j = \boldsymbol{\alpha}^T\boldsymbol{\lambda}$$

where $\boldsymbol{\lambda}$ is a point of the convex set

$$(\boldsymbol{\lambda}) = \sum_{j=1}^{k} \mu_j\mathbf{P}_j$$

then

$$f(\boldsymbol{\lambda}) = \sum_{j=1}^{k} \mu_j f(\mathbf{P}_j)$$

Proof: We can write

$$f(\boldsymbol{\lambda}) = \boldsymbol{\alpha}^T\boldsymbol{\lambda} = \boldsymbol{\alpha}^T \sum_{j=1}^{k} \mu_j\mathbf{P}_j$$

$$= \sum_{j=1}^{k} \mu_j\boldsymbol{\alpha}^T\mathbf{P}_j = \sum_{j=1}^{k} \mu_j f(\mathbf{P}_j)$$

THEOREM II: A linear function $f(\boldsymbol{\lambda})$ defined for a point $\boldsymbol{\lambda}$ of a convex set of feasible solutions has a maximum at an extreme point of the set.

Proof: In order to prove the theorem assume that the maximum is *not* assumed at an extreme point and show that a contradiction results. Therefore, assume that the maximum is assumed at a point

*Note that these vectors $\mathbf{P}_j$ are the vectors corresponding to extreme points rather than the coefficients in Eq. (4.3.1).

$$\boldsymbol{\lambda} = \sum_{j=1}^{k} \mu_j \mathbf{P}_j, \quad \mu_j \geqslant 0, \quad \sum_j \mu_j = 1$$

Therefore,

$$\max f(\boldsymbol{\lambda}) = \sum_{j=1}^{k} \mu_j f(\mathbf{P}_j)$$

Now $f(\mathbf{P}_j)$ will vary, depending upon the $\mathbf{P}_j$. Let $f(\mathbf{P}_r)$ be the largest of the $f(\mathbf{P}_j)$, $j = 1, \ldots, k$. Then

$$\max f(\boldsymbol{\lambda}) \leq \sum_{j=1}^{k} \mu_j f(\mathbf{P}_r)$$

since each term is replaced by a term which is as large or larger. Thus

$$\max f(\boldsymbol{\lambda}) \leqslant f(\mathbf{P}_r) \sum_{j=1}^{k} \mu_j = f(\mathbf{P}_r)$$

This says that the max $f(\boldsymbol{\lambda})$ for an interior point is less than or equal to the largest $f(\mathrm{P}_r)$ and hence contradicts the hypothesis that f assumed a maximum at an interior point since P_r is an extreme point. Therefore, the sign can be equals at most, and it follows that the maximum occurs at an extreme point. It follows also that if the maximum occurs at more than one extreme point the same maximum must occur in the convex set formed from these extreme points. (In the introductory problem the maximum could occur along an edge of the convex set if the slopes had been coincidentally chosen.)

This theorem is extremely important since it means that maxima need be sought only at the extreme points of the convex set of feasible solutions. It is apparent that since the convex set is determined by a finite number of hyperplanes, there will, in general, be a finite number of extreme points, hence only the feasible solutions at these extreme points need be examined. A theorem is now needed which will seek out these extreme points. In the expression

$$\sum_{j=1}^{N} \lambda_j \mathbf{P}_j = \mathbf{P}_o$$

the λ_j will be referred to as the *weights* of the $\mathbf{P}_j$.

THEOREM III: A point $\boldsymbol{\lambda} = (\lambda_1, \lambda_2, \ldots, \lambda_N)$ is an extreme point of the convex set of feasible solutions if and only if the $\mathbf{P}_j$ with positive weights ($\lambda_j > 0$) form a linearly independent set among all the $\mathbf{P}_j$, $j = 1, 2, \ldots, N$. The other $\mathbf{P}_j$ will have weights zero.

Proof: The proof must be divided into two parts: the "if" and the "only if." First assume the $\boldsymbol{\lambda}$ is an extreme point with k positive weights $\lambda_1, \lambda_2, \ldots, \lambda_k$, the remainder being zero, then

$$\sum_{j=1}^{k} \lambda_j \mathbf{P}_j = \mathbf{P}_o, \quad \lambda_j > 0 \tag{4.5.1}$$

Suppose that $\mathbf{P}_1, \mathbf{P}_2, \ldots, \mathbf{P}_k$ are linearly dependent, then there exists a set of α_j's not all zero such that

$$\sum_{j=1}^{k} \alpha_j \mathbf{P}_j = \mathbf{o} \tag{4.5.2}$$

Let c be a small positive constant, so small that

$$\lambda_j - c\alpha_j > 0 \quad \text{and} \quad \lambda_j + c\alpha_j > 0, \quad j = 1, 2, \ldots, k$$

Multiply Eq. (4.5.2) through by c and subtract from Eq. (4.5.1), then multiply Eq. (4.5.2) through c and add to Eq. (4.5.1), obtaining

$$\sum_{j=1}^{k} (\lambda_j - c\alpha_j)\mathbf{P}_j = \mathbf{P}_o, \quad \sum_{j=1}^{k} (\lambda_j + c\alpha_j)\mathbf{P}_j = \mathbf{P}_o$$

Therefore, $\boldsymbol{\lambda}_1 = \boldsymbol{\lambda} - c\boldsymbol{\alpha} \qquad \boldsymbol{\lambda}_2 = \boldsymbol{\lambda} + c\boldsymbol{\alpha}$

are each members of the convex set of feasible solutions. Adding these

$$\boldsymbol{\lambda} = \tfrac{1}{2}\boldsymbol{\lambda}_1 + \tfrac{1}{2}\boldsymbol{\lambda}_2$$

But this contradicts the hypothesis that $\boldsymbol{\lambda}$ was an extreme point and therefore the vectors $\mathbf{P}_j$, $j = 1, 2, 3, \ldots, k$, must be linearly independent. This proves that there cannot be more than m $\mathbf{P}_j$'s with positive weights, the remainder having zero weights. To prove the converse of the foregoing, let $\mathbf{P}_j$, $j = 1, 2, \ldots, k$, corresponding to the k positive weights of $\boldsymbol{\lambda}$ (belonging to the convex set of feasible solutions) be linearly independent. Suppose that $\boldsymbol{\lambda}$ is an interior point; that is, there exists a λ_1 and λ_2 such that

$$\boldsymbol{\lambda} = \mu\boldsymbol{\lambda}_1 + (1 - \mu)\boldsymbol{\lambda}_2, \quad 1 \geqslant \mu \geqslant 0$$

with $\boldsymbol{\lambda}_1$ and $\boldsymbol{\lambda}_2$ having positive weights. Then

$$\sum_{j=1}^{k} \lambda_j \mathbf{P}_j = \mathbf{P}_o, \quad \sum_{j=1}^{k} \lambda_{j1} \mathbf{P}_j = \mathbf{P}_o, \quad \sum_{j=1}^{k} \lambda_{j2} \mathbf{P}_j = \mathbf{P}_o$$

By subtraction,

$$\sum_{j=1}^{k} (\lambda_j - \lambda_{j1})\mathbf{P}_j = \mathbf{o}, \quad \sum_{j=1}^{k} (\lambda_j - \lambda_{j2})\mathbf{P}_j = \mathbf{o}$$

which are impossible, since these imply that there exists a set of constants not all zero such that $\sum_{j=1}^{k} \mu_j \mathbf{P}_j = \mathbf{o}$, that is linear dependence. This contradicts the hypothesis of linear independence and therefore λ must be an extreme point.

This theorem is important since it says there cannot be more than

$$\frac{N!}{(m)!\,(n)!}$$

extreme points since there are a maximum of N vectors $\mathbf{P}_j$ of order m and the foregoing expression is the number of possible combinations of N vectors taken m at a time ($N = n + m$).

THEOREM IV: Suppose that $\mathbf{x}$ and $\mathbf{y}$ are feasible solutions of LP and DLP, respectively, then $\mathbf{p}^T\mathbf{x} \leqslant \mathbf{b}^T\mathbf{y}$.

Proof: Since $\mathbf{x}$ is feasible

$$\mathbf{A}\mathbf{x} \leqslant \mathbf{b}$$

and taking transposes

$$\mathbf{x}^T\mathbf{A}^T \leqslant \mathbf{b}^T$$

Postmultiply by $\mathbf{y}$

$$\mathbf{x}^T\mathbf{A}^T\mathbf{y} \leqslant \mathbf{b}^T\mathbf{y}$$

Consider also

$$\mathbf{p} \leqslant \mathbf{A}^T\mathbf{y}$$

and premultiply by $\mathbf{x}^T$

$$\mathbf{x}^T\mathbf{p} \leqslant \mathbf{x}^T\mathbf{A}^T\mathbf{y}$$

then

$$\mathbf{x}^T\mathbf{p} \leqslant \mathbf{x}^T\mathbf{A}^T\mathbf{y} \leqslant \mathbf{b}^T\mathbf{y}$$

Therefore,

$$\mathbf{p}^T\mathbf{x} \leqslant \mathbf{b}^T\mathbf{y}$$

THEOREM V: If there is some $\mathbf{x}$, say $\mathbf{x}_o$, and some $\mathbf{y}$, say $\mathbf{y}_o$, for which $\mathbf{p}^T\mathbf{x}_o = \mathbf{b}^T\mathbf{y}_o$ then

$$\mathbf{p}^T\mathbf{x}_o = \max \mathbf{p}^T\mathbf{x} = \mathbf{b}^T\mathbf{y}_o = \min \mathbf{b}^T\mathbf{y}$$

Proof: Since $\mathbf{p}^T\mathbf{x} \leqslant \mathbf{b}^T\mathbf{y}$, then in particular $\mathbf{p}^T\mathbf{x} \leqslant \mathbf{b}^T\mathbf{y}_o = \mathbf{p}^T\mathbf{x}_o$ so that

$$\mathbf{p}^T\mathbf{x} \leqslant \mathbf{p}^T\mathbf{x}_o$$

therefore

$$\max(\mathbf{p}^T\mathbf{x}) = \mathbf{p}^T\mathbf{x}_o$$

and similarly,

$$\min \mathbf{b}^T\mathbf{y} = \mathbf{b}^T\mathbf{y}_o$$

and $\mathbf{x}_o$ and $\mathbf{y}_o$ are the solutions of LP and DLP.

4.6 The Simplex Method of Dantzig

The theorems in Section 4.5 enable us to solve the linear programming problem when they are combined with the methods for finding the expansion of a vector in terms of a basis discussed in Section 3.5. Theorem II guarantees that only the extreme points of the convex set of feasible solutions are candidates for the maximum of the objective function. Theorem III guarantees that, if in the expansion of $\mathbf{P}_o$ in terms of a linearly independent set of vectors the coefficients are positive, then the coefficients are the components of a vector $\boldsymbol{\lambda}$ which is an extreme point of the convex set of feasible solutions. Now when the linear program is stated as a maximum problem (not its dual which is a minimum problem), it is possible to find a feasible solution immediately by noting that the vectors $\mathbf{P}_{n+1}, \mathbf{P}_{n+2}, \ldots, \mathbf{P}_N$ are the vectors $\mathbf{e}_1, \mathbf{e}_2, \mathbf{e}_3, \ldots, \mathbf{e}_m$ and these form a basis in the m-dimensional space so that if one chooses

$$\boldsymbol{\lambda}^T = [0, 0, \ldots 0, b_1, b_2, b_3, \ldots, b_m]$$

then $\boldsymbol{\lambda}$ satisfies the set of equations

$$\mathbf{P}_o = \sum_{j=n+1}^{N} \lambda_j \mathbf{P}_j$$

since

$$\mathbf{P}_o = b_1\mathbf{e}_1 + b_2\mathbf{e}_2 + \cdots + b_m\mathbf{e}_m \tag{4.6.1}$$

(Note in the definition of our problem in Section 4.2 we have assumed that

b > **o**. This restriction will be removed later.) In this case the profit function is

$$P = \boldsymbol{\lambda}^T \mathbf{p} = [0, 0, 0, \ldots, b_1, b_2, \ldots, b_m] \begin{bmatrix} p_1 \\ p_2 \\ \cdot \\ \cdot \\ \cdot \\ p_n \\ 0 \\ 0 \\ \cdot \\ \cdot \\ \cdot \\ 0 \end{bmatrix} = 0$$

In Section 3.5, however, we devised a scheme whereby the basis could be changed one vector at a time and if we change the basis by removing one vector from it and adding one to it in a manner such that the profit function will always increase, after a finite number of changes in the basis we should arrive at the maximum value of P, since there are only a finite number of possibilities.

We know that the profit function will be a maximum for some set of vectors among the $\mathbf{P}_j$; hence it is essential to examine various combinations of the $\mathbf{P}_j$ until the maximum is reached. Suppose now that at some stage a feasible solution $\boldsymbol{\lambda}$ is at hand with a set of basic vectors $\mathbf{f}_1, \mathbf{f}_2, \ldots, \mathbf{f}_m$ where $\mathbf{f}_i$ is an $\mathbf{e}_j$, $j = 1, 2, \ldots, m$, or a $\mathbf{P}_j$, $j = 1, 2, 3, \ldots, n$. Let $\mathbf{f}_{m+1}, \ldots, \mathbf{f}_N$ then be the remainder of the original N vectors, where $\mathbf{f}_s$ is one of the latter set and $\mathbf{f}_j$ is one of the former set. Now

$$\mathbf{P}_o = \sum_{j=1}^{m} \lambda_j \mathbf{f}_j$$

and

$$\mathbf{f}_s = \sum_{j=1}^{m} \alpha_{js} \mathbf{f}_j \tag{4.6.2}$$

Define a quantity

$$z_s = \sum_{j=1}^{m} \alpha_{js} p_j$$

and let the profit function be denoted by

$$z = \sum_{j=1}^{m} \lambda_j p_j$$

where the p_j in these formulae is the appropriate p_j associated with a particular $\mathbf{f}_j$, $j = 1, 2, \ldots, n + m$. If $\mathbf{f}_j$ is a $\mathbf{P}_i$, then p_j is p_i; if, on the other hand, it is an $\mathbf{e}_k$, then $p_j = 0$.

Suppose now a vector in the set $\mathbf{f}_j$, $j = 1, 2, \ldots, m$ is replaced in the basis by a vector $\mathbf{f}_s$, then

$$P_o = \sum_{j=1}^{m} \lambda_j \mathbf{f}_j - \theta \mathbf{f}_s + \theta \mathbf{f}_s$$

Using Eq. (4.6.2),

$$\mathbf{P}_o = \sum_{j=1}^{m} \lambda_j \mathbf{f}_j - \theta \sum_{j=1}^{m} \alpha_{js} \mathbf{f}_j + \theta \mathbf{f}_s$$
$$= \sum_{j=1}^{m} (\lambda_j - \theta \alpha_{js}) \mathbf{f}_j + \theta \mathbf{f}_s$$

Hence $\mathbf{P}_o$ may be expanded in terms of the new basis—that is, the basis with a particular $\mathbf{f}_j$ removed and $\mathbf{f}_s$ added to the set—if θ is chosen as

$$\theta = \frac{\lambda_j}{\alpha_{js}}$$

Thus one term may be removed from the sum and the new expansion becomes, on changing the summation variable for convenience from j to i,

$$\mathrm{P}_o = \sum_{i=1}^{m} \left(\lambda_i - \frac{\lambda_j}{\alpha_{js}} \alpha_{is}\right) \mathbf{f}_i + \frac{\lambda_j}{\alpha_{js}} \mathbf{f}_s \tag{4.6.3}$$

In the sum the term in which $i = j$ is missing but the last term is substituted in its place.

Now suppose θ has been chosen so that

$$\frac{\lambda_j}{\alpha_{js}} > 0$$

which implies that $\alpha_{js} > 0$ and also so that

$$\lambda_i - \frac{\lambda_j}{\alpha_{js}} \alpha_{is} > 0, \quad \text{for all } i \neq j$$

then the new value of $\boldsymbol{\lambda}$ is again a feasible solution with m positive weights and it follows from Theorem III that the new set of vectors is linearly independent.

Consider now the profit function

$$z = \sum_{j=1}^{m} \lambda_j p_j$$

for the old set of $\mathbf{f}_j$. The profit function for the new set of $\mathbf{f}_j$ will be

$$z' = \sum_{i=1}^{m} \left(\lambda_i - \frac{\lambda_j}{\alpha_{js}} \alpha_{is}\right) p_i + \frac{\lambda_j}{\alpha_{js}} p_s$$
$$= \sum_{i=1}^{m} \lambda_i p_i + \frac{\lambda_j}{\alpha_{js}} p_s - \sum_{i=1}^{m} \frac{\lambda_j}{\alpha_{js}} \alpha_{is} p_i$$
$$= \sum_{i=1}^{m} \lambda_i p_i + \frac{\lambda_j}{\alpha_{js}} p_s - \frac{\lambda_j}{\alpha_{js}} \sum_{i=1}^{m} \alpha_{is} p_i$$
$$= z + \frac{\lambda_j}{\alpha_{js}} (p_s - z_s)$$

Thus z' will be increased for the new basis if $p_s - z_s > 0$. Therefore it is obvious now what the procedure should be. The vector j for removal must be such that

$$(1) \qquad \frac{\lambda_j}{\alpha_{js}} > 0$$

or

$$\alpha_{js} > 0$$

$$(2) \qquad \lambda_i - \frac{\lambda_j}{\alpha_{js}}\,\alpha_{is} > 0, \quad i \neq j$$

or λ_j/α_{js} must be the minimum value among $\alpha_{js} > 0$, and the vector s for addition is chosen so that the greatest improvement in the profit function is obtained. Therefore, the value of s is chosen for which $p_s - z_s$ is the largest.

If all these conditions can be met, a new feasible solution is obtained at an extreme point, and the process may be repeated with successive improvements in the profit function. An iterative procedure may then be used until no further improvement in the profit function is possible. The procedure will terminate certainly when all values of $p_s = z_s$ are negative or zero. It will be shown later that one has at this point an optimal solution. The procedure could also terminate if all $\alpha_{is} < 0$, as follows. Consider

$$z' = \sum_{i=1}^{m} (\lambda_i - \theta\alpha_{is})\,p_i + \theta\,p_s$$

and

$$\mathbf{P}_o = \sum_{i=1}^{m} (\lambda_i - \theta\alpha_{is})\mathbf{f}_i + \theta\mathbf{f}_s$$

In this case any $\mathbf{f}_i$ could be removed and $\mathbf{f}_s$ could be introduced with a resulting feasible vector for $\theta > 0$. Since some $p_s - z_s > 0$ it follows that

$$z' = \sum_{i=1}^{m} \lambda_i p_i + \theta(p_s - z_s)$$

can be made as large as one pleases by choosing θ large and therefore the solution is unbounded and no finite maximum exists.

The method is best described by considering the tableau below.

	$\mathbf{f}_1$	$\mathbf{f}_2$	…	$\mathbf{f}_j$	…	$\mathbf{f}_m$	…	$\mathbf{f}_s$	…	$\mathbf{f}_N$	$\mathbf{P}_o$	p_j
$\mathbf{f}_1$	1	0	…	…	…	0	…	α_{1s}	…	α_{1N}	λ_1	p_1
$\mathbf{f}_2$	0	1	…	…	…	0	…	α_{2s}	…	α_{2N}	λ_2	p_2
⋮	⋮	⋮				⋮		⋮		⋮	⋮	⋮
$\mathbf{f}_j$	0	0	…	1	…	0	…	α_{js}	…	α_{jN}	λ_j	p_j
⋮	⋮	⋮		⋮		⋮		⋮		⋮	⋮	⋮
$\mathbf{f}_m$	0	0	…	0	…	1	…	α_{ms}	…	α_{mN}	λ_m	p_m
p_s	p_1	p_2	…	p_j	…	p_m	…	p_s	…	p_N	—	—
z_s	z_1	z_2	…	z_j	…	z_m	…	z_s	…	z_N	—	—
$p_s - z_s$	$p_1 - z_1$	$p_2 - z_2$		$p_j - z_j$		$p_m - z_m$	…	$p_s - z_s$		$p_N - z_N$	—	—

Now we proceed to determine which vector should be replaced and which should be introduced into the basis.

(1) Examine the row $p_s - z_s$. Choose s so that $p_s - z_s$ is the largest positive number. This fixes s.
(2) Examine the column $\mathbf{f}_s$. Select those elements α_{is} which are greater than zero. Of these elements compute λ_i/α_{is}. Choose the smallest. This fixes the row; Call it the jth row.
(3) Use the method described in Section 3.5 for replacing vectors
   (a) Divide the jth row by α_{js} to produce a "one" in the column $\mathbf{f}_s$ at the jth row.
   (b) Obtain a zero in the ith row of the table under $\mathbf{f}_s$ by multiplying the jth row by α_{is}/α_{js} and subtracting it element by element from the ith row.
(4) Compute z_s for each column s by multiplying in the new tableau

$$z_s = \sum_{i=1}^{m} \alpha_{is} p_i$$

(5) Compute $(p_s - z_s)$ for each column.
(6) If all $p_s - z_s \leqslant 0$, the vector $\boldsymbol{\lambda}$ at this stage is optimal; if not, iterate.

Note that this procedure passes from one extreme point to another until the profit function cannot be improved further *by this method of computation.* After a numerical example, a theorem will be proved which enables one to say definitely that this is the optimal vector.

Example: Maximize

$$P = 2x_1 + 4x_2 + x_3 + x_4$$

Subject to

$$\begin{aligned} x_1 + 3x_2 + x_4 &\leqslant 4 \\ 2x_1 + x_2 \qquad &\leqslant 3 \\ x_2 + 4x_3 + x_4 &\leqslant 3 \end{aligned}$$

Let $\qquad p_1 = 2, \quad p_2 = 4, \quad p_3 = 1, \quad p_4 = 1$

$$\mathbf{P}_1 = \begin{bmatrix} 1 \\ 2 \\ 0 \end{bmatrix}, \quad \mathbf{P}_2 = \begin{bmatrix} 3 \\ 1 \\ 1 \end{bmatrix}, \quad \mathbf{P}_3 = \begin{bmatrix} 0 \\ 0 \\ 4 \end{bmatrix}, \quad \mathbf{P}_4 = \begin{bmatrix} 1 \\ 0 \\ 1 \end{bmatrix}, \quad \mathbf{P}_o = \begin{bmatrix} 4 \\ 3 \\ 3 \end{bmatrix}$$

Let $\qquad x_1 = \lambda_1, \quad x_2 = \lambda_2, \quad x_3 = \lambda_3, \quad x_4 = \lambda_4$

and introduce λ_5, λ_6, and λ_7 such that

$$\mathbf{P}_5 = \begin{bmatrix} 1 \\ 0 \\ 0 \end{bmatrix}, \quad \mathbf{P}_6 = \begin{bmatrix} 0 \\ 1 \\ 0 \end{bmatrix}, \quad \mathbf{P}_7 = \begin{bmatrix} 0 \\ 0 \\ 1 \end{bmatrix}$$

$$p_5 = 0, \quad p_6 = 0, \quad p_7 = 0$$

The tableau may then be written

I.	$\mathbf{P}_1$	$\mathbf{P}_2$	$\mathbf{P}_3$	$\mathbf{P}_4$	$\mathbf{P}_5$	$\mathbf{P}_6$	$\mathbf{P}_7$	$\mathbf{P}_0$	p_j
$\mathbf{P}_5$	1	$\textcircled{3}$	0	1	1	0	0	4	0
$\mathbf{P}_6$	2	1	0	0	0	1	0	3	0
$\mathbf{P}_7$	0	1	4	1	0	0	1	3	0
p_s	2	4	1	1	0	0	0	—	—
z_s	0	0	0	0	0	0	0	0	—
$p_s - z_s$	2	$\boxed{4}$	1	1	0	0	0	—	—

Here we note that a feasible $\boldsymbol{\lambda}$ is $\boldsymbol{\lambda}^T = [0, 0, 0, 0, 4, 3, 3]$. The quantities z_s are zero, since z_s is the product of elements of the last column by elements of the sth column and summed. It is apparent that $s = 2$. We now compute

$$\frac{\lambda_5}{\alpha_{52}}, \quad \frac{\lambda_6}{\alpha_{62}}, \quad \frac{\lambda_7}{\alpha_{72}}$$

These are $\frac{4}{3}$, $\frac{3}{1}$, $\frac{3}{1}$, respectively. Choose $\frac{4}{3}$, so replace $\mathbf{P}_5$ by $\mathbf{P}_2$

II.	$\mathbf{P}_1$	$\mathbf{P}_2$	$\mathbf{P}_3$	$\mathbf{P}_4$	$\mathbf{P}_5$	$\mathbf{P}_6$	$\mathbf{P}_7$	$\mathbf{P}_0$	p_j
$\mathbf{P}_2$	$\frac{1}{3}$	1	0	$\frac{1}{3}$	$\frac{1}{3}$	0	0	$\frac{4}{3}$	4
$\mathbf{P}_6$	$\frac{5}{3}$	0	0	$-\frac{1}{3}$	$-\frac{1}{3}$	1	0	$\frac{5}{3}$	0
$\mathbf{P}_7$	$-\frac{1}{3}$	0	$\textcircled{4}$	$\frac{2}{3}$	$-\frac{1}{3}$	0	1	$\frac{5}{3}$	0
p_s	2	4	1	1	0	0	0	—	—
z_s	$\frac{4}{3}$	4	0	$\frac{4}{3}$	$\frac{4}{3}$	0	0	$\frac{16}{3}$	—
$p_s - z_s$	$\frac{2}{3}$	0	$\boxed{1}$	$-\frac{1}{3}$	$-\frac{4}{3}$	0	0	—	—

The vector $\mathbf{P}_7$ must be replaced. Compute:

$$\frac{\lambda_7}{\alpha_{73}} \quad \text{(one choice)}$$

This is $\frac{5}{12}$

III.	$\mathbf{P}_1$	$\mathbf{P}_2$	$\mathbf{P}_3$	$\mathbf{P}_4$	$\mathbf{P}_5$	$\mathbf{P}_6$	$\mathbf{P}_7$	$\mathbf{P}_0$	p_j
$\mathbf{P}_2$	$\frac{1}{3}$	1	0	$\frac{1}{3}$	$\frac{1}{3}$	0	0	$\frac{4}{3}$	4
$\mathbf{P}_6$	$\textcircled{\frac{5}{3}}$	0	0	$-\frac{1}{3}$	$-\frac{1}{3}$	1	0	$\frac{5}{3}$	0
$\mathbf{P}_3$	$-\frac{1}{12}$	0	1	$\frac{1}{6}$	$-\frac{1}{12}$	0	$\frac{1}{4}$	$\frac{5}{12}$	1
p_s	2	4	1	1	0	0	0	—	—
z_s	$\frac{5}{4}$	4	1	$\frac{3}{2}$	$\frac{5}{4}$	0	$\frac{1}{4}$	$\frac{23}{4}$	—
$p_s - z_s$	$\boxed{\frac{3}{4}}$	0	0	$-\frac{1}{2}$	$-\frac{5}{4}$	0	$-\frac{1}{4}$	—	—

Only $\mathbf{P}_1$ may be introduced into the basis. Compute:

$$\frac{\lambda_2}{\alpha_{21}} \quad \text{and} \quad \frac{\lambda_6}{\alpha_{61}}$$

These are 4 and 1, respectively, so choose $\mathbf{P}_6$ to be removed.

IV.	$\mathbf{P}_1$	$\mathbf{P}_2$	$\mathbf{P}_3$	$\mathbf{P}_4$	$\mathbf{P}_5$	$\mathbf{P}_6$	$\mathbf{P}_7$	$\mathbf{P}_0$	p_j
$\mathbf{P}_2$	0	1	0	$\frac{2}{5}$	$\frac{2}{5}$	$-\frac{1}{5}$	0	1	4
$\mathbf{P}_1$	1	0	0	$-\frac{1}{5}$	$-\frac{1}{5}$	$\frac{3}{5}$	0	1	2
$\mathbf{P}_3$	0	0	1	$\frac{3}{20}$	$-\frac{1}{10}$	$\frac{1}{20}$	$\frac{1}{4}$	$\frac{1}{2}$	1
p_s	2	4	1	1	0	0	0	—	—
z_s	2	4	1	$\frac{27}{20}$	$\frac{11}{10}$	$\frac{9}{20}$	$\frac{1}{4}$	$\frac{13}{2}$	—
$p_s - z_s$	0	0	0	$-\frac{7}{20}$	$-\frac{11}{10}$	$-\frac{9}{20}$	$-\frac{1}{4}$	—	—

and the process can be carried no further since all $p_s - z_s$ are negative. The objective function has been carried from 0 to $5\frac{1}{3}$ to $5\frac{3}{4}$ to $6\frac{1}{2}$. The solution is

$$x_1 = 1, \quad x_2 = 1, \quad x_3 = \tfrac{1}{2}, \quad x_4 = 0$$

The solution of the dual is not straightforward, as we now show. Consider the dual to the foregoing

$$\begin{aligned} \min\,(4y_1 + 3y_2 + 3y_3) \\ y_1 + 2y_2 \geqslant 2 \\ 3y_1 + y_2 + y_3 \geqslant 4 \\ 4y_3 \geqslant 1 \\ y_1 + y_3 \geqslant 1 \\ y_j \geqslant 0 \end{aligned}$$

Now a minimum problem may be changed to a maximum problem by changing the sign of the objective function. By subtracting slack variables, the inequalities may be changed to equalities with the result that the dual may now be written as follows:

$$\begin{aligned} \max\,(-4\lambda_1 - 3\lambda_2 - 3\lambda_3) \\ \lambda_1 + 2\lambda_2 - \lambda_4 = 2 \\ 3\lambda_1 + \lambda_2 + \lambda_3 - \lambda_5 = 4 \\ 4\lambda_3 - \lambda_6 = 1 \\ \lambda_1 + \lambda_3 - \lambda_7 = 1 \\ \lambda_j \geqslant 0 \end{aligned}$$

One may now define vectors $\mathbf{P}_j$ with components as follows:

$\mathbf{P}_1$	$\mathbf{P}_2$	$\mathbf{P}_3$	$\mathbf{P}_4$	$\mathbf{P}_5$	$\mathbf{P}_6$	$\mathbf{P}_7$	$\mathbf{P}_0$
1	2	0	−1	0	0	0	2
3	1	1	0	−1	0	0	4
0	0	4	0	0	−1	0	1
1	0	1	0	0	0	−1	1

This tableau is fundamentally different from that for the original problem for the elements which appear under $\mathbf{P}_0$, $\mathbf{P}_1$, $\mathbf{P}_2$, $\mathbf{P}_3$ are not the expansions in terms of $\mathbf{P}_4$, $\mathbf{P}_5$, $\mathbf{P}_6$, $\mathbf{P}_7$, for if $\mathbf{P}_1$, $\mathbf{P}_2$, $\mathbf{P}_3$, $\mathbf{P}_o$ are expanded in terms of $\mathbf{P}_4$, $\mathbf{P}_5$, $\mathbf{P}_6$, $\mathbf{P}_7$, it is obvious that each element of $\mathbf{P}_1$, $\mathbf{P}_2$, $\mathbf{P}_3$, $\mathbf{P}_o$ will be negative so that a feasible solution will not be readily available as it was before. Therefore the first task in the dual problem is to determine a feasible solution.

4.7 Feasible Solution for the Dual Problem

As indicated in Section 4.6, the difficulty with the solution is in finding an initial feasible vector. Once a feasible solution has been found, the Simplex algorithm may be used to proceed in a step-by-step manner. Our purpose in this section is to show how an initial feasible solution for the modified minimum problem may be found. The DLP of Section 4.2 was written in the form

$$\begin{aligned} &\min(\mathbf{b}^T\mathbf{y}) \\ &\mathbf{A}^T\mathbf{y} \geqslant \mathbf{p} \\ &\quad \mathbf{y} \geqslant \mathbf{o} \end{aligned}$$

and here we assume that $\mathbf{p} > \mathbf{o}$. We shall write it in the modified form

$$\begin{aligned} &\max(-\mathbf{b}^T\mathbf{y}) \\ &\mathbf{A}^T\mathbf{y} \geqslant \mathbf{p} \\ &\quad \mathbf{y} \geqslant \mathbf{o} \end{aligned}$$

The restraints are

$$\begin{aligned} a_{11}y_1 + a_{21}y_2 + \cdots + a_{m1}y_m &\geqslant p_1 \\ a_{12}y_1 + a_{22}y_2 + \cdots + a_{m2}y_m &\geqslant p_2 \\ &\vdots \\ a_{1n}y_1 + a_{2n}y_2 + \cdots + a_{mn}y_m &\geqslant p_n \end{aligned}$$

Subtraction of slack variables changes these inequalities to equalities as

$$\begin{aligned} a_{11}\lambda_1 + a_{21}\lambda_2 + \cdots + a_{N1}\lambda_N &= p_1 \\ a_{12}\lambda_1 + a_{22}\lambda_2 + \cdots + a_{N2}\lambda_N &= p_2 \\ &\vdots \\ a_{1n}y_1 + a_{2n}y_2 + \cdots + a_{Nn}\lambda_N &= p_n \end{aligned} \tag{4.7.1}$$

where $a_{m+1,1} = -1$, $a_{m+2,2} = -1, \ldots, a_{N,n} = -1$ and the other a_{ij} which have been artificially introduced are zero. This system of equations may be written

$$\mathbf{P}_o = \sum_{j=1}^{N} \lambda_j \mathbf{P}_j \tag{4.7.2}$$

where now

$$\mathbf{P}_o^T = [p_1, p_2, \ldots, p_n] \quad \text{and} \quad \mathbf{P}_j^T = [a_{j1}, a_{j2}, \ldots, a_{jn}]$$

In Section 3.5, we developed a scheme for finding the solution of a set of simultaneous linear equations and although the example used was square it is not necessary that this be the case. For example, the set of Eq. (4.7.1) has fewer equations than unknowns. The vector $\mathbf{P}_o$ has the expansion

$$\mathbf{P}_o = \sum_{j=1}^{n} p_j \mathbf{e}_j$$

where $\mathbf{e}_j$ are the unit vectors with n components. In order to find a solution to the set of Eq. (4.7.1) or (4.7.2), we replace the $\mathbf{e}_j$ by the vectors $\{\mathbf{P}_j\}$, one at a time, until *all* are replaced. There is one problem here, however, since although we know what vectors are to be replaced, there are many vectors to serve as replacements. We can write the set of Eq. (4.7.2) in the form

$$\mathbf{P}_o = \lambda_1 \mathbf{P}_1 + \lambda_2 \mathbf{P}_2 + \cdots + \lambda_N \mathbf{P}_N + \Delta_1 \mathbf{e}_1 + \Delta_2 \mathbf{e}_2 + \cdots + \Delta_n \mathbf{e}_n \tag{4.7.3}$$

where the vector $\boldsymbol{\lambda}$ is now one with $(N + n)$ components in it. In order to find an initial feasible solution, we shall now synthesize a *new* linear programming problem by asking for the vector $\boldsymbol{\lambda}$, with $N + n$ components, which satisfies Eq. (4.7.3) and which maximizes the objective function,

$$0\lambda_1 + 0\lambda_2 + \cdots + 0\lambda_N - \Delta_1 - \Delta_2 \ldots \Delta_n$$

or

$$-\sum_{j=1}^{n} \Delta_j$$

subject in addition to the restraints

$$\lambda_j \geqslant 0, \quad j = 1, 2, 3, \ldots, N$$

$$\Delta_j \geqslant 0, \quad j = 1, 2, 3, \ldots, n$$

Now in this linear programming problem we stress that the $\{p_j\}$ in the objective function are different from those in the original dual problem, for we are at this point trying to find only *a* feasible solution. The p_j to be associated with the $\mathbf{P}_j$ are all zero and those associated with the $\mathbf{e}_j$ are -1. If there is a feasible solution to the dual problem, max $\left(-\sum_{j=1}^{n} \Delta_j\right)$ must be zero and the various steps through the algorithm must increase $-\sum_{j=1}^{n} \Delta_j$ through negative values to zero. When this objective function does become zero, we should have a feasible solution. If there is no feasible solution there must be contradictions in the restraints. For this reason the foregoing objective function is frequently called the *redundancy condition*. We now illustrate how this procedure works by showing the various steps in the Simplex method for the example at the end of Section 4.9. The successive tableaus follow.

I.	$\mathbf{P}_1$	$\mathbf{P}_2$	$\mathbf{P}_3$	$\mathbf{P}_4$	$\mathbf{P}_5$	$\mathbf{P}_6$	$\mathbf{P}_7$	$\mathbf{P}_8$	$\mathbf{P}_9$	$\mathbf{P}_{10}$	$\mathbf{P}_{11}$	$\mathbf{P}_o$	p_j
$\mathbf{P}_8$	1	2	0	−1	0	0	0	1	0	0	0	2	−1
$\mathbf{P}_9$	3	1	1	0	−1	0	0	0	1	0	0	4	−1
$\mathbf{P}_{10}$	0	0	4	0	0	−1	0	0	0	1	0	1	−1
$\mathbf{P}_{11}$	(1)	0	1	0	0	0	−1	0	0	0	1	1	−1
p_s	0	0	0	0	0	0	0	−1	−1	−1	−1	—	—
z_s	−5	−3	−6	1	1	1	1	−1	−1	−1	−1	−8	—
$p_s - z_s$	[5]	3	6	−1	−1	−1	−1	0	0	0	0	—	—

One should choose $\mathbf{P}_3$ for introduction to the basis. Suppose for interest one chooses $\mathbf{P}_1$ for introduction. Then

II.	$\mathbf{P}_1$	$\mathbf{P}_2$	$\mathbf{P}_3$	$\mathbf{P}_4$	$\mathbf{P}_5$	$\mathbf{P}_6$	$\mathbf{P}_7$	$\mathbf{P}_8$	$\mathbf{P}_9$	$\mathbf{P}_{10}$	$\mathbf{P}_{11}$	$\mathbf{P}_o$	p_j
$\mathbf{P}_8$	0	2	−1	−1	0	0	1	1	0	0	−1	1	−1
$\mathbf{P}_9$	0	1	−2	0	−1	0	(3)	0	1	0	−3	1	−1
$\mathbf{P}_{10}$	0	0	4	0	0	−1	0	0	0	1	0	1	−1
$\mathbf{P}_1$	1	0	1	0	0	0	−1	0	0	0	1	1	0
p_s	0	0	0	0	0	0	0	−1	−1	−1	−1	—	—
z_s	0	−3	−1	1	1	1	−4	−1	−1	−1	4	−3	—
$p_s - z_s$	0	3	1	−1	−1	−1	[4]	0	0	0	−5	—	—

III.	$\mathbf{P}_1$	$\mathbf{P}_2$	$\mathbf{P}_3$	$\mathbf{P}_4$	$\mathbf{P}_5$	$\mathbf{P}_6$	$\mathbf{P}_7$	$\mathbf{P}_8$	$\mathbf{P}_9$	$\mathbf{P}_{10}$	$\mathbf{P}_{11}$	$\mathbf{P}_o$	p_j
$\mathbf{P}_8$	0	$\frac{5}{3}$	$-\frac{1}{3}$	−1	$\frac{1}{3}$	0	0	1	$-\frac{1}{3}$	0	0	$\frac{2}{3}$	−1
$\mathbf{P}_7$	0	$\frac{1}{3}$	$-\frac{2}{3}$	0	$\frac{1}{3}$	0	1	0	$\frac{1}{3}$	0	−1	$\frac{1}{3}$	0
$\mathbf{P}_{10}$	0	0	(4)	0	0	−1	0	0	0	1	0	1	−1
$\mathbf{P}_1$	1	$\frac{1}{3}$	$\frac{1}{3}$	0	$-\frac{1}{3}$	0	0	0	$\frac{1}{3}$	0	0	$\frac{4}{3}$	0
p_s	0	0	0	0	0	0	0	−1	−1	−1	−1	—	—
z_s	0	$-\frac{5}{3}$	$-\frac{11}{3}$	1	$-\frac{1}{3}$	1	0	−1	$\frac{1}{3}$	−1	0	$-\frac{5}{3}$	—
$p_s - z_s$	0	$\frac{5}{3}$	[$\frac{11}{3}$]	−1	$\frac{1}{3}$	−1	0	0	$-\frac{4}{3}$	0	−1	—	—

IV.	$\mathbf{P}_1$	$\mathbf{P}_2$	$\mathbf{P}_3$	$\mathbf{P}_4$	$\mathbf{P}_5$	$\mathbf{P}_6$	$\mathbf{P}_7$	$\mathbf{P}_8$	$\mathbf{P}_9$	$\mathbf{P}_{10}$	$\mathbf{P}_{11}$	$\mathbf{P}_o$	p_j
$\mathbf{P}_8$	0	($\frac{5}{3}$)	0	−1	$\frac{1}{3}$	$-\frac{1}{12}$	0	1	$-\frac{1}{3}$	$\frac{1}{12}$	0	$\frac{3}{4}$	−1
$\mathbf{P}_7$	0	$\frac{1}{3}$	0	0	$-\frac{1}{3}$	$-\frac{1}{6}$	1	0	$\frac{1}{3}$	$\frac{1}{6}$	−1	$\frac{1}{2}$	0
$\mathbf{P}_3$	0	0	1	0	0	$-\frac{1}{4}$	0	0	0	$\frac{1}{4}$	0	$\frac{1}{4}$	0
$\mathbf{P}_1$	1	$\frac{1}{3}$	0	0	$-\frac{1}{3}$	$\frac{1}{12}$	0	0	$\frac{1}{3}$	$-\frac{1}{12}$	0	$\frac{5}{4}$	0
p_s	0	0	6	0	0	0	0	−1	−1	−1	−1	1	—
z_s	0	$-\frac{5}{3}$	0	1	$-\frac{1}{3}$	$\frac{1}{12}$	0	−1	$\frac{1}{3}$	$-\frac{1}{12}$	0	$-\frac{3}{4}$	—
$p_s - z_s$	0	[$\frac{5}{3}$]	1	−1	$\frac{1}{3}$	$-\frac{1}{12}$	0	0	$-\frac{4}{3}$	$-\frac{11}{12}$	−1	—	—

V.	$\mathbf{P}_1$	$\mathbf{P}_2$	$\mathbf{P}_3$	$\mathbf{P}_4$	$\mathbf{P}_5$	$\mathbf{P}_6$	$\mathbf{P}_7$	$\mathbf{P}_8$	$\mathbf{P}_9$	$\mathbf{P}_{10}$	$\mathbf{P}_{11}$	$\mathbf{P}_o$	p_j
$\mathbf{P}_2$	0	1	0	$-\frac{3}{5}$	$\frac{1}{5}$	$-\frac{1}{20}$	0	$\frac{3}{5}$	$-\frac{1}{5}$	$\frac{1}{20}$	0	$\frac{9}{20}$	0
$\mathbf{P}_7$	0	0	0	$\frac{1}{5}$	$-\frac{2}{5}$	$-\frac{3}{20}$	1	$-\frac{1}{5}$	$\frac{2}{5}$	$\frac{3}{20}$	-1	$\frac{7}{20}$	0
$\mathbf{P}_3$	0	0	1	0	0	$-\frac{1}{4}$	0	0	0	$\frac{1}{4}$	0	$\frac{1}{4}$	0
$\mathbf{P}_1$	1	0	0	$\frac{1}{5}$	$-\frac{2}{5}$	$\frac{1}{10}$	0	$-\frac{1}{5}$	$\frac{2}{5}$	$-\frac{1}{10}$	0	$\frac{22}{20}$	0
p_s	0	0	0	0	0	0	0	-1	-1	-1	-1	—	—
z_s	0	0	0	0	0	0	0	0	0	0	0	0	0
$p_s - z_s$	0	0	0	0	0	0	0	-1	-1	-1	-1	—	—

Therefore a feasible solution for the dual problem is

$$\boldsymbol{\lambda}^T = [\tfrac{22}{20}, \tfrac{9}{20}, \tfrac{1}{4}, 0, 0, 0, \tfrac{7}{20}]$$

Had we chosen $\mathbf{P}_3$ for introduction in the basis originally in Tableau II exactly the same result in Tableau V would be obtained and with the *same* number of steps. Thus the Simplex path just outlined is not necessarily the optimal path.

Consider then the tableau for the dual problem (the original dual problem):

VI.	$\mathbf{P}_1$	$\mathbf{P}_2$	$\mathbf{P}_3$	$\mathbf{P}_4$	$\mathbf{P}_5$	$\mathbf{P}_6$	$\mathbf{P}_7$	$\mathbf{P}_o$	p_j
$\mathbf{P}_2$	0	1	0	$-\frac{3}{5}$	$\frac{1}{5}$	$-\frac{1}{20}$	0	$\frac{9}{20}$	-3
$\mathbf{P}_7$	0	0	0	$\frac{1}{5}$	$-\frac{2}{5}$	$-\frac{3}{20}$	1	$\frac{7}{20}$	0
$\mathbf{P}_3$	0	0	1	0	0	$-\frac{1}{4}$	0	$\frac{1}{4}$	-3
$\mathbf{P}_1$	1	0	0	$\frac{1}{5}$	$-\frac{2}{5}$	$\frac{1}{10}$	0	$\frac{22}{20}$	-4
p_s	-4	-3	-3	0	0	0	0	—	—
z_s	-4	-3	-3	1	1	$\frac{1}{2}$	0	$-\frac{13}{2}$	—
$p_s - z_s$	0	0	0	-1	-1	$-\frac{1}{2}$	0	—	—

Now since all $p_s - z_s \leq 0$, it follows that the optimal vector is the same as the feasible vector. This is not always the case and, in general, the procedure of replacing vectors in the basis until an optimal vector is found must be instituted.

A *very important fact* should be noticed here. Examine Tableau IV of the original problem and Tableau VI of the dual. In Tableau IV of the LP the values of z_s for the slack variables are

$$\tfrac{11}{10}, \tfrac{9}{20}, \text{ and } \tfrac{1}{4}$$

and these are the values of λ_1, λ_2, λ_3 in the DLP. In Tableau VI of the DLP the values of z_s for the slack variables are

$$1, 1, \tfrac{1}{2}, 0$$

Hence in this linear programming problem *the solution of both the original problem and its dual is obtained by the solution of either problem.* In LP the values of z_s for the slack variables will be the solution of the DLP. In the

DLP the values of z_s for the slack variables will be the solution of LP. This result has very important consequences for the numerical computation and will be shown to be a general property of linear programs and their duals by the Simplex method.

4.8 Optimality Criterion

In the Simplex technique the procedure is carried through, replacing one vector by another, each time increasing the profit function until the profit function cannot be increased further by this scheme. The question arises, however, whether there is another combination of vectors which will produce a greater maximum.

Suppose $\boldsymbol{\lambda}$ is an optimal vector associated with the basis $\mathbf{f}_1, \mathbf{f}_2, \ldots, \mathbf{f}_m$ where $\mathbf{f}_i$ is a $\mathbf{P}_j$ or an $\mathbf{e}_j$. Then

$$\mathbf{P}_o = \sum_{j=1}^{m} \lambda_j \mathbf{f}_j$$

Then

$$z = \sum_{j=1}^{m} \lambda_j p_j$$

$$\mathbf{P}_j = \sum_{i=1}^{m} x_{ij} \mathbf{f}_i$$

$$z_j = \sum_{i=1}^{m} x_{ij} p_i$$

and in the optimal tableau $p_j - z_j \leqslant 0$.

Now suppose there is a combination of vectors $\mathbf{g}_1, \mathbf{g}_2, \ldots, \mathbf{g}_m$ with a vector $\boldsymbol{\lambda}'$ which produces a greater maximum, $z' \geqslant z$. Then

$$\mathbf{P}_o = \sum_{j=1}^{m} \lambda'_j \mathbf{g}_j$$

$$z' = \sum_{j=1}^{m} \lambda'_j p_j$$

But each $\mathbf{g}_j$ may be expanded in terms of the foregoing basis so

$$\mathbf{P}_o = \sum_{j=1}^{m} \lambda'_j \left[\sum_{i=1}^{m} x_{ij} \mathbf{f}_i \right]$$

$$= \sum_{i=1}^{m} \mathbf{f}_i \sum_{j=1} \lambda'_j x_{ij}$$

Because of the uniqueness of the expansion of $\mathbf{P}_o$ in terms of a basis, it follows that

$$\sum_{j=1}^{m} \lambda'_j x_{ij} = \lambda_i$$

Consider now the profit function

$$z' = \sum_{j=1}^{m} \lambda'_j p_j$$

But $p_j \leqslant z_j$ and λ_j is non-negative so

$$\begin{aligned} z' &\leqslant \sum_{j=1}^{m} \lambda'_j z_j \\ &\leqslant \sum_{j=1}^{m} \lambda'_j \sum_{i=1}^{m} x_{ij} p_i \\ &\leqslant \sum_{i=1}^{m} p_i \sum_{j=1}^{m} \lambda'_j x_{ij} \end{aligned}$$

but

$$\sum_{j=1}^{m} \lambda'_j x_{ij} = \lambda_i$$

so

$$z' \leqslant \sum_{i=1}^{m} p_i \lambda_i = z$$

and this leads to a contradiction. Thus there is no other combination of vectors which can lead to a higher maximum.

4.9 Duality

Consider now the original problem to find non-negative $\mathbf{x}$ such that

$$\begin{aligned} &\max \mathbf{p}^T \mathbf{x} \\ &\mathbf{A}\mathbf{x} \leqslant \mathbf{b} \\ &\mathbf{x} \geqslant \mathbf{o} \end{aligned} \qquad \text{(LP)}$$

and its dual to find non-negative $\mathbf{y}$ such that

$$\begin{aligned} &\min \mathbf{b}^T \mathbf{y} \\ &\mathbf{A}^T \mathbf{y} \geqslant \mathbf{p} \\ &\mathbf{y} \geqslant \mathbf{o} \end{aligned} \qquad \text{(DLP)}$$

We shall now show that, if LP has a solution, then DLP has a solution and it is the same as LP in the sense that

$$\max \mathbf{p}^T \mathbf{x} = \min \mathbf{b}^T \mathbf{y}$$

and that the slack variables in LP are the components of the optimal solution $\mathbf{y}$ in DLP and conversely.

As before, let $x_1 = \lambda_1, x_2 = \lambda_2, \ldots, x_n = \lambda_n$ and let the slack variables be $\lambda_{n+1}, \lambda_{n+2}, \ldots, \lambda_{n+m}$ so that

$$\mathbf{A}\mathbf{x} + \mathbf{I}\mathbf{s} = \mathbf{b}$$

where

$$\mathbf{s} = \begin{bmatrix} \lambda_{n+1} \\ \lambda_{n+2} \\ \cdot \\ \cdot \\ \cdot \\ \lambda_{n+m} \end{bmatrix}$$

Then a new vector may be written

$$\boldsymbol{\xi} = \begin{bmatrix} \mathbf{X} \\ \mathbf{s} \end{bmatrix} = \begin{bmatrix} \lambda_1 \\ \lambda_2 \\ \cdot \\ \cdot \\ \cdot \\ \lambda_{n+1} \\ \cdot \\ \cdot \\ \cdot \\ \lambda_{n+m} \end{bmatrix}$$

and with the partitioned matrix $\{\mathbf{A}, \mathbf{I}\}$ Eq. (4.9.1) may be written

$$[\mathbf{A}, \mathbf{I}]\boldsymbol{\xi} = \mathbf{b}$$

where the partitioned matrix is the matrix

$$\begin{bmatrix} a_{11} & a_{12} & \cdots & a_{1n} & 1 & 0 & 0 & 0 & 0 \\ a_{21} & a_{22} & \cdots & a_{2n} & 0 & 1 & & \cdots & 0 \\ \cdot & & & & \cdot & \cdot & & & \\ \cdot & & & & \cdot & \cdot & & & \\ \cdot & & & & \cdot & \cdot & & & \\ a_{m1} & a_{m2} & \cdots & a_{mn} & \underbrace{0 \quad 0 \quad \cdots \quad \cdots \quad 1}_{m \text{ columns}} \end{bmatrix}$$

or

$$[\mathbf{P}_1 \mathbf{P}_2 \ldots \mathbf{P}_n \mathbf{e}_1 \mathbf{e}_2 \ldots \mathbf{e}_m] \tag{4.9.2}$$

Suppose now we consider some basis made up of m vectors $\mathbf{f}_1, \mathbf{f}_2, \ldots, \mathbf{f}_m$ such that

$$\mathbf{P}_j = \sum_{i=1}^{m} x_{ij}\mathbf{f}_i$$

$$\mathbf{e}_j = \sum_{i=1}^{m} x'_{ij}\mathbf{f}_i$$

then these equations may be written as a single matrix equation

$$[\mathbf{A}, \mathbf{I}] = [\mathbf{f}_1 \mathbf{f}_2 \ldots \mathbf{f}_m] \begin{bmatrix} x_{11} & x_{12} & \cdots & x_{1n} & x'_{11} & x'_{12} & \cdots & x'_{1m} \\ x_{21} & x_{22} & \cdots & x_{2n} & x'_{21} & x'_{22} & \cdots & x'_{2m} \\ \cdot & & & & & & & \\ \cdot & & & & & & & \\ \cdot & & & & & & & \\ x_{m1} & x_{m2} & \cdots & x_{mn} & x'_{m1} & x'_{m2} & \cdots & x'_{mm} \end{bmatrix}$$

or

$$\begin{aligned} [\mathbf{A}, \mathbf{I}] &= [\mathbf{f}_1 \mathbf{f}_2 \ldots \mathbf{f}_m][\mathbf{X}, \mathbf{Y}] \\ &= \mathbf{B}[\mathbf{X}, \mathbf{Y}] \end{aligned}$$

where $\mathbf{B}$ is the matrix of the basis vectors and where

$$\mathbf{X} = \begin{bmatrix} x_{11} & x_{12} & \cdots & x_{1n} \\ x_{21} & x_{22} & \cdots & x_{2n} \\ \cdot \\ \cdot \\ \cdot \\ x_{m1} & x_{m2} & \cdots & x_{mn} \end{bmatrix}$$

and

$$\mathbf{Y} = \begin{bmatrix} x'_{11} & x'_{12} & \cdots & x'_{1m} \\ x'_{21} & x'_{22} & \cdots & x'_{2m} \\ \cdot \\ \cdot \\ \cdot \\ x'_{m1} & x'_{m2} & \cdots & x'_{mm} \end{bmatrix}$$

Thus,

$$\mathbf{A} = \mathbf{BX} \tag{4.9.3}$$

$$\mathbf{I} = \mathbf{BY} \tag{4.9.4}$$

so that

$$\mathbf{X} = \mathbf{B}^{-1}\mathbf{A} \tag{4.9.5}$$

$$\mathbf{Y} = \mathbf{B}^{-1}\mathbf{I} = \mathbf{B}^{-1} \tag{4.9.6}$$

Now consider the partitioned matrix which is really the first tableau in the Simplex method (written in slightly different form than our previous tableaus).

$$\begin{array}{cccc} \mathbf{P}_o & \mathbf{P}_1\,\mathbf{P}_2 \ldots \mathbf{P}_n & \mathbf{e}_1\,\mathbf{e}_2 \ldots \mathbf{e}_m \\ [\mathbf{b}] & [\quad \mathbf{A} \quad] & [\quad \mathbf{I} \quad] \\ [0] & [\quad \mathbf{p}^T \quad] & [\quad \mathbf{o}^T \quad] \end{array}$$

The zero in the lower left-hand corner is a scalar whereas $\mathbf{o}^T$ in the lower right-hand corner is the transpose of the zero vector with m components. This tableau is the partitioned matrix $\mathbf{E}$ on which certain operations are performed:

$$\mathbf{E} = \begin{bmatrix} \mathbf{b} & \mathbf{A} & \mathbf{I} \\ 0 & \mathbf{p}^T & \mathbf{o}^T \end{bmatrix}$$

At any stage of the Simplex method the tableau has the form

$$\begin{array}{cccc} \mathbf{P}_o & \mathbf{P}_1\,\mathbf{P}_2 \ldots \mathbf{P}_n & \mathbf{e}_1\,\mathbf{e}_2 \ldots \mathbf{e}_m \\ [\boldsymbol{\lambda}^*] & [\quad \mathbf{X} \quad] & [\quad \mathbf{Y} \quad] \\ [\,z\,] & [\quad \boldsymbol{\gamma}^T \quad] & [\quad -\mathbf{w}^T \quad] \end{array}$$

where z is the profit function

$$z = \sum_{j=1}^{m} \lambda_j^* p_j^*$$

and where λ_j^* is the jth component of $\boldsymbol{\lambda}^*$ and is the particular λ_j associated with the vector $\mathbf{f}_j$ in the basis and p_j^* is the profit factor associated with that λ_j. The elements in the vector $\boldsymbol{\gamma}$ are the $z_s - p_s$ for the particular $\mathbf{P}_s$, and $\mathbf{w}$ has as components z_s. Now consider the vectors $\boldsymbol{\gamma}$ and $\mathbf{w}$ in detail.

$$\begin{aligned}\boldsymbol{\gamma}^T &= [z_1 - p_1, z_2 - p_2, \ldots, z_n - p_n] \\ &= \mathbf{z}^T - \mathbf{p}^T \end{aligned} \tag{4.9.7}$$

The value of z_s is

$$z_s = \sum_{i=1}^{m} x_{is} p_i^* = [p_1^* \, p_2^* \, \ldots \, p_m^*] \begin{bmatrix} x_{1s} \\ x_{2s} \\ \cdot \\ \cdot \\ \cdot \\ x_{ms} \end{bmatrix}$$

$$\mathbf{z}^T = [p_1^* \, p_2^* \, \ldots \, p_m^*] \begin{bmatrix} x_{11} & x_{12} & \ldots & x_{1n} \\ x_{21} & x_{22} & \ldots & \\ \cdot & \cdot & & \cdot \\ \cdot & \cdot & & \cdot \\ \cdot & \cdot & & \cdot \\ x_{m1} & x_{m2} & \ldots & x_{mn} \end{bmatrix}$$

$$\mathbf{z}^T = \mathbf{p}^{*T}\mathbf{X}$$

Also

$$-\mathbf{w}^T = [z_{n+1} z_{n+2} \, \ldots \, z_{n+n}]$$

or

$$-\mathbf{w}^T = \mathbf{p}^{*T}\mathbf{Y}$$

Thus we have a method of computation for each of the matrices in the partitioned matrix, which is a tableau at some stage of the Simplex method. Let

$$\mathbf{F} = \begin{bmatrix} \boldsymbol{\lambda}^* & \mathbf{X} & \mathbf{Y} \\ z & \boldsymbol{\gamma}^T & -\mathbf{w}^T \end{bmatrix}$$

We might ask the question: Is it possible to find a matrix $\mathbf{J}$ such that

$$\mathbf{E} = \mathbf{JF}$$

where $\mathbf{J}$ may be partitioned: First, since $\mathbf{E}$ and $\mathbf{F}$ each has two rows and three columns, $\mathbf{J}$ must have two rows and two columns. Let

$$\mathbf{J} = \begin{bmatrix} a & b \\ c & d \end{bmatrix}$$

so

$$\begin{bmatrix} \mathbf{b} & \mathbf{A} & \mathbf{I} \\ 0 & \mathbf{p}^T & \mathbf{o}^T \end{bmatrix} \begin{bmatrix} a & b \\ c & d \end{bmatrix} \begin{bmatrix} \boldsymbol{\lambda}^* & \mathbf{X} & \mathbf{Y} \\ z & \boldsymbol{\gamma}^T & -\mathbf{w}^T \end{bmatrix}$$

Since

$$\mathbf{b} = a\boldsymbol{\lambda}^* + bz \qquad (1)$$

$$\mathbf{A} = a\mathbf{X} + b\boldsymbol{\gamma}^T \qquad (2)$$

$$\mathbf{I} = a\mathbf{Y} - b\mathbf{w}^T \qquad (3)$$

$$0 = c\boldsymbol{\lambda}^* + dz \qquad (4)$$

$$\mathbf{p}^T = c\mathbf{X} + d\boldsymbol{\gamma}^T \qquad (5)$$
$$\mathbf{o}^T = c\mathbf{Y} - d\mathbf{w}^T \qquad (6)$$

some conclusions may be drawn concerning a, b, c, d.

From (4), z and 0 are scalars so d must be a scalar and c must be the transpose of a vector with m elements. Therefore a must be a matrix and b must be a vector with the same number of rows as a. Since a is a matrix, it must have the same number of columns as $\boldsymbol{\lambda}^*$ has rows, that is, m. From (3) since $\mathbf{I}$ is square (m by m), a must be square and have m rows. Therefore,

a is an m by m matrix
b is an m by 1 matrix (vector)
c is a 1 by m matrix (vector transpose)
d is a scalar

Therefore a matrix (partitioned) $\mathbf{J}$ may exist. From Eq. (4.9.3) and (2)

$$a = \mathbf{B}, \quad b = \mathbf{o}$$

This agrees with (1) and (3) since

$$\mathbf{b} = \mathbf{B}\boldsymbol{\lambda}^*$$
$$\mathbf{I} = \mathbf{B}\mathbf{Y} = \mathbf{B}\mathbf{B}^{-1}$$

Consider Eq. (4.9.7)

$$\boldsymbol{\gamma}^T = -\mathbf{p}^T + \mathbf{z}^T = -\mathbf{p}^T - \mathbf{p}^{*T}\mathbf{X}$$

or

$$\mathbf{p}^T = \mathbf{p}^{*T}\mathbf{X} - \boldsymbol{\gamma}^T \tag{4.9.9}$$

From (5)

$$c = \mathbf{p}^{*T}, \quad d = -1 \text{ (scalar)}$$

Note that this agrees with (4) and (6), since

$$z = \mathbf{p}^{*T}\boldsymbol{\lambda}^*$$

and this is the definition of the profit function, and

$$-\mathbf{w}^T = \mathbf{p}^{*T}\mathbf{Y} = \mathbf{p}^{*T}\mathbf{B}^{-1}$$

Thus

$$\begin{bmatrix} \mathbf{b} & \mathbf{A} & \mathbf{I} \\ 0 & \mathbf{p}^T & \mathbf{o}^T \end{bmatrix} = \begin{bmatrix} \mathbf{B} & \mathbf{o} \\ \mathbf{p}^{*T} & -1 \end{bmatrix} \begin{bmatrix} \boldsymbol{\lambda}^* & \mathbf{X} & \mathbf{B}^{-1} \\ z & \boldsymbol{\gamma}^T & -\mathbf{w}^T \end{bmatrix}$$

$\mathbf{J}$ is nonsingular, for its determinant is

$$-|\mathbf{B}|$$

the determinant of $\mathbf{B}$ (which is nonsingular since the rank of $\mathbf{B}$ is m and therefore the rank of $\mathbf{J}$ is $m + 1$) and it has an inverse; therefore,

$$\begin{bmatrix} \boldsymbol{\lambda}^* & \mathbf{X} & \mathbf{B}^{-1} \\ z & \boldsymbol{\gamma}^T & -\mathbf{w}^T \end{bmatrix} = \begin{bmatrix} \mathbf{B} & \mathbf{o} \\ \mathbf{p}^{*T} & -1 \end{bmatrix}^{-1} \begin{bmatrix} \mathbf{b} & \mathbf{A} & \mathbf{I} \\ 0 & \mathbf{p}^T & \mathbf{o}^T \end{bmatrix}$$

Hence one could go to any basis by means of this matrix transformation.

Note that the elements in $\mathbf{p}^{*T}$ will be those profit factors which occur in the chosen basis $\mathbf{B}$.

Now consider the vector $\mathbf{w}^T$ and suppose that LP has a solution where $\mathbf{x}$ is an optimal vector such that $p_s - z_s \leqslant 0$ for all s. The two relations

$$-\mathbf{w}^T = \mathbf{p}^{*T}\mathbf{Y} \tag{4.9.8}$$

$$\mathbf{p}^T = \mathbf{p}^{*T}\mathbf{X} - \boldsymbol{\gamma}^T$$

may be used as follows. Since $\mathbf{Y} = \mathbf{B}^{-1}$

$$\mathbf{w}^T = -\mathbf{p}^{*T}\mathbf{B}^{-1}$$

and

$$\mathbf{w}^T\mathbf{A} = -\mathbf{p}^{*T}\mathbf{B}^{-1}\mathbf{A} \tag{4.9.10}$$

Also

$$\mathbf{p}^T = \mathbf{p}^{*T}\mathbf{X} - \boldsymbol{\gamma}^T = \mathbf{p}^{*T}\mathbf{B}^{-1}\mathbf{A} - \boldsymbol{\gamma}^T \tag{4.9.11}$$

from Eq. (4.9.5). Combine Eq. (4.9.11) and Eq. (4.9.10), obtaining

$$-\mathbf{p}^T = +\,\mathbf{w}^T\mathbf{A} + \boldsymbol{\gamma}^T$$

or

$$-\mathbf{p}^T \geqslant \mathbf{w}^T\mathbf{A}$$

since $\boldsymbol{\gamma}^T$ is a non-negative vector (each component is greater than or equal to zero). Therefore

$$-\mathbf{w}^T\mathbf{A} \geqslant \mathbf{p}^T$$

and taking the transpose of each side

$$\mathbf{A}^T(-\mathbf{w}) \geqslant \mathbf{p} \tag{4.9.12}$$

Consider now the vector $\mathbf{w}$. It may be written

$$-\mathbf{w}^T = [z_{n+1} - p_{n+1},\quad z_{n+2} - p_{n+2},\quad \ldots,\quad z_{n+m} - p_{n+m}]$$

since $p_{n+1}, p_{n+2}, \ldots, p_{n+m}$ are all zero, so

$$-\mathbf{w} \geqslant 0$$

since in an optimal tableau

$$z_s - p_s \geqslant 0 \tag{4.9.13}$$

Thus the vector $-\mathbf{w}$ *is a feasible solution to the DLP.*

We now show that it is an optimum feasible vector. Consider the relation given in Eq. (4.9.8):

$$-\mathbf{w}^T = \mathbf{p}^{*T}\mathbf{B}^{-1}$$

Then

$$-\mathbf{w}^T\mathbf{b} = \mathbf{p}^{*T}\mathbf{B}^{-1}\mathbf{b}$$

But

$$\mathbf{b} = \mathbf{B}\boldsymbol{\lambda}^*$$

for $\boldsymbol{\lambda}^*$ an optimal solution of LP. Thus

$$-\mathbf{w}^T\mathbf{b} = \mathbf{p}^{*T}\mathbf{B}^{-1}\mathbf{B}\boldsymbol{\lambda}^*$$

$$= \mathbf{p}^{*T}\boldsymbol{\lambda}^*$$

or taking the transpose of the left-hand side

$$\mathbf{b}^T(-\mathbf{w}) = \mathbf{p}^{*T}\boldsymbol{\lambda}^*$$

since these are scalars. Note that the right-hand side is max $(\mathbf{p}^T\boldsymbol{\lambda})$. Thus we have shown that there exists a vector $(-\mathbf{w})$ which is a feasible solution of the DLP, and has for a value the value of LP. But from Theorems IV and V of Section 4.5 it follows that this vector $-\mathbf{w}$ is the optimal solution of the dual problem. Hence we have shown that if LP, the direct problem, has a solution, its dual also has a solution, and is given by the vector

$$\mathbf{y}^T = [z_{n+1}, \quad z_{n+2}, \quad \ldots, \quad z_{n+m}]$$

where these are the z_s's calculated for the slack variables of the direct problem.

It will be shown that the same holds for the DLP. The dual problem will be put in the maximum form. It then becomes a problem to determine a non-negative vector $\mathbf{y}$ such that

$$\max\,(-\mathbf{b}^T\mathbf{y})$$
$$\mathbf{A}^T\mathbf{y} \geqslant \mathbf{p}$$
$$\mathbf{y} \geqslant \mathrm{o}$$

The new matrix transformation is of the form

$$\begin{bmatrix} \mathbf{p} & \mathbf{A}^T & -\mathbf{I} \\ 0 & -\mathbf{b}^T & \mathbf{o} \end{bmatrix} = \begin{bmatrix} \mathbf{B} & \mathbf{0} \\ -\mathbf{b}^{*T} & -1 \end{bmatrix} \begin{bmatrix} \boldsymbol{\lambda}^* & \mathbf{X} & -\mathbf{B}^{-1} \\ z & \boldsymbol{\gamma}^T & -\mathbf{w}^T \end{bmatrix}$$

where, although the same letters are used, the quantities **I**, **X**, **B**, etc., are matrices with n rows whereas in LP they each had m rows. Then

$$\mathbf{p} = \mathbf{B}\boldsymbol{\lambda}^*$$
$$\mathbf{A}^T = \mathbf{B}\mathbf{X}$$
$$\mathbf{I} = \mathbf{B}\mathbf{B}^{-1}$$
$$z = -\mathbf{b}^{*T}\boldsymbol{\lambda}^*$$
$$\mathbf{b}^T = \mathbf{b}^{*T}\mathbf{X} + \boldsymbol{\gamma}^T$$
$$\mathbf{w}^T = -\mathbf{b}^{*T}\mathbf{B}^{-1}$$

Then $$\mathbf{w}^T\mathbf{p} = -\mathbf{b}^{*T}\mathbf{B}^{-1}\mathbf{p} = -\mathbf{b}^{*T}\mathbf{B}^{-1}\mathbf{B}\boldsymbol{\lambda}^* = -\mathbf{b}^{*T}\boldsymbol{\lambda}^*$$

and $$-\mathbf{w}^T\mathbf{p} = \min\,(\mathbf{b}^T\boldsymbol{\lambda}) \tag{4.9.14}$$

Since $b_{m+1}, b_{m+2}, \ldots, b_{m+n}$ are all zero,

$$-\mathbf{w}^T = [z_{m+1} + b_{m+1}, \quad z_{m+2} + b_{m+2}, \quad \ldots, \quad z_{m+n} + b_{m+n}]$$

and because $z_{m+j} + b_{m+j} \geqslant 0$, $j = 1$ to n, since they are in an optimal tableau, it follows that

$$-\mathbf{w}^T \geqslant \mathbf{o}^T \tag{4.9.15}$$

Consider now

$$\mathbf{b}^T = \mathbf{b}^{*T}\mathbf{X} + \boldsymbol{\gamma}^T = \mathbf{b}^{*T}\mathbf{B}^{-1}\mathbf{A}^T + \boldsymbol{\gamma}^T$$
$$= -\mathbf{w}^T\mathbf{A}^T + \boldsymbol{\gamma}^T$$

Since $\gamma^T \geq 0$

$$\mathbf{b}^T \geqslant -\mathbf{w}^T \mathbf{A}^T$$

Taking the transpose of each side

$$\mathbf{A}^T(-\mathbf{w}) \leqslant \mathbf{b} \tag{4.9.16}$$

From Eq. (4.9.15) and (4.9.16) **w** is a feasible vector for LP; from Theorems IV and V, Eq. (4.9.14) indicates that (−**w**) is an optimal vector for LP. Hence it has been shown that, if either LP or DLP has an optimal solution, the other does; and the solution of both problems is obtained from the solution of either, the values of z_s for the slack variables being the values of the original variables in the soultion of the dual problem.

4.10 The General Problem of Linear Programming

In the problems considered up to this point the inequalities appearing in the constraints have been unidirectional. In more general formulations, this will not be the case, the restraints appearing as equalities and inequalities in both directions. Suppose now the restraints are written so that $b_j \geqslant 0$, which is always possible by a trivial change in sign. The general problem will then have the form: to find a non-negative vector **x** such that

$$\max(\mathbf{p}^T\mathbf{x})$$

$$a_{j1}x_1 + a_{j2}x_2 + \cdots + a_{jn}x_n = \begin{cases} \leqslant b_j, & 1 \leqslant j \leqslant m_1 \\ = b_j, & m_1 + 1 \leqslant j \leqslant m_2 \\ \geqslant b_j, & m_2 + 1 \leqslant j \leqslant m \end{cases}$$

$$x_j \geqslant 0, \quad j = 1 \text{ to } n$$

It will then be necessary to append slack variables to the first m_1 inequations and to the last $(m - m_2)$ which adds m_1 unit vectors to the system and $(m - m_2)$ negative unit vectors. It is then necessary to add $(m_2 - m_1)$ and $(m - m_2)$ auxiliary variables which will introduce the same number of unit vectors. The system now contains m unit vectors which will form a basis. If the auxiliary variables are denoted by Δ_j then the original problem will have a solution, provided that the maximum value of

$$-\sum_{j=m_1+1}^{m} \Delta_j$$

is zero; that is, that the auxiliary variables can be removed from the system. In this problem the conventional Simplex method is used with profit factors −1 associated with each of the auxiliary variables and zero with the slack and regular variables. If this problem has a solution *a* feasible vector will be generated which may then be used as a starting point for the problem posed above.

In most problems it is necessary to generate such a feasible vector by this scheme since the restraints may be of mixed type as previously cited. It is usually useful to do this anyway since redundancies and incompatibilities in the system will manifest themselves in the inability to generate such a vector.

In computations it is clear that one has a choice in the problem to solve numerically, either LP or DLP, for one problem might be considerably easier than the other. It is said that the number of iterations in a typical program is of order $2m$ and this would indicate that the problem with the minimum number of rows should be solved. If a series of problems must be solved in which the restraints b_j are changed, then it is apparent that the dual problem should be solved since after the first solution a feasible vector is available for succeeding problems. Changes in the profit function do not affect the feasibility of a vector.

4.11 Matrix Solution of the Linear Programming Problem

In Section 4.9 it was shown that the Simplex method could be characterized by a certain matrix transformation

$$\mathbf{E} = \mathbf{J}\mathbf{F} \tag{4.11.1}$$

This formula has in it an implicit method for the computation which is useful and is similar to the modified Simplex method of Dantzig.

Suppose $\mathbf{J}_+$ is the transformation matrix defined for some other basis $\mathbf{B}_+$ which increases the profit function, then

$$\mathbf{E} = \mathbf{J}_+\mathbf{F}_+ \tag{4.11.2}$$

where

$$\mathbf{J}_+ = \begin{bmatrix} \mathbf{B}_+ & \mathbf{o} \\ \mathbf{p}_+^{*T} & -1 \end{bmatrix}$$

and

$$\mathbf{F}_+ = \begin{bmatrix} \boldsymbol{\lambda}_+^* & \mathbf{X}_+ & \mathbf{B}_+^{-1} \\ z_+ & \boldsymbol{\gamma}_+^T & -\mathbf{w}_+^T \end{bmatrix}$$

If one combines Eq. (4.11.1) and (4.11.2), it follows that

$$\mathbf{J}\mathbf{F} = \mathbf{J}_+\mathbf{F}_+$$

or

$$\mathbf{F}_+ = \mathbf{J}_+^{-1}\mathbf{J}\mathbf{F}$$

Hence, given a tableau in matrix form $\mathbf{F}$, the transformation

$$\mathbf{J}_+^{-1}\mathbf{J}$$

will transform $\mathbf{F}$ to $\mathbf{F}_+$. Consider now the calculation of $\mathbf{J}_+^{-1}$

$$\begin{bmatrix} \mathbf{B}_+ & \mathbf{o} \\ \mathbf{p}_+^{*T} & -1 \end{bmatrix} \begin{bmatrix} a & b \\ c & d \end{bmatrix} = \begin{bmatrix} \mathbf{I}_m & \mathbf{0} \\ \mathbf{o}^T & \mathbf{I}_1 \end{bmatrix}$$

where $\mathbf{I}_m$ is the idem matrix of order m, then

$$\begin{bmatrix} a & b \\ c & d \end{bmatrix}$$

is the matrix $\mathbf{J}_+^{-1}$ and

$$\mathbf{B}_+ a = \mathbf{I}_m$$

$$\mathbf{B}_+ b = \mathbf{0}$$

$$\mathbf{p}_+^{*T} a - c = \mathbf{o}^T$$

$$\mathbf{p}_+^{*T} b - d = \mathbf{I}_1 = 1$$

Therefore

$$a = \mathbf{B}_+^{-1}$$

$$b = \mathbf{o} \text{ (since } \mathbf{B}_+ \text{ is nonsingular)}$$

$$c = \mathbf{p}_+^{*T}\mathbf{B}_+^{-1}$$

$$d = -1$$

and

$$\mathbf{J}_+^{-1} = \begin{bmatrix} \mathbf{B}_+^{-1} & \mathbf{o} \\ \mathbf{p}_+^{*T}\mathbf{B}_+^{-1} & -1 \end{bmatrix}$$

Thus

$$\mathbf{J}_+^{-1}\mathbf{J} = \begin{bmatrix} \mathbf{B}_+^{-1} & \mathbf{o} \\ \mathbf{p}_+^{*T}\mathbf{B}^{-1} & -1 \end{bmatrix} \begin{bmatrix} \mathbf{B} & \mathbf{o} \\ \mathbf{p}^{*T} & -1 \end{bmatrix}$$

$$= \begin{bmatrix} \mathbf{B}_+^{-1}\mathbf{B} & \mathbf{0} \\ \mathbf{p}_+^{*T}\mathbf{B}_+^{-1}\mathbf{B} - \mathbf{p}^{*T} & 1 \end{bmatrix}$$

$\mathbf{F}_+$ then will have as elements

$$\boldsymbol{\lambda}_+^* = \mathbf{B}_+^{-1}\mathbf{B}\boldsymbol{\lambda}^* \tag{4.11.3}$$

$$\mathbf{X}_+ = \mathbf{B}_+^{-1}\mathbf{B}\mathbf{X}$$

$$\mathbf{B}_+^{-1} = \mathbf{B}_+^{-1}\mathbf{B}\mathbf{B}^{-1} = \mathbf{B}_+^{-1}$$

$$z_+ = [\mathbf{p}_+^{*T}\mathbf{B}_+^{-1}\mathbf{B} - \mathbf{p}^{*T}]\boldsymbol{\lambda}^* + z \tag{4.11.4}$$

$$\gamma_+^T = (\mathbf{p}_+^{*T}\mathbf{B}_+^{-1}\mathbf{B} - \mathbf{p}^{*T})\mathbf{X} + \boldsymbol{\gamma}^T \tag{4.11.5}$$

$$-\mathbf{w}_+^T = (\mathbf{p}_+^{*T}\mathbf{B}_+^{-1}\mathbf{B} - \mathbf{p}^{*T})\mathbf{B}^{-1} - \mathbf{w}^T \tag{4.11.6}$$

In these formulae it is apparent that the quantity $\mathbf{B}_+^{-1}\mathbf{B}$ is the important one. Consider now the matrix $\mathbf{F}$. Examine the vector $\boldsymbol{\gamma}^T$ and let the index of that component which is largest numerically and negative be s. This will designate a column vector not in $\mathbf{B}$ but certainly in $\mathbf{X}$ and will have the effect of increasing the profit function. Examine the elements of this vector and eliminate those that are negative. Form the ratio of corresponding elements of $\boldsymbol{\lambda}^*$ and positive elements x_{is} of the s vector above and designate the subscript of the component of $\boldsymbol{\lambda}^*$ such that λ_i/x_{is} is a minimum as r. Suppose this λ_r corresponds to the jth column of **B**. Then the new basis $\mathbf{B}_+$ is the same as the old basis except that the jth column has been replaced by the s vector (which was not in **B** but which was in **X**).

Let **B** be a matrix $[\mathbf{x}_{ij}]$ with m rows and m columns with $\mathbf{P}_j$ as the jth column. Now the jth column may be replaced by a column vector

$$\mathbf{v} = \begin{bmatrix} c_1 \\ c_2 \\ \cdot \\ \cdot \\ \cdot \\ c_m \end{bmatrix}$$

by the operation

$$\mathbf{B} + (\mathbf{v} - \mathbf{p}_j)\mathbf{e}_j^T = \mathbf{B}_+$$

since $(\mathbf{v} - \mathbf{P}_j)\,\mathbf{e}_j^T$ is a matrix having as elements in the jth column $c_i - x_{ij}$, $i = 1$ to m, and zeroes in all other positions.

Now in Householder* the inverse of $\mathbf{B}_+$ may be written

$$\begin{aligned} \mathbf{B}_+^{-1} &= [\mathbf{B} + (\mathbf{v} - \mathbf{p}_j)\mathbf{e}_j^T]^{-1} \\ &= \mathbf{B}^{-1} - [\mathbf{B}^{-1}(\mathbf{v} - \mathbf{p}_j)][\mathbf{e}_j^T\mathbf{B}^{-1}](1 + \mathbf{e}_j^T\mathbf{B}^{-1}(\mathbf{v} - \mathbf{p}_j))^{-1} \end{aligned}$$

The last term in parentheses is a scalar, $1 + \beta$, where β is the jth element of $\mathbf{B}^{-1}\,(\mathbf{v} - \mathbf{P}_j)$. But $\mathbf{B}^{-1}\,\mathbf{P}_j$ is the jth column of the unit matrix and the jth element is one, that is,

$$\mathbf{e}_j^T\mathbf{B}^{-1}\mathbf{P}_j = 1$$

so

$$1 + \mathbf{e}_j^T\mathbf{B}^{-1}(\mathbf{v} - \mathbf{p}_j) = \mathbf{e}_j^T\mathbf{B}^{-1}\mathbf{v} = \Delta$$

and

$$\mathbf{B}_+^{-1} = \mathbf{B}^{-1} - [\mathbf{B}^{-1}(\mathbf{v} - \mathbf{p}_j)][\mathbf{e}_j^T\mathbf{B}^{-1}]\Delta^{-1}$$

Now suppose the vector to be introduced into the basis is known and that to be removed is also known. Let $\mathbf{P}_j$ be removed and $\mathbf{v}$ introduced. Since **B** is known,

(1) Compute $\mathbf{B}^{-1}$
(2) Form the row vector $\mathbf{e}_j^T\mathbf{B}^{-1}$
(3) Form the vector $(\mathbf{v} - \mathbf{P}_j)$
(4) Compute the matrix $(\mathbf{v} - \mathbf{P}_j)\,[\mathbf{e}_j^T\mathbf{B}^{-1}]$
(5) Compute the matrix $\mathbf{B}^{-1}\,(\mathbf{v} - \mathbf{P}_j)\,(\mathbf{e}_j^T\mathbf{B}^{-1}) = \mathbf{D}$ and multiply by Δ^{-1}
(6) Compute $\mathbf{B}_+^{-1} = \mathbf{B}^{-1} - \mathbf{D}\,\Delta^{-1}$

Thus from Eq. (4.11.3) $\boldsymbol{\lambda}_+^*$ may be easily calculated as well as $\mathbf{X}_+$. From Eq. (4.11.3),

$$z_+ = \mathbf{p}_+^*\boldsymbol{\lambda}_+^*$$

From Eq. (4.11.5),

$$\boldsymbol{\gamma}_+^T = \mathbf{p}_+^{*T}\mathbf{X}_+ - \mathbf{p}^T$$

*Alton S. Householder, *Principles of Numerical Analysis*, New York: McGraw-Hill, 1953, p. 79.

and from Eq. (4.11.6)

$$-\mathbf{w}_{+}^{T} = \mathbf{p}_{+}^{*T}\mathbf{B}_{+}^{-1}$$

Hence one can use a complete matrix method to determine the optimal solution.

4.12 Degeneracy and Its Solution

Degeneracy arises in a problem when the vector P_o may be expressed at any stage in terms of less than m linear independent vectors. If at some stage there are $m - 1$ non-zero λ_j's or less then in the next stage the profit function will not be increased since the value of θ in this next stage will certainly be zero. In the previous cases, that is, those in which there were m λ_j's, the profit function always increased and thus one could be sure that the iterative procedure never returned to a previous stage because of the uniqueness of the expansion in terms of the $\mathbf{P}_j$. If the profit function does not increase, no such assurance is possible and examples have been constructed in which it is possible to get into a computational loop.

The technique for rectifying this difficulty is to perturb the vector $\mathbf{P}_o$ in such a way that there are always m positive λ_j's during the course of the calculation. Let the perturbed vector be $\mathbf{P}_o(\epsilon)$ where ϵ is the perturbation parameter, and where $\mathbf{P}_o(\epsilon)$ is defined as

$$\mathbf{P}_o(\epsilon) = \mathbf{P}_o + \sum_{j=1}^{N} \epsilon^j \mathbf{P}_j$$

It will be shown during the course of the computation that there are always m non-zero λ's and, further, that using a perturbation of this form has computational advantages not present in other possible perturbations. It will be shown also that ϵ is a "phantom" variable in the sense that it really never needs to be specified or written down.

Consider the following example. To maximize

$$x_1 - x_2$$

subject to

$$2x_1 - x_2 \leqslant 4$$
$$x_1 - 2x_2 \leqslant 2$$
$$x_1 + x_2 \leqslant 5$$

Adding slack variables x_3, x_4, x_5, the problem becomes

$$2x_1 - x_2 + x_3 = 4$$
$$x_1 - 2x_2 + x_4 = 2$$
$$x_1 + x_2 + x_5 = 5$$

Then $p_1 = 1$, $p_2 = -1$, $p_3 = p_4 = p_5 = 0$. The vectors are

$\mathbf{P}_0$	$\mathbf{P}_1$	$\mathbf{P}_2$	$\mathbf{P}_3$	$\mathbf{P}_4$	$\mathbf{P}_5$
4	2	−1	1	0	0
2	1	−2	0	1	0
5	1	1	0	0	1

where we have varied our original tableaus by writing the vector $\mathbf{P}_0$ at the left. The vector $\mathbf{P}_0(\epsilon)$ has components

$$4 + 2\epsilon - \epsilon^2 + \epsilon^3$$
$$2 + \epsilon - 2\epsilon^2 + \epsilon^4$$
$$5 + \epsilon + \epsilon^2 + \epsilon^5$$

and the new perturbed tableau is

I.	$\mathbf{P}_0(\epsilon)$	$\mathbf{P}_1$	$\mathbf{P}_2$	$\mathbf{P}_3$	$\mathbf{P}_4$	$\mathbf{P}_5$	p_j
$\mathbf{P}_3$	$4 + 2\epsilon - \epsilon^2 + \epsilon^3$	2	−1	1	0	0	0
$\mathbf{P}_4$	$2 + \epsilon - 2\epsilon^2 + \epsilon^4$	①	−2	0	1	0	0
$\mathbf{P}_5$	$5 + \epsilon + \epsilon^2 + \epsilon^5$	1	1	0	0	1	0
p_s	—	1	−1	0	0	0	—
z_s	0	0	0	0	0	0	—
$p_s - z_s$	—	$\boxed{1}$	−1	0	0	0	—

From this tableau the vector $\mathbf{P}_1$ must be introduced. To determine which vector is to be removed one sees first that all of x_{31}, x_{41}, x_{51} are positive so that one computes the ratios

$$\frac{4 + 2\epsilon - \epsilon^2 + \epsilon^3}{2} = 2 + \epsilon - \frac{\epsilon^2}{2} + \frac{\epsilon^3}{2}$$

$$\frac{2 + \epsilon - 2\epsilon^2 + \epsilon^4}{1} = 2 + \epsilon - 2\epsilon^2 + \epsilon^4$$

$$\frac{5 + \epsilon + \epsilon^2 + \epsilon^5}{1} = 5 + \epsilon + \epsilon^2 + \epsilon^5$$

and considers which is the smallest. If one thinks of ϵ as a small positive number, then the second is the smallest and $\mathbf{P}_4$ is removed. The new tableau is

II.	$\mathbf{P}_0(\epsilon)$	$\mathbf{P}_1$	$\mathbf{P}_2$	$\mathbf{P}_3$	$\mathbf{P}_4$	$\mathbf{P}_5$	p
$\mathbf{P}_3$	$3\epsilon^2 + \epsilon^3 - 2\epsilon^4$	0	③	1	−2	0	0
$\mathbf{P}_1$	$2 + \epsilon - 2\epsilon^2 + \epsilon^4$	1	−2	0	1	0	1
$\mathbf{P}_5$	$3 + 3\epsilon^2 - \epsilon^4 + \epsilon^5$	0	3	0	−1	1	0
p_s	—	1	−1	0	0	0	—
z_s	$2 + \epsilon - 2\epsilon^2 + \epsilon^4$	1	−2	0	1	0	—
$p_s - z_s$	$-2 - \epsilon + 2\epsilon^2 - \epsilon^4$	0	$\boxed{1}$	0	−1	0	—

The vector to be removed is $\mathbf{P}_3$ since

$$\epsilon^2 + \frac{\epsilon^3}{3} - \frac{2\epsilon^4}{3}$$

is smaller than

$$3 + 3\epsilon^2 - \epsilon^4 + \epsilon^5$$

The new tableau is

III.	$\mathbf{P}_o(\epsilon)$	$\mathbf{P}_1$	$\mathbf{P}_2$	$\mathbf{P}_3$	$\mathbf{P}_4$	$\mathbf{P}_5$	p_s
$\mathbf{P}_2$	$\epsilon^2 + \frac{\epsilon^3}{3} - \frac{2}{3}\epsilon$	0	1	$\frac{1}{3}$	$-\frac{2}{3}$	0	-1
$\mathbf{P}_1$	$2 + \epsilon + \frac{2}{3}\epsilon^3 - \frac{1}{3}\epsilon^4$	1	0	$\frac{2}{3}$	$-\frac{1}{3}$	0	1
$\mathbf{P}_5$	$3 - \epsilon^3 + {}^4 + \epsilon^5$	0	0	-1	1	1	0
p_s	—	1	-1	0	0	0	—
z_s	$2 + \epsilon - \epsilon^2 + \frac{1}{3}\epsilon^3 + \frac{1}{3}\epsilon^4$	1	-1	$\frac{1}{3}$	$\frac{1}{3}$	0	—
$p_s - z_s$	—	0	0	$-\frac{1}{3}$	$-\frac{1}{3}$	0	—

Since the last row $(p_s - z_s)$ is negative or zero the optimal solution has been reached and

$$x_1 = 2, \quad x_2 = 0, \quad x_5 = 3, \quad x_3 = x_4 = 0$$

and max $(x_1 - x_2) = 2$. Now there are many things to be noted in these three tableaus. First, the components of $\mathbf{P}_o(\epsilon)$ can never become zero during the course of a computation if one starts with a basis, since each component of $\mathbf{P}_o(\epsilon)$ initially contains a power of ϵ not present in any other component. Another way of saying this is that the components of $\mathbf{P}_o(\epsilon)$ are a set of polynomials in ϵ which are linearly independent. Second, the coefficients of the polynomials at any stage are precisely those numbers which appear in the same row of the tableau. For example, in III opposite $\mathbf{P}_1$ the polynomial is $2 + \epsilon + \frac{2}{3}\epsilon^3 - \frac{1}{3}\epsilon^4$ and the row is 2, 1, 0, $\frac{2}{3}$, $-\frac{1}{3}$, 0 where $\lambda_2 = 2$ at this stage. With this in mind, it is clear that the following procedure may be followed to determine which vector is to be replaced. Suppose that at some stage of a regular tableau there are two candidates for removal from the basis because the ratios of λ_i/x_{ij} and λ_k/x_{kj} are the same. Then compare the corresponding values of

$$\frac{x_{ir}}{x_{ij}} \quad \text{for fixed } i \text{ and } r = 1, 2, \ldots, n$$

with

$$\frac{x_{kt}}{x_{kj}} \quad \text{for fixed } k \text{ and } t = 1, 2, \ldots, n$$

where $\mathbf{P}_i$ and $\mathbf{P}_k$ are the candidates for removal from the basis and $\mathbf{P}_j$ is the vector to be introduced. Let x_{io} be λ_i. Then the ratios are

$$\frac{x_{io}}{x_{ij}}, \frac{x_{i1}}{x_{ij}}, \frac{x_{i2}}{x_{ij}}, \ldots, 1, \ldots, \frac{x_{in}}{x_{ij}}$$

$$\frac{x_{ko}}{x_{kj}}, \frac{x_{k1}}{x_{kj}}, \frac{x_{k2}}{x_{kj}}, \ldots, 1, \ldots, \frac{x_{kn}}{x_{kj}}$$

Compare the values in a column; the row which has an algebraically smallest value in a column first is then the vector to be removed. In Tableau I the foregoing values are

$$\mathbf{P}_3; \ \frac{4}{2}, \ \frac{2}{2}, \ \frac{-1}{2}, \ \frac{1}{2}, \ \frac{0}{2}, \ \frac{0}{2}$$

$$\mathbf{P}_4; \ \frac{2}{1}, \ \frac{1}{1}, \ \frac{-2}{1}, \ \frac{0}{1}, \ \frac{1}{1}, \ \frac{0}{1}$$

The first two are the same so choose the third since $-\frac{2}{1} < -\frac{1}{2}$ and so $\mathbf{P}_4$ must be removed.

Therefore in the computations the value of the profit function will be increased at each stage since no λ_j in $\mathbf{P}_o(\epsilon)$ is ever zero and therefore with the ϵ perturbation it is not possible to return to an earlier basis and the possibility of a loop in the calculations is impossible. As just pointed out, however, it is not necessary in the computations to write the ϵ polynomials since a comparison of the ratios of terms in a row with those in another row will indicate the vector to be removed uniquely. This will have the result of increasing the perturbed profit function even though the effect of the perturbation is not spelled out in detail in the tableaus.

In general terms we may proceed as follows: Suppose at some stage we have a basic feasible solution $\lambda_1, \lambda_2, \ldots, \lambda_m$ with $\mathbf{P}_1, \mathbf{P}_2, \ldots, \mathbf{P}_m$ where a renumbering of the variables has been made. Then

$$\mathbf{P}_o = \sum_{j=1}^{m} \lambda_j \mathbf{P}_j$$

$$z = \sum_{j=1}^{m} \lambda_j p_j$$

$$\mathbf{P}_k = \sum_{j=1}^{m} x_{jk} \mathbf{P}_j$$

Define

$$z_s = \sum_{j=1}^{m} x_{js} p_j$$

$$\begin{aligned}\mathbf{P}_o(\epsilon) &= \mathbf{P}_o + \sum_{k=1}^{n} \epsilon^k \mathbf{P}_k \\ &= \sum_{j=1}^{m} \lambda_j \mathbf{P}_j + \sum_{k=1}^{n} \epsilon^k \sum_{j=1}^{m} x_{jk} \mathbf{P}_j \\ &= \sum_{j=1}^{m} \left[\lambda_j + \sum_{k=1}^{n} \epsilon^k x_{jk} \right] \mathbf{P}_j\end{aligned}$$

Therefore the perturbed $\mathbf{P}_o$ is $\mathbf{P}_o(\epsilon)$ with $\boldsymbol{\lambda}$ having elements

$$\lambda_j + \sum_{k=1}^{n} \epsilon^k x_{jk}$$

and this can never be zero if ϵ is sufficiently small. Consider now the perturbed profit function

$$z' = \sum_{j=1}^{m}\left(\lambda_j + \sum_{k=1}^{n}\epsilon^k x_{jk}\right)p_j$$
$$= \sum_{j=1}^{m}\lambda_j p_j + \sum_{k=1}^{n}\epsilon^k \sum_{j=1}^{m} x_{jk}p_j$$
$$= \sum_{j=1}^{m}\lambda_j p_j + \sum_{k=1}^{n}\epsilon^k z_k$$

Suppose now the basis is changed in the usual way

$$\mathbf{P}_o(\epsilon) = \sum_{j=1}^{m}\left(\lambda_j + \sum_{k=1}^{n}\epsilon^k x_{jk}\right)\mathbf{P}_j - \theta\mathbf{P}_s + \theta\mathbf{P}_s$$
$$= \sum_{j=1}^{m}\left(\lambda_j + \sum_{k=1}^{n}\epsilon^k x_{jk} - \theta x_{js}\right)\mathbf{P}_j + \theta\mathbf{P}_s$$

and the associated profit function is

$$z'' = \sum_{j=1}^{m}\left(\lambda_j + \sum_{k=1}^{n}\epsilon^k x_{jk} - \theta x_{js}\right)p_j + \theta p_s$$
$$= z' + \theta(p_s - z_s)$$

As before, choose s so $p_s - z_s$ is largest and positive. Choose

$$\theta = \min_j \frac{\lambda_j + \sum_{k=1}^{n}\epsilon^k x_{jk}}{x_{js}}$$

among those j for which $x_{js} > 0$ and for fixed s. Now there is always a unique θ since the polynomials in ϵ corresponding to the vectors in the basis by the definition of $\mathbf{P}_o(\epsilon)$ are always linearly independent and in fact each one contains a power of ϵ not present in the others. Therefore only one vector can be replaced by a new vector.

4.12 Optimal Solutions

In the optimal tableau there are some $p_s - z_s = 0$ and thus if these vectors are put into the basis no change in the profit function will take place. It is possible therefore by a method of exhaustion to determine all those extreme points at which the maximum is assumed. Let $\boldsymbol{\lambda}_1$ and $\boldsymbol{\lambda}_2$ be two extreme points at which the maximum is assumed, then

$$z = \mathbf{p}^T\boldsymbol{\lambda}_1 = \mathbf{p}^T\boldsymbol{\lambda}_2$$

then $$z = \mathbf{p}^T(\alpha\boldsymbol{\lambda}_1 + \beta\boldsymbol{\lambda}_2) \qquad \alpha + \beta = 1; \qquad \alpha, \beta \; 0$$

Since the $\boldsymbol{\lambda}$ form a convex set, that is,

$$\boldsymbol{\lambda} = \alpha\boldsymbol{\lambda}_1 + \beta\boldsymbol{\lambda}_2$$

then
$$z = \mathbf{p}^T \boldsymbol{\lambda}$$
is a solution and therefore every point in the set of points determined by the optimal points (convex hull) is an optimal solution. This may be very important in economic problems since it gives a choice of operating conditions which may be more convenient in one case than another.

EXERCISES

1. Solve the problem of Acrivos in Section 4.1. First derive the expressions for the restrictions.

2. Using a geometrical argument find

(a) $\max\,(x_1 + x_2 - 1)$

(b) $\min\,(3x_1 - x_2 + 6)$

(c) $\max\,(-2x_1 - 2x_2 + 2)$

with the following restraints:

$$\begin{aligned} 2x_2 - x_1 &\leqslant 8 \\ 3x_2 + x_1 &\leqslant 22 \\ x_1 + x_2 &\leqslant 12 \\ 10x_2 + x_1 &\geqslant 10 \\ x_1, x_2 &\geqslant 0 \end{aligned}$$

3. In a blending problem to produce a certain product, two blending stocks are available and three minimum specifications must be met. The product must be produced at minimum cost. The corresponding linear programming problem is the following:

$$\begin{aligned} x_1 + 3x_2 &\geqslant 8 \\ 3x_1 + 4x_2 &\geqslant 19 \\ 3x_1 + x_2 &\geqslant 7 \\ \min(50x_1 &+ 25x_2) \\ x_1, x_2 &\geqslant 0 \end{aligned}$$

Write the dual problem and solve it.

4. Solve the problem as stated in Problem 3 by the method of Section 4.7. (Note, as shown on p. 100, that it was not necessary to solve this problem. Illustrate.)

Eigenvalues and Eigenvectors 5

5.1 Introduction

Up to this point, we have used matrices in order to make the description of certain problems more compact and elegant, but one has the feeling that their introduction was more or less unnecessary. From this point on, however, many of the deeper properties of matrices will be introduced and their applications exploited. It will be shown that the treatment of first-order differential equations of linear type, for example, can be developed in a systematic way otherwise not easily done. We shall introduce a number of new terms such as *eigenvalue*, *eigenvector*, and *eigenrow* and prove two important theorems upon which most of the advanced work on matrices depends, the Hamilton-Cayley theorem and Sylvester's formula. Based on these a whole theory of functions of matrices may be developed which will be treated only superficially here but which will have wide application at some future time.

As an introduction, consider the system of equations

$$\mathbf{A}\mathbf{x} = \lambda\mathbf{x} \tag{5.1.1}$$

where $\mathbf{A}$ is a square matrix with real elements and λ is an as yet unspecified parameter. In most of what follows all matrices will be square. Equation

(5.1.1) is a set of homogeneous algebraic equations as is obvious when written in the form

$$(\mathbf{A} - \lambda\mathbf{I})\mathbf{x} = \begin{bmatrix} a_{11} - \lambda & a_{12} & \cdots & a_{1n} \\ a_{21} & a_{22} - \lambda & \cdots & a_{2n} \\ \cdot & & & \\ \cdot & & & \\ \cdot & & & \\ a_{n1} & a_{n2} & \cdots & a_{nn} - \lambda \end{bmatrix} \begin{bmatrix} x_1 \\ x_2 \\ \cdot \\ \cdot \\ \cdot \\ x_n \end{bmatrix} = \mathbf{0}$$

In general there will be no solution other than the trivial solution but we ask the question: Are there values of λ which will produce nontrivial solutions? In this work it will generally be assumed that $\mathbf{A}$ is nonsingular. The necessary and sufficient condition that there will be nontrivial solutions is that the determinant of the coefficients be zero or that

$$\det(A - \lambda\mathbf{I}) = \begin{vmatrix} a_{11} - \lambda & a_{12} & \cdots & a_{1n} \\ a_{21} & a_{22} - \lambda & \cdots & a_{2n} \\ \cdot & \cdot & & \cdot \\ \cdot & \cdot & & \cdot \\ \cdot & \cdot & & \cdot \\ a_{n1} & a_{n2} & \cdots & a_{nn} \end{vmatrix} = 0$$

This determinant when expanded will be a polynomial of the nth degree in λ of the form

$$P_n(\lambda) = (-\lambda)^n + a_1(-\lambda)^{n-1} + a_2(-\lambda)^{n-2} + \cdots + a_{n-1}(-\lambda) + a_n$$

and
$$\det(\mathbf{A} - \lambda\mathbf{I}) = P_n(\lambda) = 0$$

is the condition for nontrivial solutions. From the fundamental theorem of algebra it is known that a polynomial of the nth degree has n zeroes; that is, there are n values λ_i such that

$$P_n(\lambda_i) = 0, \quad i = 1, 2, \ldots, n$$

The $\{\lambda_i\}$ which satisfy this equation may be complex, since an algebraic equation with real coefficients may have roots in complex conjugate pairs, or they may be complex since the polynomial may have complex coefficients if the matrix $\mathbf{A}$ has nonreal elements. *The values of* λ_i *which satisfy this equation are said to be the eigenvalues of the matrix* $\mathbf{A}$. $\mathbf{A} - \lambda\mathbf{I}$ *is called the characteristic matrix, det* $(\mathbf{A} - \lambda\mathbf{I})$ *is called the characteristic determinant, and det* $(\mathbf{A} - \lambda\mathbf{I}) = 0$ *is called the characteristic equation. In the remainder of this book it will be assumed that all the eigenvalues are simple unless otherwise stated explicitly to the contrary.* Then

$$P_n(\lambda_i) = 0, \quad P'_n(\lambda_i) \neq 0$$

5.2 Eigenvectors

In the set of equations

$$(\mathbf{A} - \lambda_j \mathbf{I})\,\mathbf{x} = \mathbf{o} \tag{5.2.1}$$

where λ_j is an eigenvalue of $\mathbf{A}$, it is known that there will be nontrivial solutions since the rank of $(\mathbf{A} - \lambda_j\mathbf{I})$ is less than n. Since the eigenvalues are simple its rank is $n - 1$. In this case from Section 2.10 it is known that for a particular λ_j there will be a solution vector $\mathbf{x}_j$ which is determined only up to an arbitrary multiplicative factor and is any non-zero column of adj $(\mathbf{A} - \lambda_j\mathbf{I})$. We have previously shown that the rank of the matrix adj $(\mathbf{A} - \lambda_j\mathbf{I})$ is one and therefore each column is a constant multiple of every other column. Some care must be used in specifying a column, however, for it may happen that there are whole columns of zeroes. *The solution* $\mathbf{x}_j$ *which corresponds to a given* λ_j *is called an eigenvector and is said to belong to* λ_j. For each λ_j there is an eigenvector and these eigenvectors possess two important properties; for simple λ_j the set of eigenvectors $\{\mathbf{x}_j\}$ forms a linearly independent set, and the eigenvectors possess an orthogonality property (in general, a biorthogonality property).

If $\mathbf{x}_j$ and $\mathbf{x}_j$ are two eigenvectors which belong to λ_i and λ_j, respectively, then one cannot be a constant multiple of the other, for if it were, then

$$\mathbf{x}_j = c\mathbf{x}_i$$

and

$$\mathbf{A}\mathbf{x}_j = c\mathbf{A}\mathbf{x}_i$$

$$\lambda_j\mathbf{x}_j = c\lambda_i\mathbf{x}_i$$

and

$$\lambda_j c\mathbf{x}_i = c\lambda_i\mathbf{x}_i$$

and this can be true only if $\lambda_i = \lambda_j$, which is contrary to assumption. Therefore, for different eigenvalues the corresponding eigenvectors do not stand in simple proportion. Suppose now of the n eigenvectors there are only r linearly independent ones, and let these be the first r. Then any of the other $n - r$ vectors may be expressed as a linear combination of the first r

$$\mathbf{x}_j = c_1\mathbf{x}_1 + c_2\mathbf{x}_2 + \cdots + c_r\mathbf{x}_r$$

where j is one of the numbers from $r + 1$ to n and the $\{c_i\}$ are specific to that j. Then

$$\mathbf{A}\mathbf{x}_j = c_1\mathbf{A}\mathbf{x}_1 + c_2\mathbf{A}\mathbf{x}_2 + \cdots + c_r\mathbf{A}\mathbf{x}_r$$

or

$$\lambda_j\mathbf{x}_j = c_1\lambda_1\mathbf{x}_1 + c_2\lambda_2\mathbf{x}_2 + \cdots + c_r\lambda_r\mathbf{x}_r$$

But

$$\lambda_j\mathbf{x}_j = c_1\lambda_j\mathbf{x}_1 + c_2\lambda_j\mathbf{x}_2 + \cdots + c_r\lambda_j\mathbf{x}_r$$

by simple multiplication of the vector $\mathbf{x}_j$ and its expansion by λ_j. If these two expressions for $\lambda_j \mathbf{x}_j$ are subtracted

$$\mathbf{o} = c_1(\lambda_1 - \lambda_j)\mathbf{x}_1 + c_2(\lambda_2 - \lambda_j)\mathbf{x}_2 + \cdots + c_r(\lambda_r - \lambda_j)\mathbf{x}_r$$

this expression implies that there exists a set of constants not all zero such that the vectors $\mathbf{x}_1, \mathbf{x}_2, \ldots, \mathbf{x}_r$ are linearly dependent, clearly contrary to hypothesis. Therefore the set of n vectors must be linearly independent.

5.3 The Biorthogonality Property of Distinct Eigenvectors

Another eigenvalue problem may also be formed from the same matrix $\mathbf{A}$ as

$$\mathbf{y}^T\mathbf{A} = \eta\mathbf{y}^T$$

and by using the reversal rule for transposes we have

$$\mathbf{A}^T\mathbf{y} = \eta\mathbf{y}$$

We ask the same question as previously: What is the condition that the equivalent system

$$(\mathbf{A}^T - \eta\mathbf{I})\mathbf{y} = \mathbf{o}$$

have nontrivial solutions in y, and the answer is the same; that is, η must be a solution of the equation

$$\det(\mathbf{A}^T - \eta\mathbf{I}) = 0$$

But

$$\det(\mathbf{A}^T - \eta\mathbf{I}) = \begin{vmatrix} a_{11} - \eta & a_{21} & \cdots & a_{n1} \\ a_{12} & a_{22} - \eta & \cdots & a_{n2} \\ \cdot & & & \\ \cdot & & & \\ \cdot & & & \\ a_{1n} & a_{2n} & \cdots & a_{nn} - \eta \end{vmatrix} = 0$$

and from the fact that a determinant is unchanged if rows and columns are interchanged gives

$$\det(\mathbf{A} - \lambda\mathbf{I}) = \det(\mathbf{A}^T - \lambda\mathbf{I})$$

It follows that the two eigenvalue problems

$$\mathbf{A}\mathbf{x} = \lambda\mathbf{x} \quad \text{and} \quad \mathbf{y}^T\mathbf{A} = \eta\mathbf{y}^T$$

have the same eigenvalues, $\lambda_1, \lambda_2, \ldots, \lambda_n$. The eigenvectors, however, are not the same for the nontrivial solutions of $(\mathbf{A}^T - \lambda_j\mathbf{I})\mathbf{y} = 0$ will be non-zero columns of

$$\operatorname{adj}(\mathbf{A}^T - \lambda_j\mathbf{I})$$

for all j, whereas the nontrivial solutions of $(\mathbf{A} - \lambda_j\mathbf{I})\,\mathbf{x} = \mathbf{o}$ were non-zero columns of

$$\text{adj}\,(\mathbf{A} - \lambda_j\mathbf{I})$$

and these are clearly not the same unless the matrix $\mathbf{A}$ is symmetric (in this case $\mathbf{A} = \mathbf{A}^T$). But the columns of adj $(\mathbf{A}^T - \lambda_j\mathbf{I})$ are the rows of adj $(\mathbf{A} - \lambda_j\mathbf{I})$, and therefore a nontrivial solution of $\mathbf{A}^T\mathbf{y} = \eta\mathbf{y}$ will be one of the non-zero rows of adj $(\mathbf{A} - \lambda_j\mathbf{I})$. This is a very fortunate circumstance, for it follows that the two eigenvectors, one for each problem, may be computed by one calculation; that is, a non-zero column of the adjoint of the characteristic matrix will be a nontrivial solution for one problem, whereas a non-zero row of the same adjoint matrix will be a nontrivial solution for the transposed problem.

Hence two sets of linearly independent vectors are generated, a set $\{\mathbf{x}_j\}$ and a set $\{\mathbf{y}_j^T\}$ We now show that these two sets of vectors possess a biorthogonality property; each member of one set is orthogonal to each member of the other set except for the one that has a common eigenvalue. For, write

$$\mathbf{A}\mathbf{x}_j = \lambda_j\mathbf{x}_j$$

$$\mathbf{y}_i^T\mathbf{A} = \lambda_i\mathbf{y}_i^T, \quad i \neq j$$

and therefore $\lambda_i \neq \lambda_j$. Premultiply the first by $\mathbf{y}_i^T$ and postmultiply the second by $\mathbf{x}_j$

$$\mathbf{y}_i^T\mathbf{A}\mathbf{x}_j = \lambda_j\mathbf{y}_i^T\mathbf{x}_j$$

$$\mathbf{y}_i^T\mathbf{A}\mathbf{x}_j = \lambda_i\mathbf{y}_i^T\mathbf{x}_j$$

and subtract

$$(\lambda_j - \lambda_i)\mathbf{y}_i^T\mathbf{x}_j = 0$$

Since $\lambda_j \neq \lambda_i$ it follows that

$$\mathbf{y}_i^T\mathbf{x}_j = 0, \quad i \neq j$$

and since this is a scalar

$$\mathbf{x}_j^T\mathbf{y}_i = 0, \quad i \neq j$$

The term *orthogonality* is used here as a generalization from three-dimensional analytic geometry where, if two lines have direction cosines $\cos\alpha_1$, $\cos\alpha_2$, $\cos\alpha_3$ and $\cos\beta_1$, $\cos\beta_2$, $\cos\beta_3$ where the α_j and β_j are the direction angles with the coordinate axes, the two lines are perpendicular, or orthogonal, if

$$\cos\alpha_1\cos\beta_1 + \cos\alpha_2\cos\beta_2 + \cos\alpha_3\cos\beta_3 = 0$$

If the lines have direction numbers n_1, n_2, n_3 and m_1, m_2, m_3, then the normalized direction numbers are the direction cosines, or

$$\cos\alpha_j = \frac{n_j}{\sqrt{\sum_j n_j^2}}, \quad \cos\beta_j = \frac{m_j}{\sqrt{\sum_j m_j^2}}$$

and the vectors $\mathbf{n}^T = (n_1\ n_2\ n_3)$ and $\mathbf{m}^T = (m_1\ m_2\ m_3)$ satisfy

$$\mathbf{n}^T\mathbf{m} = n_1m_1 + n_2m_2 + n_3m_3 = 0$$

for two orthogonal lines. The direction numbers of a line through the origin to the point (n_1, n_2, n_3) are obviously n_1, n_2, n_3.

Let us show now that

$$\mathbf{y}_i^T\mathbf{x}_i \neq 0$$

To show this we will prove that given a set of n linearly independent vectors with n components there is no vector orthogonal to all members of the set, except the zero vector. If there were, then if $\mathbf{z}$ is the vector, $\mathbf{z}^T\mathbf{x}_j = 0$, for all j, or

$$\begin{aligned} z_1x_{11} + z_2x_{21} + \cdots + z_nx_{n1} &= 0 \\ z_1x_{12} + z_2z_{22} + \cdots + z_nx_{n2} &= 0 \\ &\vdots \\ z_1x_{1n} + z_2x_{2n} + \cdots + z_nx_{nn} &= 0 \end{aligned}$$

and this set of equations can have only the trivial solution in $z_1, z_2, \ldots, z_n$ since the determinant of the coefficients is not zero. Therefore the only vector orthogonal to all members of the set is the zero vector.

5.4 The Symmetric Case

If $\mathbf{A}$ is a real symmetric matrix, the situation is considerably simplified, for as mentioned in Section 5.3 for $\mathbf{A} = \mathbf{A}^T$

$$\text{adj}\,(\mathbf{A} - \lambda_j\mathbf{I}) = \text{adj}\,(\mathbf{A}^T - \lambda_j\mathbf{I})$$

so that this matrix is symmetric. Thus the vectors $\{\mathbf{x}_j\}$ and $\{\mathbf{y}_j\}$ are identical and

$$\mathbf{x}_j^T\mathbf{x}_i = 0, \quad i \neq j$$

and we have an orthogonal set of vectors $\{\mathbf{x}_j\}$; that is, each member of the set is orthogonal to each of the other members of the set. A second simplification also results since in this case it can be shown that the eigenvalues are always real numbers, whereas in the previous case they could be complex. Therefore the characteristic polynomial $P_n(\lambda)$ has real coefficients for a real symmetric matrix and if there is a zero of the form $\alpha + i\beta$, there must also be one of the form $\alpha - i\beta$, its complex conjugate. Suppose there are two complex eigenvalues of the form

$$\lambda_+ = \alpha + i\beta$$

$$\lambda_- = \alpha - i\beta$$

then $\mathbf{x}_+$ and $\mathbf{x}_-$ are the corresponding eigenvectors. Then these two eigenvectors are orthogonal so that $\mathbf{x}_+^T\mathbf{x}_- = 0$. The two vectors belonging to the complex conjugate eigenvalues differ only in that where there is an i in the one the other will have $-i$ so that the vectors $\mathbf{x}_+$ and $\mathbf{x}_-$ must have corresponding elements which are complex conjugates. Therefore

$$\mathbf{x}_+^* = \mathbf{x}_-, \qquad \mathbf{x}_-^* = \mathbf{x}_+$$

where the asterisk indicates the complex conjugate, so that the product $\mathbf{x}_+^T\mathbf{x}_-$ is a sum of squares of real numbers and hence never zero unless each of the eigenvectors is identically zero, which contradicts the hypothesis that an eigenvector is a non-zero vector. Thus there can be no complex eigenvalues.

5.5 The Expansion of an Arbitrary Vector

As mentioned earlier a matrix $\mathbf{A}$ may now be considered to generate a set, or two sets, of linearly independent vectors through the eigenvalue problem associated with the matrix. Given a matrix $\mathbf{A}$ there are two eigenvalue problems associated with it

$$\mathbf{A}\mathbf{x} = \lambda\mathbf{x} \quad \text{and} \quad \mathbf{A}^T\mathbf{y} = \eta\mathbf{y}$$

and the eigenvalues of these two problems are the same since the characteristic equations are identical. There are two sets of vectors $\{\mathbf{x}_j\}$ and $\{\mathbf{y}_j\}$ corresponding to $\{\lambda_j\}$ and each member of each set is obtained as a non-zero column and a non-zero row, respectively, of the matrix

$$\text{adj}\,(\mathbf{A} - \lambda_j\mathbf{I})$$

Further $\{\mathbf{x}_j\}$ and $\{\mathbf{y}_j\}$ are each linearly independent sets and

$$\mathbf{y}_i^T\mathbf{x}_j \begin{cases} = 0, & i \neq j \\ \neq 0, & i = j \end{cases}$$

If we now consider an arbitrary vector $\mathbf{z}$ with the same number of components as the vectors $\{\mathbf{x}_j\}$ its expansion in terms of the vectors $\{\mathbf{x}_j\}$ is simple and straightforward, provided that the set of vectors $\{\mathbf{x}_j\}$ is closed with respect to $\mathbf{z}$. For if

$$\mathbf{z} = \alpha_1\mathbf{x}_1 + \alpha_2\mathbf{x}_2 + \cdots + \alpha_n\mathbf{x}_n = \sum_j \alpha_j\mathbf{x}_j$$

then multiplication on the left by $\mathbf{y}_i^T$, a member of $\{\mathbf{y}_j\}$, gives

$$\mathbf{y}_i^T\mathbf{z} = \sum_j \alpha_j\mathbf{y}_i^T\mathbf{x}_j$$

Each member on the right is zero except $\mathbf{y}_i^T\mathbf{x}_i$ so

$$\mathbf{y}_i^T\mathbf{z} = \alpha_i\mathbf{y}_i^T\mathbf{x}_i$$

and

$$\alpha_i = \frac{\mathbf{y}_i^T\mathbf{z}}{\mathbf{y}_i^T\mathbf{x}_i}$$

Thus,

$$\mathbf{z} = \sum_j \frac{\mathbf{y}_j^T\mathbf{z}}{\mathbf{y}_j^T\mathbf{x}_j}\mathbf{x}_j$$

In the special case that **A** is symmetric

$$\mathbf{z} = \sum_j \beta_j\mathbf{x}_j$$

then

$$\mathbf{z} = \sum_j \frac{\mathbf{x}_j^T\mathbf{z}}{\mathbf{x}_j^T\mathbf{x}_j}\mathbf{x}_j$$

Applications of these expansions occur later in this work and they will be of one or the other of two forms. It seems that we always refer to linear algebraic equations and we will do so again. To find the solution of

$$\mathbf{Ax} = \mathbf{b}$$

we have a new technique. We bring this up at this point since the methods used in finding solutions of linear differential equations are analogous to the present methods, and in fact, more advanced works show that the whole subject may be unified into an all-embracing theory. The subject of *spectral theory* is beyond our present ambitions and intent. In the preceding equation assume a solution of the set of algebraic equations in the form

$$\mathbf{x} = \sum_j \alpha_j\mathbf{x}_j$$

where the $\{\mathbf{x}_j\}$ are the eigenvectors belonging to the eigenvalues $\{\lambda_j\}$ of the matrix **A**. Then if **b** has a representation

$$\mathbf{b} = \sum_j \beta_j\mathbf{x}_j$$

substitution of this into the equation $\mathbf{Ax} = \mathbf{b}$ gives

$$\mathbf{A}(\sum_j \alpha_j\mathbf{x}_j) = \sum_j \alpha_j\mathbf{Ax}_j = \sum_j \alpha_j\lambda_j\mathbf{x}_j = \sum_j \beta_j\mathbf{x}_j$$

or

$$\sum_j (\alpha_j\lambda_j - \beta_j)\,\mathbf{x}_j = \mathbf{o}$$

and because the $\{\mathbf{x}_j\}$ form a linearly independent set

$$\alpha_j\lambda_j - \beta_j = 0 \quad \text{or} \quad \alpha_j = \frac{\beta_j}{\lambda_j}$$

and

$$\mathbf{x} = \sum_j \frac{\beta_j}{\lambda_j}\mathbf{x}_j$$

where λ_j cannot be zero since we are assuming that **A** is nonsingular.

If one wishes a solution of

$$\mathbf{Ax} - \gamma\mathbf{x} = \mathbf{b}$$

then the same procedure may be used to give

$$\mathbf{x} = \sum_j \frac{\beta_j}{\lambda_j - \gamma}\mathbf{x}_j$$

and so this system has a solution only if γ is *not* an eigenvalue of **A**.

5.6 Examples

Before proceeding further it will be instructive to present a few examples.

(a) Symmetric matrix

$$\mathbf{A} = \begin{bmatrix} 1 & -1 & 2 \\ -1 & 2 & 0 \\ 2 & 0 & -1 \end{bmatrix}$$

$$\mathbf{A} - \lambda\mathbf{I} = \begin{bmatrix} 1-\lambda & -1 & 2 \\ -1 & 2-\lambda & 0 \\ 2 & 0 & -1-\lambda \end{bmatrix}$$

$$\det(\mathbf{A} - \lambda\mathbf{I}) = \begin{vmatrix} 1-\lambda & -1 & 2 \\ -1 & 2-\lambda & 0 \\ 2 & 0 & -1-\lambda \end{vmatrix} = -\lambda^3 + 2\lambda^2 + 6\lambda - 9$$

$$= (\lambda - 3)(\lambda^2 + \lambda - 3)$$

$$\lambda_1 = 3, \quad \lambda_2 = \tfrac{1}{2}(-1 + \sqrt{13}), \quad \lambda_3 = \tfrac{1}{2}(-1 - \sqrt{13})$$

for $\lambda_1 = 3$

$$\operatorname{adj}(\mathbf{A} - \lambda_1\mathbf{I}) = \operatorname{adj}\begin{bmatrix} -2 & -1 & 2 \\ -1 & -1 & 0 \\ 2 & 0 & -4 \end{bmatrix} = \begin{bmatrix} 4 & -4 & +2 \\ -4 & 4 & -2 \\ 2 & -2 & 1 \end{bmatrix}; \qquad \mathbf{x}_1 = \begin{bmatrix} +2 \\ -2 \\ 1 \end{bmatrix}$$

for $\lambda_2 = \frac{1}{2}(-1 + \sqrt{13})$

$$\operatorname{adj}(\mathbf{A} - \lambda_2\mathbf{I}) = \operatorname{adj}\begin{bmatrix} +\frac{3}{2} - \frac{1}{2}\sqrt{13} & -1 & 2 \\ -1 & \frac{5}{2} - \frac{1}{2}\sqrt{13} & 0 \\ 2 & 0 & -\frac{1}{2} - \frac{1}{2}\sqrt{13} \end{bmatrix}$$

$$= \begin{bmatrix} 2 - \sqrt{13} & -\frac{1}{2} - \frac{1}{2}\sqrt{13} & -5 + \sqrt{13} \\ -\frac{1}{2} - \frac{1}{2}\sqrt{13} & -\frac{3}{2} - \frac{1}{2}\sqrt{13} & -2 \\ -5 + \sqrt{13} & -2 & +6 - 2\sqrt{13} \end{bmatrix}; \begin{bmatrix} -5 + \sqrt{13} \\ -2 \\ 6 - 2\sqrt{13} \end{bmatrix} = \mathbf{x}_2$$

for $\lambda_3 = \frac{1}{2}(-1 - \sqrt{13})$

$$\operatorname{adj}(\mathbf{A} - \lambda_3\mathbf{I}) = \operatorname{adj}\begin{bmatrix} \frac{3}{2} + \frac{1}{2}\sqrt{13} & -1 & 2 \\ -1 & \frac{5}{2} + \frac{1}{2}\sqrt{13} & 0 \\ 2 & 0 & -\frac{1}{2} + \frac{1}{2}\sqrt{13} \end{bmatrix}$$

$$= \begin{bmatrix} 2 + \sqrt{13} & -\frac{1}{2} + \frac{1}{2}\sqrt{13} & -5 - \sqrt{13} \\ -\frac{1}{2} + \frac{1}{2}\sqrt{13} & -\frac{3}{2} + \frac{1}{2}\sqrt{13} & -2 \\ -5 - \sqrt{13} & -2 & 6 + 2\sqrt{13} \end{bmatrix}; \qquad \mathbf{x}_3 = \begin{bmatrix} -5 - \sqrt{13} \\ -2 \\ 6 + 2\sqrt{13} \end{bmatrix}$$

$$\mathbf{x}_1^T\mathbf{x}_2 = (2)(-5 + \sqrt{13}) + (-2)(-2) + (1)(6-2\sqrt{13}) = 0$$

$$\mathbf{x}_2^T\mathbf{x}_3 = (-5 + \sqrt{13})(-5 + \sqrt{13}) + (-2)(-2) + (6 - 2\sqrt{13})(6+2\sqrt{13}) = 0$$

$$\mathbf{x}_3^T\mathbf{x}_1 = (-5 - \sqrt{13})(2) + (-2)(-2) + (6 + 2\sqrt{13})(1) = 0$$

(b) Unsymmetric matrix

$$\mathbf{A} = \begin{bmatrix} 6 & -4 & 7 \\ 0 & 2 & -3 \\ -1 & 1 & 1 \end{bmatrix}$$

$$\mathbf{A} - \lambda\mathbf{I} = \begin{bmatrix} 6-\lambda & -4 & 7 \\ 0 & 2-\lambda & -3 \\ -1 & 1 & 1-\lambda \end{bmatrix}$$

$$\det(\mathbf{A} - \lambda\mathbf{I}) = -\lambda^3 + 9\lambda^2 - 30\lambda + 32 = 0$$

$$\lambda_1 = 2, \quad \lambda_2 = \frac{7}{2} + \frac{\sqrt{15}}{2}i, \quad \lambda_3 = \frac{7}{2} - \frac{\sqrt{15}}{2}i$$

for $\lambda_1 = 2$

$$\operatorname{adj}(\mathbf{A} - \lambda_1\mathbf{I}) = \operatorname{adj}\begin{bmatrix} 4 & -4 & 7 \\ 0 & 0 & -3 \\ -1 & 1 & -1 \end{bmatrix}$$

$$= \begin{bmatrix} 3 & 3 & 12 \\ 3 & 3 & 12 \\ 0 & 0 & 0 \end{bmatrix}; \qquad \mathbf{x}_1 = \begin{bmatrix} 3 \\ 3 \\ 0 \end{bmatrix}; \qquad \mathbf{y}_1 = \begin{bmatrix} 1 \\ 1 \\ 4 \end{bmatrix}$$

for $\lambda_2 = \frac{7}{2} + \frac{\sqrt{15}}{2}i$

$$\operatorname{adj}(\mathbf{A} - \lambda_2\mathbf{I}) = \operatorname{adj}\begin{bmatrix} \frac{5}{2} - \frac{\sqrt{15}}{2}i & -4 & 7 \\ 0 & -\frac{3}{2} - \frac{\sqrt{15}}{2}i & -3 \\ -1 & 1 & -\frac{5}{2} - \frac{\sqrt{15}}{2}i \end{bmatrix}$$

$$= \begin{bmatrix} 3 + 2\sqrt{15}\,i & -3 - 2\sqrt{15}\,i & \frac{45}{2} + \frac{7}{2}\sqrt{15}\,i \\ 3 & -3 & \frac{15}{2} - 3\frac{\sqrt{15}}{2}\,i \\ -\frac{3}{2} - \frac{\sqrt{15}}{2}i & +\frac{3}{2} + \frac{\sqrt{15}}{2}i & -\frac{15}{2} - \frac{\sqrt{15}}{2}i \end{bmatrix}$$

$$\mathbf{x}_2 = \begin{bmatrix} 3 + 2\sqrt{15}\,i \\ 3 \\ -\dfrac{3}{2} - \dfrac{\sqrt{15}}{2}\,i \end{bmatrix}; \qquad \mathbf{y}_2 = \begin{bmatrix} 1 \\ -1 \\ \dfrac{5}{2} - \dfrac{\sqrt{15}}{2}\,i \end{bmatrix}$$

for $\lambda_3 = \dfrac{7}{2} = \dfrac{\sqrt{15}}{2}\,i$

$$\operatorname{adj}(\mathbf{A} - \lambda_3\mathbf{I}) = \begin{bmatrix} 3 - 2\sqrt{15}\,i & -3 + 1\sqrt{15}\,i & \dfrac{45}{2} - \dfrac{7}{2}\sqrt{15}\,i \\ 3 & -3 & \dfrac{15}{2} + 3\dfrac{\sqrt{15}}{3}\,i \\ -\dfrac{3}{2} + \dfrac{\sqrt{15}}{2}\,i & \dfrac{3}{2} - \dfrac{\sqrt{15}}{2}\,i & -\dfrac{15}{2} + \dfrac{\sqrt{15}}{2}\,i \end{bmatrix}$$

$$\mathbf{x}_3 = \begin{bmatrix} 3 - 2\sqrt{15}\,i \\ 3 \\ -\dfrac{3}{2} + \dfrac{\sqrt{15}}{2}\,i \end{bmatrix}; \qquad \mathbf{y}_3 = \begin{bmatrix} 3 \\ -3 \\ \dfrac{15}{2} + 3\dfrac{\sqrt{15}}{2}\,i \end{bmatrix}$$

A simple calculation will verify that

$$\mathbf{y}_i^T\mathbf{x}_j = 0; \qquad i \neq j$$

5.7 Application to a System of First-Order Differential Equations

The methods of treatment of linear simultaneous algebraic equations described in Section 5.5 may be extended to treat linear simultaneous first-order differential equations with constant coefficients. Consider the system

$$\frac{dx_1}{dt} = a_{11}x_1 + a_{12}x_2 + \cdots + a_{1n}x_n + b_1$$

$$\frac{dx_2}{dt} = a_{21}x_1 + a_{22}x_2 + \cdots + a_{2n}x_n + b_2$$

$$\vdots$$

$$\frac{dx_n}{dt} = a_{n1}x_1 + a_{n2}x_2 + \cdots + a_{nn}a_n + b_n$$

which may be written in the compact notation of vectors and matrices

$$\frac{d\mathbf{x}}{dt} = \mathbf{Ax} + \mathbf{b} \tag{5.7.1}$$

It is well known that such a system of equations has a solution passing through a prescribed point

$$\left.\begin{aligned} x_1 &= x_1^i \\ x_2 &= x_2^i \\ &\cdot \\ &\cdot \\ &\cdot \\ x_n &= x_n^i \end{aligned}\right\} \text{ when } \quad t = 0 \tag{5.7.2}$$

or

$$\mathbf{x} = \mathbf{x}^i \quad \text{when} \quad t = 0$$

As is usual in seeking the solutions of differential equations, the procedure appears to be arbitrary and so once one thinks he has a solution it must be shown that it satisfies the differential equation and the auxiliary conditions, in this case Eq. (5.7.1) and (5.7.2). Consider first the equation

$$\frac{d\mathbf{x}}{dt} = \mathbf{Ax} \tag{5.7.3}$$

the homogeneous equation and assume it has a solution of the form

$$\mathbf{x} = \mathbf{z}e^{\lambda t} \tag{5.7.4}$$

where $\mathbf{z}$ is a vector of constants and λ is an unknown parameter. If it is to satisfy Eq. (5.7.3)

$$\mathbf{z}\lambda e^{\lambda t} = \mathbf{Az}e^{\lambda t}$$

Since $e^{\lambda t}$ cannot be zero

$$\mathbf{Az} = \lambda\mathbf{z}$$

and because we desire vectors $\mathbf{z}$ which are not the zero vector the condition that Eq. (5.7.4) be a solution is that λ be an eigenvalue of $\mathbf{A}$ and $\mathbf{z}$ be an eigenvector belonging to the eigenvalue. Therefore let λ_j be an eigenvalue which satisfies

$$\det(\mathbf{A} - \lambda_j\mathbf{I}) = 0$$

and $\mathbf{z}_j$ be a column of

$$\operatorname{adj}(\mathbf{A} - \lambda_j\mathbf{I})$$

Then there are n solutions

$$\mathbf{x} = \mathbf{z}_j e^{\lambda_j t}$$

with each $\mathbf{z}_j$ determined only up to a multiplicative constant.

One can find a particular solution of the nonhomogeneous equation (5.7.1) easily, for if one assumes that there is a constant vector $\mathbf{K}$ which is a solution, then

$$\mathbf{o} = \mathbf{AK} + \mathbf{b}$$

and since $\mathbf{A}$ is assumed to be nonsingular

$$\mathbf{K} = -\mathbf{A}^{-1}\mathbf{b}$$

Therefore the general solution, that is, one containing n linearly independent solutions, is

$$\mathbf{x} = \sum_j c_j \mathbf{z}_j e^{\lambda_j t} - \mathbf{A}^{-1}\mathbf{b} \tag{5.7.5}$$

where the arbitrary constants enter in a natural way because of the nature of $\mathbf{z}_j$. If Eq (5.7.5) is to satisfy the auxiliary condition, when $t = 0$,

$$\mathbf{x}^i = \sum_j c_j \mathbf{z}_j - \mathbf{A}^{-1}\mathbf{b} \tag{5.7.6}$$

and the c_j must be determined from this equation. In reality this is a set of linear simultaneous algebraic equations for the $\{\mathbf{c}_j\}$ since $\mathbf{x}^i$ and $\mathbf{A}^{-1}\,\mathbf{b}$ are known vectors and the rank of the system is n because of the linear independence of the n vectors $\{\mathbf{z}_j\}$. However, associated with the vector set $\{\mathbf{z}_j\}$ is the biorthogonal set of eigenrows of $\mathbf{A}$, $\{\mathbf{w}_j^T\}$, and

$$\mathbf{w}_k^T \mathbf{z}_j = 0, \quad k \neq j$$

Hence,

$$\mathbf{w}_k^T(\mathbf{x}^i + \mathbf{A}^{-1}\mathbf{b}) = \sum_j c_j \mathbf{w}_k^T \mathbf{z}_j$$

and therefore

$$c_j = \frac{\mathbf{w}_j^T(\mathbf{x}^i + \mathbf{A}^{-1}\mathbf{b})}{\mathbf{w}_j^T \mathbf{z}_j}$$

and the desired solution is

$$\mathbf{x} = \sum_j \frac{\mathbf{w}_j^T(\mathbf{x}^i + \mathbf{A}^{-1}\mathbf{b})}{\mathbf{w}_j^T \mathbf{z}_j} \mathbf{z}_j e^{\lambda_j t} - \mathbf{A}^{-1}\mathbf{b} \tag{5.7.7}$$

A direct substitution will show that it is a solution. If $\mathbf{b} = \mathbf{o}$, then

$$\mathbf{x} = \sum_j \frac{\mathbf{w}_j^T \mathbf{x}^i}{\mathbf{w}_j^T \mathbf{z}_j} \mathbf{z}_j e^{\lambda_j t}$$

In the special case the matrix $\mathbf{A}$ is symmetric

$$\mathbf{x} = \sum_j \frac{\mathbf{z}_j^T(\mathbf{x}^i + \mathbf{A}^{-1}\mathbf{b})}{\mathbf{z}_j^T \mathbf{z}_j} \mathbf{z}_j e^{\lambda_j t} - \mathbf{A}^{-1}\mathbf{b} \tag{5.7.8}$$

It was shown in Section 5.4 that the eigenvalues of a symmetric matrix were always real so that the solution Eq. (5.7.8) cannot have oscillatory solutions. In the non-symmetric case, however, the eigenvalues of $\mathbf{A}$ may be complex, but if the elements of $\mathbf{A}$ are real, the complex eigenvalues must occur in pairs of complex conjugates and in this case the general situation may be damped or growing oscillations or in case there are eigenvalues which are pure imaginary numbers there will be sustained oscillations. Consider these three cases:

(a) $\mathbf{A}$ symmetric. Consider the set of equations

$$\frac{dx_1}{dt} = x_1 - x_2 + 2x_3 + a_1; \qquad x_1 = x_1^{(i)}, \quad t = 0$$

$$\frac{dx^2}{dt} = -x_1 + 2x_2 \qquad + a_2; \qquad x_2 = x_2^{(i)}, \quad t = 0$$

$$\frac{dx_3}{dt} = 2x_1 \qquad -x_3 + a_3; \qquad x_3 = x_3^{(i)}, \quad t = 0$$

where the matrix **A** is the same as the one considered in Example (a) in Section 5.6. The details are left as an exercise for the reader to complete.

(b) **A** unsymmetric. Consider the set of equations

$$\frac{dx_1}{dt} = 6x_1 - 4x_2 + 7x_3; \qquad x_1 = x_1^{(i)}, \quad t = 0$$

$$\frac{dx_2}{dt} = 2x_2 - 3x_3; \qquad x_2 = x_2^{(i)}, \quad t = 0$$

$$\frac{dx_3}{dt} = -x_1 + x_2 + x_3; \qquad x_3 = x_3^{(i)}, \quad t = 0$$

where the matrix **A** is the same as the one considered in Example (b) in Section 5.6. The details are left to the reader to fill in and to show that the solution will be real and have the character of a growing oscillation.

(c) **A** unsymmetric. Consider the set of equations

$$\frac{dx_1}{dt} = -x_1 - x_2 + 2x_3; \qquad x_1 = x_1^{(i)}, \quad t = 0$$

$$\frac{dx_2}{dt} = +x_1 - x_3; \qquad x_2 = x_2^{(i)}, \quad t = 0$$

$$\frac{dx_3}{dt} = -x_2 - x_3; \qquad x_3 = x_3^{(i)}, \quad t = 0$$

Here

$$\det(\mathbf{A} - \lambda\mathbf{I}) = \begin{vmatrix} -1-\lambda & -1 & 2 \\ 1 & -\lambda & -1 \\ 0 & -1 & -1-\lambda \end{vmatrix} = -\lambda^3 - 2\lambda^2 - \lambda - 2$$

$$= -(\lambda + 2)(\lambda^2 + 1); \qquad \lambda_1 = -2, \lambda_2 = i, \lambda_3 = -i$$

The reader should show that the solution consists of one part which damps off to zero for large values of the time and one part which represents a sustained oscillation.

5.8 Air-Water Interaction Process

In an air-water interaction process in countercurrent flow in a cooling tower, the following three equations may be written after considerable simplification of the mathematical and physical model of the process. (For a derivation see any standard text.)

$$\frac{dH}{dh} = \frac{Ka}{W}(H_w - H)$$

$$\frac{dt}{dh} = \frac{Ua}{WS}(T - t)$$

$$\frac{dT}{dh} = \frac{\lambda W}{Lc_p}\frac{dH}{dh} + \frac{WS}{Lc_p}\frac{dt}{dh}$$

where h = tower running variable height,
T = temperature of the water, T_i outlet water temperature
t = temperature of the air, t_i inlet air temperature
H = humidity of the air
λ = latent heat
L = water flow rate
W = air rate
Ka = mass transfer coefficient
Ua = heat transfer coefficient
S = humid heat
H_w = saturated humidity at local water temperature
p = vapor pressure of water
c_p = specific heat of water

The data* in the problem are

Inlet water temperature = 90°F
L = 3000 #/hr
W = 2000 #/hr
H (entering air) = 0.0030 lb H_2O/lb air
Ua = 35.6 BTU/hr °F cu ft
Ka = 105 lb/hr cu ft unit humidity difference
λ = 1000 BTU/lb
S = 0.26 BTU(°F) #air
t_i = 70°F
T_i = 75°F
total pressure = 760 mm

The relation between H_w, the staturated humidity, and the vapor pressure of water is given by the relation

$$H_w = \frac{18}{29}\frac{p}{760 - p}$$

with p in millimeters of mercury. In the range of interest, 75°–90°F, this function may be approximated by the straight line

$$H_w = 0.000827\,T - 0.0432$$

It is desired to obtain the solutions of the differential equations listed above subject to the conditions

$$\left.\begin{aligned} T &= T_i \\ t &= t_i \\ H &= H_i \end{aligned}\right\} \text{ at } \quad h = 0$$

and to determine the tower height to cool water from 90° to 75°.

*T. K. Sherwood and Charles E. Reed, *Applied Mathematics in Chemical Engineering*, New York: McGraw-Hill, 1939, p. 134. Used with permission of publisher.

Solution: One's first inclination is to write the system of differential equations in the form:

$$\frac{dH}{dh} = \frac{Ka}{W}(H_w - H) \quad \text{(water transfer)}$$

$$\frac{dt}{dh} = \frac{Ua}{WS}(T - t) \quad \text{(heat transfer)}$$

$$\frac{dT}{dh} = \frac{\lambda Ka}{Lc_p}(H_w - H) + \frac{Ua}{Lc_p}(T - t) \quad \text{(energy)}$$

and treat it as a system of three simultaneous differential equations in the three dependent variables H, t, and T. Since, however, H_w may be written

$$H_w = AT - B$$

$$\frac{dH}{dh} = -\frac{Ka}{W}H + 0t + \frac{Ka}{W}AT - \frac{Ka}{W}B$$

$$\frac{dt}{dh} = 0H - \frac{Ua}{WS}t + \frac{Ua}{WS}T$$

$$\frac{dT}{dh} = -\frac{\lambda Ka}{Lc_p}H - \frac{Ua}{Lc_p}t + \left[\frac{Ua}{Lc_p} + \frac{\lambda Ka}{Lc_p}A\right]T - \frac{\lambda Ka}{Lc_p}B$$

the matrix

$$\begin{bmatrix} -\frac{Ka}{W} & 0 & \frac{Ka}{W}A \\ 0 & -\frac{Ua}{WS} & \frac{Ua}{WS} \\ -\frac{\lambda Ka}{Lc_p} & -\frac{Ua}{Lc_p} & \frac{Ua}{Lc_p} + \frac{\lambda Ka}{Lc_p}A \end{bmatrix}$$

has rank two since the third row is a linear combination of the first two rows. Hence this matrix is singular and it will not be possible to find the particular solution of the nonhomogeneous equation as described earlier since the inverse of the coefficient matrix does not exist. The third differential equation may be integrated directly to give

$$T - T_i = \frac{\lambda W}{Lc_p}(H - H_i) + \frac{WS}{Lc_p}(t - t_i)$$

and this value may be used in the other two equations to produce two equations and two unknowns. Thus,

$$T = \left(T_i - \frac{\lambda W}{Lc_p}H_i - \frac{WS}{Lc_p}t_i\right) + \frac{\lambda W}{Lc_p}H + \frac{WS}{Lc_p}t$$

$$= a' + bH + ct$$

$$H_w = Aa' - B + AbH + cAt$$

and

$$\frac{dH}{dh} = \frac{Ka}{W}(-1 + Ab)H + \frac{Ka}{W}cAt - \frac{BKa}{W} + \frac{Ka}{W}Aa'$$

$$\frac{dh}{dt} = \frac{Ua}{WS}bH + \frac{Ua}{WS}(c - 1)t + \frac{Uaa'}{WS}$$

These equations are now in a form suitable for solution. (This is left as an exercise for the reader.)

5.9 Hamilton-Cayley Theorem

As mentioned in the introductory paragraph, most of the analysis before this chapter did not involve sophisticated properties of matrices; from this point on, however, the character of the subject changes abruptly because of the Hamilton-Cayley theorem, a theorem at the base of almost all advanced matrix theory. First we state the theorem and illustrate it with a numerical example; Then, because of its great importance, the proof will be sketched.

Given a square matrix **A** and its characteristic equation

$$P_n(\lambda) = (-\lambda)^n + a_1(-\lambda)^{n-1} + a_2(-\lambda)^{n-2} + \cdots + a_{n-1}(-\lambda) + a_n = 0$$

the Hamilton-Cayley theorem states: Every square matrix satisfies its own characteristic equation; that is,

$$P_n(\mathbf{A}) = (-\mathbf{A})^n + a_1(-\mathbf{A})^{n-1} + a_2(-\mathbf{A})^{n-2} + \cdots + a_{n-1}(-\mathbf{A}) + a_n\mathbf{I} = \mathbf{O}$$

The left-hand side of this equation is a matrix polynomial and from Section 2.3 we know that such polynomials exist. In order to fix ideas let us consider the matrix

$$\mathbf{A} = \begin{bmatrix} 6 & -4 & 7 \\ 0 & 2 & -3 \\ -1 & 1 & 1 \end{bmatrix}$$

and its characteristic equation

$$P_n(\lambda) = -\lambda^3 + 9\lambda^2 - 30\lambda + 32 = 0$$

A direct calculation will show that

$$-\begin{bmatrix} 6 & -4 & 7 \\ 0 & 2 & -3 \\ -1 & 1 & 1 \end{bmatrix}^3 + 9\begin{bmatrix} 6 & -4 & 7 \\ 0 & 2 & -3 \\ -1 & 1 & 1 \end{bmatrix}^2 - 30\begin{bmatrix} 6 & -4 & 7 \\ 0 & 2 & -3 \\ -1 & 1 & 1 \end{bmatrix}$$
$$+ 32\begin{bmatrix} 1 & 0 & 0 \\ 0 & 1 & 0 \\ 0 & 0 & 1 \end{bmatrix} = \begin{bmatrix} 0 & 0 & 0 \\ 0 & 0 & 0 \\ 0 & 0 & 0 \end{bmatrix}$$

5.10 A Useful Lemma

In the proof of the Hamilton-Cayley theorem use is made of the fact that if

$$\mathbf{Ab} = \mathbf{o}$$

for any vector $\mathbf{b}$ then $\mathbf{A} = \mathbf{O}$. Since it is true for any vector, choose n linearly independent vectors $\{\mathbf{b}_j\}$. Then $\mathbf{Ab}_j = \mathbf{o}$ is equivalent to

$$\begin{aligned} a_{11}b_{1j} + a_{12}b_{2j} + \cdots + a_{1n}b_{nj} &= 0 \\ a_{21}b_{1j} + a_{22}b_{2j} + \cdots + a_{2n}b_{nj} &= 0 \\ &\vdots \\ a_{n1}b_{1j} + a_{n2}b_{2j} + \cdots + a_{nn}b_{nj} &= 0 \end{aligned}$$

Each of these equations holds for $j = 1, 2, 3, \ldots, n$. The first equation is then a system of n equations

$$\begin{aligned} a_{11}b_{11} + a_{12}b_{21} + \cdots + a_{1n}b_{n1} &= 0 \\ a_{11}b_{12} + a_{12}b_{22} + \cdots + a_{1n}b_{n2} &= 0 \\ &\vdots \\ a_{11}b_{1n} + a_{12}b_{2n} + \cdots + a_{1n}b_{nn} &= 0 \end{aligned}$$

in which the unknowns are $a_{11}, a_{12}, \ldots, a_{1n}$ and since the determinant of their coefficients is not zero the only solution is the trivial solution. One can repeat the same analysis with each of the other sets of equations and hence the only matrix $\mathbf{A}$ such that

$$\mathbf{Ab} = \mathbf{o}$$

for any vector $\mathbf{b}$ is the zero matrix itself. Note that in this proof it is not necessary that $\mathbf{A}$ be square.

5.11 Proof of Hamilton-Cayley Theorem (H-C)

If $\mathbf{A}$ is a square matrix, its characteristic equation may be written in the form,

$$\lambda^n + b_1\lambda^{n-1} + b_2\lambda^{n-2} + \cdots + b_{n-1}\lambda + b_n = 0$$

If λ_s is an eigenvalue and $\mathbf{x}_s$ the corresponding eigenvector

$$\mathbf{Ax}_s = \lambda_s\mathbf{x}_s$$

then

$$\mathbf{AAx}_s = \mathbf{A}^2\mathbf{x}_s = \lambda_s\mathbf{Ax}_s = \lambda_s^2\mathbf{x}_s$$

and, in general,

$$\mathbf{A}^m\mathbf{x}_s = \lambda_s^m\mathbf{x}_s$$

so that if λ_s is an eigenvalue of $\mathbf{A}$, λ_s^m is an eigenvalue of $\mathbf{A}^m$ with the same eigenvector.

If $Q(x)$ is any scalar polynomial

$$Q(\lambda_s) = c_0\lambda_s^m + c_1\lambda_s^{m-1} + c_2\lambda_s^{m-2} + \cdots + c_{n-1}\lambda_s + c_n$$

then multiplying by $\mathbf{x}_s$ to make it a vector identity

$$Q(\lambda_s)\mathbf{x}_s = c_0\lambda_s^m\mathbf{x}_s + c_1\lambda_s^{m-1}\mathbf{x}_s + c_2\lambda_s^{m-1}\mathbf{x}_s + \cdots + c_{n-1}\lambda_s\mathbf{x}_s + c_n\mathbf{x}_s$$

But $\mathbf{A}^p\mathbf{x}_s = \lambda_s^p\mathbf{x}_s$ for p an integer so that

$$\begin{aligned} Q(\lambda_s)\mathbf{x}_s &= c_0\mathbf{A}^m\mathbf{x}_s + c_1\mathbf{A}^{m-1}\mathbf{x}_s + c_2\mathbf{A}^{m-2}\mathbf{x}_s + \cdots + c_{n-1}\mathbf{A}\mathbf{x}_s + c_n\mathbf{I}\mathbf{x}_s \\ &= (c_0\mathbf{A}^m + c_1\mathbf{A}^{m-1} + c_2\mathbf{A}^{m-2} + \cdots + c_{n-1}\mathbf{A} + c_n\mathbf{I})\mathbf{x}_s \end{aligned}$$

and therefore

$$Q(\mathbf{A})\mathbf{x}_s = Q(\lambda_s)\mathbf{x}_s \tag{5.11.1}$$

Thus if λ is an eigenvalue of $\mathbf{A}$ and $Q(\mathbf{A})$ is any polynomial defined on $\mathbf{A}$, then $Q(\lambda_s)$ is an eigenvalue of $Q(\mathbf{A})$. Although it is not pertinent to this section it should be noted also that

$$\mathbf{A}^{-1}\mathbf{x}_s = \lambda_s^{-1}\mathbf{x}_s$$

so that

$$\mathbf{A}^{-p}\mathbf{x}_s = \lambda_s^{-p}\mathbf{x}_s$$

and therefore, given any expression of the form

$$F(\mathbf{A}) = \sum_{j=-p}^{j=q} c_j\mathbf{A}^j$$

then

$$F(\mathbf{A})\mathbf{x}_s = F(\lambda_s)\mathbf{x}_s$$

Now for the proof. If $P_n(\lambda_s) = 0$ is the characteristic equation, then

$$P_n(\mathbf{A})\mathbf{x}_s = P_n(\lambda_s)\mathbf{x}_s$$

and therefore

$$P_n(\mathbf{A})\mathbf{x}_s = 0$$

But *any* vector, of order n, may be represented as a linear combination of eigenvectors

$$\mathbf{x} = \sum_s \alpha_s\mathbf{x}_s$$

so that

$$(\mathbf{A}^n + b_1\mathbf{A}^{n-1} + b_2\mathbf{A}^{n-2} + \cdots + b_{n-1}\mathbf{A} + b_n\mathbf{I})\mathbf{x}_s$$

may be multiplied by α_s and summed from $s = 1$ to n to give

$$(\mathbf{A}^n + b_1\mathbf{A}^{n-1} + \cdots + b_{n-1}\mathbf{A} + b_n\mathbf{I})\sum_s \alpha_s\mathbf{x}_s = 0$$

and since $\sum_s \alpha_s\mathbf{x}_s$ is *any* vector, from the lemma,

$$\mathbf{A}^n + b_1\mathbf{A}^{n-1} + b_2\mathbf{A}^{n-2} + \cdots + b_{n-1}\mathbf{A} + b_n\mathbf{I} = \mathbf{0}$$

which is the statement that $\mathbf{A}$ satisfies its own characteristic equation.

5.12 Some Deductions from the Hamilton-Cayley Theorem

Some very interesting deductions may be drawn from the Hamilton-Cayley theorem. For example, if the characteristic equation is written in the form,

$$P_n(\lambda) = \lambda^n + b_1\lambda^{n-1} + b_2\lambda^{n-2} + \cdots + b_{n-1}\lambda + b_n = 0$$

then

$$\mathbf{A}^n = -b_1\mathbf{A}^{n-1} - b_2\mathbf{A}^{n-2} + \cdots + b_{n-1}\mathbf{A} - b_n\mathbf{I} \tag{5.12.1}$$

and the nth power of an nth-order matrix may be expressed as a linear combination of the first n powers of $\mathbf{A}$; that is, $\mathbf{A}^0$, $\mathbf{A}^1$, $\mathbf{A}^2$, . . . , $\mathbf{A}^{n-1}$. If this equation is multiplied through by $\mathbf{A}$

$$\begin{aligned}\mathbf{A}^{n+1} &= -b_1\mathbf{A}^n - b_2\mathbf{A}^{n-1} + \cdots - b_{n-1}\mathbf{A}^2 - b_n\mathbf{A}\\ &= -b_1[-b_1\mathbf{A}^{n-1} - b_2\mathbf{A}^{n-1} + \cdots - b_{n-1}\mathbf{A} - b_n\mathbf{I}]\\ &\quad -b_2\mathbf{A}^{n-1} - b_3\mathbf{A}^{n-2} + \cdots - b_{n-1}\mathbf{A}^2 - b_n\mathbf{A}\\ &= c_1\mathbf{A}^{n-1} + c_2\mathbf{A}^{n-2} + \cdots + c_{n-1}\mathbf{A} + c_n\mathbf{I}\end{aligned}$$

This multiplication may be repeated and the term A^n may be substituted for again to give an expression of the form

$$\mathbf{A}^{n+2} = d_1\mathbf{A}^{n-1} + d_2\mathbf{A}^{n-2} + \cdots + d_{n-1}\mathbf{A} + d_n\mathbf{I}$$

Successive multiplications by $\mathbf{A}$ and substitutions for $\mathbf{A}^n$ will give

$$\mathbf{A}^p = \alpha_1\mathbf{A}^{n-1} + \alpha_2\mathbf{A}^{n-2} + \cdots + \alpha_{n-1}\mathbf{A} + \alpha_n\mathbf{I}$$

Thus if $\mathbf{A}$ is a square matrix of order n, any integral power p of the matrix $\mathbf{A}$ may be expressed as a linear combination of the first n powers of that matrix. Hence if one considers a polynomial of arbitrary degree q defined on a matrix, $\mathbf{A}$, of nth order with scalar coefficients, say,

$$\sum_{j=0}^{q} \beta_j\mathbf{A}^j$$

then that polynomial may be expressed as

$$\sum_{j=0}^{n-1} \gamma_j\mathbf{A}^j$$

where $\{\gamma_j\}$ may be computed from the $\{\beta_j\}$ and the $\{b_j\}$ of the characteristic equation.

Note also that Eq. (5.12.1) can be written in the form

$$\mathbf{I} = -\frac{1}{b_n}\left[\mathbf{A}^n + b_1\mathbf{A}^{n-1} + b_2\mathbf{A}^{n-2} + \cdots + b_{n-1}\mathbf{A}\right]$$

If this is multiplied by $\mathbf{A}^{-1}$, one obtains

$$\mathbf{A}^{-1} = -\frac{1}{b_n}\left[\mathbf{A}^{n-1} + b_1\mathbf{A}^{n-2} + b_2\mathbf{A}^{n-3} + \cdots + b_{n-2}\mathbf{A} + b_{n-1}\mathbf{I}\right]$$

so that the inverse of a nonsingular square matrix may be expressed as a linear combination of the first n powers of that matrix. A procedure analogous to the preceding one may now be applied, for if this equation is multiplied through by $\mathbf{A}^{-1}$, the left-hand side will become $\mathbf{A}^{-2}$ and the last term on the right will be $\mathbf{A}^{-1}$. The foregoing expression can then be substituted for $\mathbf{A}^{-1}$, giving a formula of the form,

$$\mathbf{A}^{-2} = g_1\mathbf{A}^{n-1} + g_2\mathbf{A}^{n-2} + \cdots + g_{n-1}\mathbf{A} + g_n\mathbf{I}$$

Repetition of this procedure certainly gives

$$\mathbf{A}^{-q} = h_1\mathbf{A}^{n-1} + h_2\mathbf{A}^{n-2} + \cdots + h_{n-1}\mathbf{A} + h_n\mathbf{I}$$

where q is any integer. Thus any expression of the form

$$\sum_{j=-q_1}^{q_2} \beta_j\mathbf{A}^j, \quad q_1 > 0, \quad q_2 > 0$$

may be expressed as

$$\sum_{j=0}^{n-1} \epsilon_j\mathbf{A}^j$$

that is, a polynomial in $\mathbf{A}$ and $\mathbf{A}^{-1}$ may be expressed as a linear combination of the first n powers of $\mathbf{A}$. Thus the set of polynomials defined over an nth-order matrix $\mathbf{A}$ and $\mathbf{A}^{-1}$ are of maximum degree $n - 1$. This certainly is a result that could not have been anticipated.

In order for this procedure to be generally useful one needs the expression for the characteristic equation and this involves tedious computations to obtain the coefficients $\{\mathbf{b}_j\}$. In general this is a nontrivial computation. It may be shown that if the sum of the diagonal elements of a square matrix is called its *trace* (abbreviated Tr), with

$$T_1 = \text{Tr}\,\mathbf{A}$$

$$T_2 = \text{Tr}\,\mathbf{A}^2$$

and, in general,

$$T_j = \text{Tr}\,\mathbf{A}^j$$

then

$$b_1 = -T_1$$

$$b_2 = -\frac{1}{2}(b_1T_1 + T_2)$$

$$b_3 = -\frac{1}{3}(b_2T_1 + b_1T_2 + T_3)$$

$$\vdots$$

$$b_n = -\frac{1}{n}(b_{n-1}T_1 + b_{n-2}T_2 + \cdots + b_1T_{n-1} + T_n)$$

where the $\{\mathbf{b}_j\}$ are the coefficients in the characteristic equation.

As an illustration of the application of the Hamilton-Cayley theorem, consider the numerical matrix

$$\mathbf{A} = \begin{bmatrix} 6 & -4 & 7 \\ 0 & 2 & -3 \\ -1 & 1 & 1 \end{bmatrix}$$

where the characteristic equation is

$$P_n(\lambda) = \lambda^3 - 9\lambda^2 + 30\lambda - 32 = 0$$

and so

$$\begin{bmatrix} 6 & -4 & 7 \\ 0 & 2 & -3 \\ -1 & 1 & 1 \end{bmatrix}^3 = 9\begin{bmatrix} 6 & -4 & 7 \\ 0 & 2 & -3 \\ -1 & 1 & 1 \end{bmatrix}^2 - 30\begin{bmatrix} 6 & -4 & 7 \\ 0 & 2 & -3 \\ -1 & 1 & 1 \end{bmatrix} + 32\begin{bmatrix} 1 & 0 & 0 \\ 0 & 1 & 0 \\ 0 & 0 & 1 \end{bmatrix}$$

which may be directly verified. Also

$$\begin{bmatrix} 6 & -4 & 7 \\ 0 & 2 & -3 \\ -1 & 1 & 1 \end{bmatrix}^{-1} = \tfrac{1}{32}\left\{\begin{bmatrix} 6 & -4 & -7 \\ 0 & 2 & -3 \\ -1 & 1 & 1 \end{bmatrix}^2 - 9\begin{bmatrix} 6 & -4 & 7 \\ 0 & 2 & -3 \\ -1 & 1 & 1 \end{bmatrix}^1 + 30\begin{bmatrix} 1 & 0 & 0 \\ 0 & 1 & 0 \\ 0 & 0 & 1 \end{bmatrix}\right\}$$

$$= \tfrac{1}{32}\begin{bmatrix} 5 & 11 & -2 \\ 3 & 13 & 18 \\ 2 & -2 & 12 \end{bmatrix}$$

5.13 Sylvester's Theorem

A second theorem which builds on the Hamilton-Cayley theorem and which is important theoretically and very useful in computations is Sylvester's theorem. Let us consider a scalar polynomial $Q(x)$ of degree $n - 1$ written in the Lagrangian form where the $\{\alpha_j\}$ are, for the moment, any n different scalars. Then

$$\begin{aligned} Q(x) = {} & c_1(x - \alpha_2)(x - \alpha_3)(x - \alpha_4) \ldots (x - \alpha_n) \\ & + c_2(x - \alpha_1)(x - \alpha_3)(x - \alpha_4) \ldots (x - \alpha_n) \\ & + c_3(x - \alpha_1)(x - \alpha_2)(x - \alpha_4) \ldots (x - \alpha_n) \\ & \quad \vdots \\ & + c_n(x - \alpha_1)(x - \alpha_2)(x - \alpha_3) \ldots (x - \alpha_{n-1}) \end{aligned}$$

Note that the jth product in the sum omits the factor $(x - \alpha_j)$. A matrix polynomial defined on the matrix $\mathbf{A}$ may be written

$$\begin{aligned} Q(\mathbf{A}) = {} & c_1(\mathbf{A} - \alpha_2\mathbf{I})(\mathbf{A} - \alpha_3\mathbf{I})(\mathbf{A} - \alpha_4\mathbf{I}) \ldots (\mathbf{A} - \alpha_n\mathbf{I}) \\ & + c_2(\mathbf{A} - \alpha_1\mathbf{I})(\mathbf{A} - \alpha_3\mathbf{I})(\mathbf{A} - \alpha_4\mathbf{I}) \ldots (\mathbf{A} - \alpha_n\mathbf{I}) \\ & \quad \vdots \\ & + c_n(\mathbf{A} - \alpha_1\mathbf{I})(\mathbf{A} - \alpha_2\mathbf{I}) \ldots (\mathbf{A} - \alpha_{n-1}\mathbf{I}) \end{aligned} \tag{5.13.1}$$

Before proceeding, it should be recognized that the order of terms in a product is immaterial since

$$(\mathbf{A} - \alpha_i\mathbf{I})(\mathbf{A} - \alpha_j\mathbf{I}) = \mathbf{A}^2 - (\alpha_i + \alpha_j)\mathbf{A} + \alpha_i\alpha_j\mathbf{I} = (\mathbf{A} - \alpha_j\mathbf{I})(\mathbf{A} - \alpha_i\mathbf{I})$$

that is, the factors in a product of this form may be written in any order. If one supposes the $\{\alpha_j\}$ are given, then $Q(\mathbf{A})$ is a polynomial of degree $(n - 1)$ and contains n constants $\{c_j\}$. It is clear also that if the various terms in the sums are multiplied out, we obtain a polynomial of degree $(n - 1)$, whose coefficients will be a linear combination of the $\{c_j\}$. For example, the coefficient of $\mathbf{A}^{n-1}$ will be the sum

$$c_1 + c_2 + c_3 + \cdots + c_n$$

the coefficient of $\mathbf{A}^{n-2}$ will be

$$\begin{aligned}
&- c_1(\alpha_2 + \alpha_3 + \alpha_4 + \cdots + \alpha_n) \\
&- c_2(\alpha_1 + \alpha_3 + \alpha_4 + \cdots + \alpha_n) \\
&- c_3(\alpha_1 + \alpha_2 + \alpha_4 + \cdots + \alpha_n) \\
&\quad\vdots \\
&- c_n(\alpha_1 + \alpha_2 + \alpha_3 + \cdots + \alpha_{n-1})
\end{aligned}$$

and it can be shown that the coefficient of $\mathbf{A}^{n-3}$ will be a linear combination of the $\{c_j\}$ in which the coefficients will be the sum of all possible products of α_j taken two at a time, omitting the appropriate α_j in each sum, and the coefficient of $\mathbf{I}$ will be

$$(-1)^{n-1}[(\alpha_2\alpha_3 \ldots \alpha_n)c_1 + (\alpha_1\alpha_3 \ldots \alpha_n)c_2 + (\alpha_1\alpha_2\alpha_4 \ldots \alpha_n)c_3 \\ \ldots (\alpha_1\alpha_2 \ldots \alpha_{n-1})c_n]$$

Thus if a polynomial of degree $n - 1$ in $\mathbf{A}$ is written

$$d_1\mathbf{A}^{n-1} + d_2\mathbf{A}^{n-2} + d_3\mathbf{A}^{n-3} + \cdots + d_{n-1}\mathbf{A} + d_n\mathbf{I}$$

then the $\{d_j\}$ will be related to the $\{c_j\}$ by

$$\begin{aligned}
d_1 &= c_1 + c_2 + c_2 + \cdots + c_n \\
d_2 &= \beta_1c_1 + \beta_2c_2 + \cdots + \beta_nc_n \\
d_3 &= \gamma_1c_1 + \gamma_2c_2 + \cdots + \gamma_nc_n \\
&\;\;\vdots \\
d_n &= \nu_1c_1 + \nu_2c_2 + \cdots + \nu_vc_n
\end{aligned} \tag{5.13.2}$$

where the $\{\beta_j\}$, $\{\gamma_j\}$, . . . , $\{\nu_j\}$ are related to the $\{\alpha_j\}$ as previously outlined. If the α_j are specified, then the $\{c_j\}$ may be computed; hence any matrix polynomial of degree $n - 1$ may be written in the Lagrangian form, assuming that the $\{\alpha_j\}$ may be chosen so that the determinant of the coefficients of the $\{c_j\}$ in Eq. (5.13.2) is not zero.

Consider now Eq. (5.13.1) and postmultiply each side by $\mathbf{x}_j$ and at the same time let $\alpha_j = \lambda_j$ where λ_j is an eigenvalue of $\mathbf{A}$. Equation (5.13.1) may be written

$$\begin{aligned} Q(\mathbf{A})\mathbf{x}_j = c_1 \prod_{i=2}^{n} (\mathbf{A} - \lambda_i \mathbf{I})\mathbf{x}_j \\ + c_2 \prod_{i=1,2}^{n} (\mathbf{A} - \lambda_i \mathbf{I})\mathbf{x}_j \\ + c_2 \prod_{i=1,3}^{n} (\mathbf{A} - \lambda_i \mathbf{I})\mathbf{x}_j \\ \cdot \\ \cdot \\ \cdot \\ + c_n \prod_{i=1}^{n-1} (\mathbf{A} - \lambda_i \mathbf{I})\mathbf{x}_j \end{aligned}$$

where the notation $\prod_{i=1,k}^{n}$ indicates the product of all the factors from one to n but omitting $i = k$. But

$$(\mathbf{A} - \lambda_i \mathbf{I})\mathbf{x}_j = \mathbf{A}\mathbf{x}_j - \lambda_i \mathbf{x}_j = \lambda_j \mathbf{x}_j - \lambda_i \mathbf{x}_j = (\lambda_j - \lambda_i)\mathbf{x}_j$$

Each of the products contains a factor $(\mathbf{A} - \lambda_j \mathbf{I})$ except the one whose coefficient is c_j, so each term except that one will produce a factor $\lambda_j - \lambda_j = 0$. The jth term in the sum will be

$$Q(\mathbf{A})\mathbf{x}_j = c_j \prod_{i=1,j}^{n} (\lambda_j - \lambda_i)\mathbf{x}_j$$

From Eq. (5.11.1),

$$Q(\mathbf{A})\mathbf{x}_j = Q(\lambda_j)\mathbf{x}_j$$

so

$$Q(\lambda_j)\mathbf{x}_j = c_j \prod_{i=1,j}^{n} (\lambda_j - \lambda_i)\mathbf{x}_j$$

and

$$c_j = \frac{Q(\lambda_j)}{\prod_{i=1,j}^{n} (\lambda_j - \lambda_i)}$$

thus

$$Q(\mathbf{A}) = \sum_{j=1}^{n} Q(\lambda_j) \frac{\prod_{i=1,j}^{n} (\mathbf{A} - \lambda_i \mathbf{I})}{\prod_{i=1,j}^{n} (\lambda_j - \lambda_i)} \tag{5.13.3}$$

which is Sylvester's formula.

As Eq. (5.13.3) stands, it evidently holds only for matrix polynomials of degree $n - 1$. Remember, however, that an arbitrary polynomial $P(\mathbf{A})$ in $\mathbf{A}$ and $\mathbf{A}^{-1}$ may be expressed as a polynomial of degree $n - 1$ in $\mathbf{A}$. Also, one can perform the same operations on the characteristic equation in λ_s that one performed on the characteristic equation with $\mathbf{A}$ so that

$$\mathbf{A}^n = -b_1 \mathbf{A}^{n-1} - b_2 \mathbf{A}^{n-2} - b_3 \mathbf{A}^{n-3} \ldots - b_{n-1}\mathbf{A} - b_n \mathbf{I}$$

$$\lambda_s^n = -b_1 \lambda_s^{n-1} - b_2 \lambda_s^{n-2} - b_3 \lambda_s^{n-3} \ldots - b_{n-1}\lambda_s - b_n$$

and thus

$P(\mathbf{A})$ will reduce to $Q(\mathbf{A})$ (a polynomial of degree $n-1$ in $\mathbf{A}$)
$P(\lambda_s)$ will reduce to $Q(\lambda_s)$ (the same polynomial in λ_s)

and therefore *Eq. (5.13.3), Sylvester's formula, holds for an arbitrary polynomial in A and A^{-1}.*

5.14 An Alternative Form for Eq. (5.13.3)

Sylvester's formula may be written in another form which is frequently more convenient for computations. Suppose $f(y)$ is a scalar polynomial of degree n

$$f(y) = y^n + h_1 y^{n-1} + h_2 y^{n-2} + \cdots + h_{n-1} y + h_n$$

then

$$\begin{aligned} f(y) - f(x) &= (y^n - x^n) + h_1(y^{n-1} - x^{n-1}) \\ &\quad + h_2(y^{n-2} - x^{n-2}) + \cdots + h_{n-1}(y - x) \\ &= (y - x)\phi(y, x) \end{aligned} \tag{5.14.1}$$

where $\phi(y, x)$ is a polynomial of degree $(n - 1)$ in y and x. If λ is a scalar, then the foregoing algebraic identity may be written as a matrix identity

$$f(\lambda\mathbf{I}) - f(\mathbf{A}) = (\lambda\mathbf{I} - \mathbf{A})\phi(\lambda\mathbf{I}, \mathbf{A}) \tag{5.14.2}$$

If we let

$$f(\lambda) = \det(\lambda\mathbf{I} - \mathbf{A})$$

then $f(\mathbf{A}) = \mathbf{0}$, since $f(\lambda)$ is the negative of the characteristic determinant, and

$$\begin{aligned} f(\lambda\mathbf{I}) = (\lambda\mathbf{I} - \mathbf{A})\phi(\lambda\mathbf{I}, \mathbf{A}) &= \mathbf{I}f(\lambda) \\ &= (\lambda\mathbf{I} - \mathbf{A}) \operatorname{adj}(\lambda\mathbf{I} - \mathbf{A}) \end{aligned}$$

the latter statement following from Eq. (2.7.2). So

$$\phi(\lambda\mathbf{I}, \mathbf{A}) = \operatorname{adj}(\lambda\mathbf{I} - \mathbf{A}) \tag{5.14.3}$$

Consider again Eq. (5.14.1). This is an identity in y and x. If $y = \lambda_r$, an eigenvalue of $\mathbf{A}$ so that $f(\lambda_r) = 0$, and $x = \lambda$, then

$$f(\lambda) = (\lambda - \lambda_r)\phi(\lambda_r, \lambda)$$

and therefore

$$\phi(\lambda_r, \lambda) = \prod_{j=1, r}^{n} (\lambda - \lambda_j)$$

a polynomial identity in λ. Therefore it can be written as a matrix polynomial identity

$$\phi(\lambda_r\mathbf{I}, \mathbf{A}) = \prod_{j=1, r}^{n} (\mathbf{A} - \lambda_j\mathbf{I})$$

It is apparent that Eq. (5.14.3) also holds for $\lambda = \lambda_r$ and therefore

$$\prod_{j=1,r}^{n} (\mathbf{A} - \lambda_j \mathbf{I}) = \operatorname{adj}(\mathbf{A} - \lambda_r \mathbf{I})$$

Sylvester's formula may then be written

$$Q(\mathbf{A}) = \sum_{j=1}^{n} Q(\lambda_j) \frac{\operatorname{adj}(\lambda_j \mathbf{I} - \mathbf{A})}{\prod_{i=1,j}^{n} (\lambda_j - \lambda_i)} \tag{5.14.4}$$

As an example consider the computation of the inverse of

$$\mathbf{A} = \begin{bmatrix} 1 & -3 \\ -2 & 2 \end{bmatrix}$$

$$\det(\mathbf{A} - \lambda \mathbf{I}) = \lambda^2 - 3\lambda - 4 = (\lambda - 4)(\lambda + 1); \qquad \lambda_1 = 4, \quad \lambda_2 = -1$$

then

$$\begin{bmatrix} 1 & -3 \\ -2 & 2 \end{bmatrix}^{-1} = \frac{1}{\lambda_1} \frac{\operatorname{adj}(\lambda_1 \mathbf{I} - \mathbf{A})}{\lambda_1 - \lambda_2} + \frac{1}{\lambda_2} \frac{\operatorname{adj}(\lambda_2 \mathbf{I} - \mathbf{A})}{\lambda_2 - \lambda_1}$$

$$= \frac{1}{4} \frac{\begin{bmatrix} 2 & -3 \\ -2 & 3 \end{bmatrix}}{5} + \frac{1}{-1} \frac{\begin{bmatrix} -3 & -3 \\ -2 & -2 \end{bmatrix}}{-5}$$

$$= \begin{bmatrix} -\frac{1}{2} & -\frac{3}{4} \\ -\frac{1}{2} & -\frac{1}{4} \end{bmatrix}$$

For the same example compute $\mathbf{A}^4 + 3\mathbf{A}^2 + 6\mathbf{A} = \mathbf{B}$

$$\mathbf{B} = \tfrac{328}{5} \begin{bmatrix} 2 & -3 \\ -2 & 3 \end{bmatrix} + \tfrac{2}{5} \begin{bmatrix} -3 & -3 \\ -2 & -2 \end{bmatrix} = \tfrac{1}{5} \begin{bmatrix} 130 & -198 \\ -132 & 196 \end{bmatrix}$$

Now Eq. (5.14.4) was derived for a polynomial of degree $n - 1$ and it then followed by the use of the Hamilton-Cayley theorem that it was also valid for an arbitrary polynomial in $\mathbf{A}$ and $\mathbf{A}^{-1}$, if $\mathbf{A}$ is nonsingular. It is also true for polynomials of degree less than $n - 1$ since these are polynomials of degree $n - 1$ with certain coefficients equal to zero. Thus, for example,

$$\mathbf{A} = \sum_{j=1}^{n} \lambda_j \frac{\operatorname{adj}(\lambda_j \mathbf{I} - \mathbf{A})}{\prod_{i=1,j}^{n} (\lambda_j - \lambda_i)} \tag{5.14.5}$$

or *any* power of $\mathbf{A}$ may be expressed as

$$\mathbf{A}^q = \sum_{j=1}^{n} \lambda_j{}^q \frac{\operatorname{adj}(\lambda_j \mathbf{I} - \mathbf{A})}{\prod_{i=1,j}^{n} (\lambda_j - \lambda_i)}$$

for q, a positive or negative integer. It would be tempting to use such a formula for q, a proper fraction, but this is in general not true and this topic will not be treated in this book.

It is interesting that Sylvester's formula is really an expansion theorem and, as such, has some interesting properties. If Eq. (5.14.5) is multiplied through on the left by

$$\frac{\operatorname{adj}(\lambda_r \mathbf{I} - \mathbf{A})}{\prod_{i=1,r}^{n} (\lambda_r - \lambda_i)}$$

then

$$\mathbf{A}\frac{\operatorname{adj}(\lambda_r \mathbf{I} - \mathbf{A})}{\prod_{i=1,r}^{n} (\lambda_r - \lambda_i)} = \sum_{j=1}^{n} \lambda_j \frac{\operatorname{adj}(\lambda_j \mathbf{I} - \mathbf{A})}{\prod_{i=1,j}^{n} (\lambda_j - \lambda_i)} \frac{\operatorname{adj}(\lambda_r \mathbf{I} - \mathbf{A})}{\prod_{i=1,r}^{n} (\lambda_r - \lambda_i)}$$

Because

$$(\lambda_r \mathbf{I} - \mathbf{A}) \operatorname{adj}(\lambda_r \mathbf{I} - \mathbf{A}) = \mathbf{0}$$

it follows that

$$\lambda_r \frac{\operatorname{adj}(\lambda_r \mathbf{I} - \mathbf{A})}{\prod_{i=1,r}^{n} (\lambda_r - \lambda_i)} = \sum_{j=1}^{n} \lambda_j \frac{\operatorname{adj}(\lambda_j \mathbf{I} - \mathbf{A})}{\prod_{i=1,j}^{n} (\lambda_j - \lambda_i)} \frac{\operatorname{adj}(\lambda_r \mathbf{I} - \mathbf{A})}{\prod_{i=1,r}^{n} (\lambda_r - \lambda_i)}$$

and thus defining

$$\mathbf{S}_r = \frac{\operatorname{adj}(\lambda_r \mathbf{I} - \mathbf{A})}{\prod_{i=1,r}^{n} (\lambda_r - \lambda_i)}$$

there results

$$\mathbf{S}_r \mathbf{S}_j = 0; \qquad r \neq j$$

and

$$\mathbf{S}_r^2 = \mathbf{S}_r$$

and in general

$$\mathbf{S}_r^m = \mathbf{S}_r$$

Thus the matrices $\mathbf{S}_r$ play the role of unity matrices which have no inverses. Equation (5.14.4) may be written for $\mathbf{I}$ also so that

$$\mathbf{I} = \sum_{r=1}^{n} \mathbf{S}_r$$

The idea of an expansion formula then becomes more apparent when Eq. (5.14.4) is written in the form

$$Q(\mathbf{A}) = \sum_{r=1}^{n} Q(\lambda_r) \mathbf{S}_r$$

where the $Q(\lambda_r)$ may be considered as coefficients determined by the character of a particular function defined on a given matrix $\mathbf{A}$, and the $\mathbf{S}_r$ are always matrices of the same form but with different numerical values depending only upon the eigenvalues of $\mathbf{A}$ but not depending upon Q.

5.15 Gram-Schmidt Orthogonalization Process

In Chapters 3 and 4 we discussed the expansion of a vector in terms of a complete set of vectors. In Chapter 3, in particular, we were concerned with a finite number of vectors and we devised an algorithm for the systematic expansion of a vector in terms of a basis. In Chapter 4, this procedure was exploited in its application to linear programming. In preceding sections of this chapter it was shown that, if one had a complete set of vectors which was orthogonal, the expansion of a vector in terms of a basis was expedited considerably. This leads to a natural problem of how one might form an orthogonal basis from a nonorthogonal basis, the solution of which is called the Gram-Schmidt orthogonalization procedure.

Let there be given a set of n real vectors $\{\mathbf{x}_i\}$ which form a basis and suppose, for convenience only, the vectors have been normalized by dividing each by the square root of the sum of the squares of its components. Let $\mathbf{x}_1$ be the first vector and call it $\mathbf{u}_1$. We now want to choose a second vector $\mathbf{u}_2'$ which is orthogonal to $\mathbf{u}_1$. Form

$$\mathbf{u}_2' = \mathbf{x}_2 - c_o\mathbf{u}_1$$

If $\mathbf{u}_2'$ is to be orthogonal to $\mathbf{u}_1$, then, premultiplying by $\mathbf{u}_1^T$,

$$0 = \mathbf{u}_1^T\mathbf{u}_2' = \mathbf{u}_1^T\mathbf{x}_2 - c_o$$

or

$$c_o = \mathbf{u}_1^T\mathbf{x}_2$$

so

$$\mathbf{u}_2' = \mathbf{x}_2 - (\mathbf{u}_1^T\mathbf{x}_2)\mathbf{u}_1$$

and when it is normalized call it $\mathbf{u}_2$. Consider now a third vector $\mathbf{u}_3'$ which is to be orthogonal to $\mathbf{u}_1$ and $\mathbf{u}_2$. Let

$$\mathbf{u}_3' = \mathbf{x}_3 - c_1\mathbf{u}_1 - c_2\mathbf{u}_2$$

Multiply by $\mathbf{u}_1^T$ and $\mathbf{u}_2^T$, separately, obtaining

$$c_1 = \mathbf{u}_1^T\mathbf{x}_3$$

$$c_2 = \mathbf{u}_2^T\mathbf{x}_3$$

so

$$\mathbf{u}_3' = \mathbf{x}_3 - (\mathbf{u}_1^T\mathbf{x}_3)\mathbf{u}_1 - (\mathbf{u}_2^T\mathbf{x}_3)\mathbf{u}_2$$

Let the vector $\mathbf{u}_3'$ when normalized be $\mathbf{u}_3$. This process may be repeated and

$$\mathbf{u}_4' = \mathbf{x}_4 - (\mathbf{u}_1^T\mathbf{x}_4)\mathbf{u}_1 - (\mathbf{u}_2^T\mathbf{x}_4)\mathbf{u}_2 - (\mathbf{u}_3^T\mathbf{x}_4)\mathbf{u}_3$$

and if $\mathbf{u}_4'$ is normalized we call it $\mathbf{u}_4$. In general,

$$\mathbf{u}_s' = \mathbf{x}_s - \sum_{j=1}^{s-1} (\mathbf{u}_j^T\mathbf{x}_s)\mathbf{u}_j$$

Obviously this procedure cannot fail to produce an "orthonormal" set of vectors since, if $\mathbf{u}_s'$ is zero for any s, then $\mathbf{x}_s$ must be a linear combination of the $\mathbf{u}_j$ which is impossible since the $\{\mathbf{u}_j\}$ are a linear combination of the $\{\mathbf{x}_j\}$. This is a useful procedure and is used in the approximate computation of eigenvectors and eigenvalues in the next section.

5.16 The Calculation of Eigenvalues and Eigenvetors

A number of examples have been used in the previous sections in which eigenvalues and eigenvectors have been computed. These calculations were made for matrices of very small order and for matrix elements which were small integers. Hence various determinants could be easily computed and, in particular, the characteristic equation could be obtained with ease. Obviously, for large systems, less tedious and more direct methods must be devised as well as methods of approximation. At the end of Section 5.12, a method was presented for obtaining the characteristic equation from the characteristic matrix which required the computation of the various powers of the matrix itself. Once the characteristic equation is obtained, the eigenvalues may be obtained by a variety of numerical procedures and the eigenvectors obtained by a straightforward application of previously discussed methods for the solution of homogeneous equations. Although the purpose of this work is mathematical rather than numerical and although there are extensive treatises which treat numerical methods, it may not be out of place to discuss some of the methods used very briefly to obtain the characteristic equation, eigenvalues, and eigenvectors. Most of the methods which have been developed for the computation of eigenvalues and eigenrows are applicable to special matrices and general methods for arbitrary non-symmetric matrices will not be presented here. Consider for the moment symmetric matrices. It is interesting and useful to realize that, if an approximate eigenvector is known, with an approximation of order ϵ, the corresponding approximate eigenvalue is determined to the second order in ϵ. This may be seen as follows: If

$$\mathbf{A}\mathbf{x}_j = \lambda_j \mathbf{x}_j$$

where $\mathbf{x}_j$ and λ_j are exact eigenvectors and eigenvalues of a symmetric matrix then

$$\mathbf{x}_j^T \mathbf{A}\mathbf{x}_j = \lambda_j \mathbf{x}_j^T \mathbf{x}_j$$

or

$$\lambda_j = \frac{\mathbf{x}_j^T \mathbf{A}\mathbf{x}_j}{\mathbf{x}_j^T \mathbf{x}_j}$$

The question then arises as to what one can say about the value λ' obtained from the same ratio if a vector $\mathbf{z}$ is substituted for $\mathbf{x}_j$,

$$\lambda' = \frac{\mathbf{z}^T \mathbf{A}\mathbf{z}}{\mathbf{z}^T \mathbf{z}} \tag{5.16.1}$$

Suppose that

$$\mathbf{z} = \mathbf{x}_j + \boldsymbol{\epsilon}$$

where $\boldsymbol{\epsilon}$ is a small vector, where *small* means that each component is small. Now

$$\begin{aligned}
\lambda' \mathbf{z}^T\mathbf{z} &= \mathbf{z}^T\mathbf{A}\mathbf{z} \\
&= (\mathbf{x}_j^T + \boldsymbol{\epsilon}^T)(\mathbf{A})(\mathbf{x}_j + \boldsymbol{\epsilon}) \\
&= \mathbf{x}_j^T\mathbf{A}\mathbf{x}_j + \mathbf{x}_j^T\mathbf{A}\boldsymbol{\epsilon} + \boldsymbol{\epsilon}^T\mathbf{A}\mathbf{x}_j + \boldsymbol{\epsilon}^T\mathbf{A}\boldsymbol{\epsilon} \\
&= \lambda_j\mathbf{x}_j^T\mathbf{x}_j + \mathbf{x}_j^T\mathbf{A}\boldsymbol{\epsilon} + \lambda_j\boldsymbol{\epsilon}^T\mathbf{x}_j + \boldsymbol{\epsilon}^T\mathbf{A}\boldsymbol{\epsilon}
\end{aligned}$$

Form

$$(\lambda' - \lambda_j)\mathbf{z}^T\mathbf{z} = \mathbf{x}_j^T\mathbf{A}\boldsymbol{\epsilon} + \boldsymbol{\epsilon}^T\mathbf{A}\boldsymbol{\epsilon} - \lambda_j\mathbf{x}_j^T\boldsymbol{\epsilon} - \lambda_j\boldsymbol{\epsilon}^T\boldsymbol{\epsilon}$$

and if $\mathbf{A}$ is symmetric

$$\mathbf{x}_j^T\mathbf{A}\boldsymbol{\epsilon} = \lambda_j\mathbf{x}_j^T\boldsymbol{\epsilon}$$

so

$$\lambda' - \lambda_j = \frac{\boldsymbol{\epsilon}^T\mathbf{A}\boldsymbol{\epsilon} - \lambda_j\boldsymbol{\epsilon}^T\boldsymbol{\epsilon}}{\mathbf{z}^T\mathbf{z}}$$

and therefore $\lambda' - \lambda$ is second order in $\boldsymbol{\epsilon}$ if $\mathbf{z} - \mathbf{x}_j$ is first order in $\boldsymbol{\epsilon}$. This has an important bearing in calculating eigenvalues, for it means there is a good possibility that the successive substitution method for calculating eigenvectors and eigenrows should converge. That is, if one assumes a value for $\mathbf{x}_j$, say $\mathbf{z}^{(1)}$, which is near to the exact value of $\mathbf{x}_j$ an approximate eigenvalue $\lambda_j^{(1)}$ will be obtained. The new approximate eigenvector $\mathbf{z}^{(2)}$ may be obtained* from

$$(\mathbf{A} - \lambda_j^{(1)}\mathbf{I})\mathbf{z}^{(2)} = 0$$

Substitution of $\mathbf{z}^{(2)}$ into Eq. (5.16.1) will produce a new $\lambda_j^{(2)}$, etc. Then hopefully

$$|\lambda_j^{(m)} - \lambda_j^{(m+1)}| < \eta$$

where $\eta \longrightarrow 0$ as m increases.

Suppose that a matrix has real eigenvalues and suppose that the eigenvalues are distinct and let them be ordered so that

$$|\lambda_1| > |\lambda_2| > |\lambda_3| \ldots > |\lambda_n|$$

and consider

$$\mathbf{y} = \mathbf{A}\mathbf{x}$$

Now assume a vector $\mathbf{x}$, call it $\mathbf{x}^{(0)}$ and let it be such that its largest element is unity, then a $\mathbf{y}^{(1)}$ is generated

$$\mathbf{y}^{(1)} = \mathbf{A}\mathbf{x}^{(0)}$$

*Since $\lambda_j^{(1)}$ is known only approximately, $\mathbf{z}^{(2)}$ must be obtained by omitting one of the equations from the homogeneous set of equations, otherwise the only solution is the trivial one.

Now repeat the operation but instead of substituting $\mathbf{y}^{(1)}$ into $\mathbf{y} = \mathbf{Ax}$ for $\mathbf{x}$ directly modify it by dividing the vector by its largest element (in absolute value) thus making the largest element unity. Call this $k^{(1)}$ so

$$\mathbf{y}^{(1)} = k^{(1)}\mathbf{z}^{(1)}$$

then

$$\mathbf{y}^{(2)} = \mathbf{Az}^{(1)}$$

and let

$$\mathbf{y}^{(2)} = k^{(2)}\mathbf{z}^{(2)}$$

where $k^{(2)}$ is the largest element in absolute value in $\mathbf{y}^{(2)}$. Repetition of this procedure will produce

$$\mathbf{y}^{(p+1)} = \mathbf{Az}^{(p)}$$

$$\mathbf{y}^{(p+1)} = k^{(p+1)}\mathbf{z}^{(p+1)}$$

where

$$k^{(p+1)} \longrightarrow \lambda$$

an eigenvalue and $\mathbf{y}^{(p+1)}$ will be the corresponding eigenvector. It will be shown that this procedure always leads to the eigenvalue which is largest in absolute value for a symmetric matrix. If the smallest eigenvalue is desired, then it is obvious that one should start with the equation

$$\mathbf{A}^{-1}\mathbf{x} = \lambda^{-1}\mathbf{x}$$

or

$$\mathbf{y} = \mathbf{A}^{-1}\mathbf{x}$$

and repeat the same analysis. The eigenvalue will then be the largest λ^{-1} in absolute value or the smallest λ in absolute value.

As an example consider

$$\mathbf{A} = \begin{bmatrix} 1 & 1 \\ 2 & 1 \end{bmatrix}; \qquad \lambda_1 = 2.414, \quad \lambda_2 = -0.414$$

and assume as a first guess

$$\begin{bmatrix} 1 \\ 0 \end{bmatrix}$$

then the successive eigenvalues and eigenvector approximations are

$$2, 2, \tfrac{5}{2}, \tfrac{12}{5}, \tfrac{29}{12}, \tfrac{70}{29}, \tfrac{169}{70}, \ldots$$

and $\frac{169}{70} = 2.414$. The vectors are

$$\begin{bmatrix} 1 \\ 2 \end{bmatrix}, \begin{bmatrix} \frac{3}{2} \\ 2 \end{bmatrix}, \begin{bmatrix} \frac{7}{4} \\ \frac{5}{2} \end{bmatrix}, \begin{bmatrix} \frac{17}{10} \\ \frac{12}{5} \end{bmatrix}, \begin{bmatrix} \frac{41}{24} \\ \frac{29}{12} \end{bmatrix}, \begin{bmatrix} \frac{99}{58} \\ \frac{70}{29} \end{bmatrix}, \begin{bmatrix} \frac{239}{140} \\ \frac{169}{70} \end{bmatrix}, \ldots$$

Also

$$\mathbf{A}^{-1} = \begin{bmatrix} -1 & 1 \\ 2 & -1 \end{bmatrix}$$

and with the same first guess, the successive reciprocal of the eigenvalues are

$$2, -2, -\tfrac{5}{2}, -\tfrac{12}{5}, -\tfrac{29}{12}, -\tfrac{70}{29}, -\tfrac{169}{70}, \ldots$$

and $-\frac{70}{169} = 0.414$. The corresponding eigenvectors are

$$\begin{bmatrix} -1 \\ 2 \end{bmatrix}, \begin{bmatrix} \frac{3}{2} \\ -2 \end{bmatrix}, \begin{bmatrix} \frac{7}{4} \\ -\frac{5}{2} \end{bmatrix}, \begin{bmatrix} \frac{17}{10} \\ -\frac{12}{5} \end{bmatrix}, \begin{bmatrix} \frac{41}{24} \\ -\frac{29}{12} \end{bmatrix}, \begin{bmatrix} \frac{99}{58} \\ -\frac{70}{29} \end{bmatrix}, \begin{bmatrix} \frac{239}{140} \\ -\frac{169}{70} \end{bmatrix}, \ldots$$

This technique will not work, in general, for complex eigenvalues.

If the matrix is a symmetric one, then *any* vector, say, $\mathbf{x}^{(0)}$ may be expanded in terms of the eigenvectors

$$\mathbf{x}^{(0)} = \sum_j \alpha_j \mathbf{x}_j$$

and

$$\mathbf{y}^{(1)} = \mathbf{A}\mathbf{x}^{(0)} = \sum_j \alpha_j \lambda_j \mathbf{x}_j$$

and since $\mathbf{y}^{(1)} = k^{(1)}\mathbf{z}^{(1)}$ it follows that

$$\mathbf{y}^{(2)} = \mathbf{A}\frac{\mathbf{y}^{(1)}}{k^{(1)}} = \frac{1}{k^{(1)}} \sum_j \alpha_j \lambda_j^2 \mathbf{x}_j$$

Continued substitutions will give, using $\mathbf{y}^{(p-1)} = k^{(p-1)}\mathbf{z}^{(p-1)}$ and $\mathbf{y}^{(p+1)} = \mathbf{A}\mathbf{z}^p$,

$$\mathbf{y}^{(p)} = \frac{1}{\prod_{j=1}^{p-1} k^{(j)}} \sum_j \alpha_j \lambda_j^p \mathbf{x}_j$$

$$\mathbf{y}^{(p+1)} = \frac{1}{\prod_{j=1}^{p} k^{(j)}} \sum_j \alpha_j \lambda_j^{p+1} \mathbf{x}_j$$

Now if λ_1 is the largest eigenvalue in absolute value,

$$\mathbf{y}^{(p)} = \frac{\lambda_1^p}{\prod_{j=1}^{p-1} k^{(j)}} \sum_j \alpha_j \left(\frac{\lambda_j}{\lambda_1}\right)^p \mathbf{x}_j$$

so that for p large

$$\mathbf{y}^{(p)} = \frac{\lambda_1^p}{\prod_{j=1}^{p-1} k^{(j)}} \alpha_1 \mathbf{x}_1$$

$$\mathbf{y}^{(p+1)} = \frac{\lambda_1^{p+1}}{\prod_{j=1}^{p} k^{(j)}} \alpha_1 \mathbf{x}_1$$

Thus when the difference $\mathbf{y}^{(p+1)} - \mathbf{y}^{(p)}$ is small

$$k^{(p)} \cong \lambda_1$$

for large p.

We have a procedure for obtaining the largest eigenvalue and corresponding eigenvector and (as pointed out if $\mathbf{A}^{-1}$ is used instead of $\mathbf{A}$) the smallest eigenvalue and corresponding eigenvector; and some attention should be given to finding the remaining eigenvalues. In order to find the second largest eigenvalue in absolute value, one method is to choose a vector

$\mathbf{x}^{o'}$ initially which is orthogonal to the vector $\mathbf{y}^{(p+1)}$. The representation of $\mathbf{x}^{o'}$ in terms of the basic set of vectors will not contain a term in $\mathbf{x}_1$, that is,

$$\mathbf{x}^{o'} = \sum_{j=2}^{n} \beta_j \mathbf{x}_j$$

and the foregoing procedure should then, since $\mathbf{x}_1$ is *not* a member of the expansion, give convergence to λ_2. In actual fact, however, the computations carried out are never exact, and it is found that convergence is again to λ_1, but usually very slowly. This occurs since the errors introduced cause more and more of the $\mathbf{x}_1$ component to be introduced into the calculations. This difficulty can be corrected as follows: We arrange the calculations so that at each stage of the iterative procedure the calculated vector is orthogonal to the vector or vectors already known. Let $\mathbf{x}^o$ be an arbitrary vector and form

$$k_2^{(1)}\mathbf{z}^{(1)} = \mathbf{y}^{(1)} = \mathbf{A}\mathbf{x}^{(o)} - \alpha_1 \mathbf{x}_1$$

where $\mathbf{x}_1$ has been previously obtained and corresponds to λ_1 and where $k_2^{(1)}$ is a scalar so that $\mathbf{z}^{(1)}$ has a largest component of unity. α_1 is determined so that $\mathbf{y}^{(1)}$ is orthogonal to $\mathbf{x}_1$

$$\mathbf{x}_1^T \mathbf{A}\mathbf{x}^{(o)} - \alpha_1 \mathbf{x}_1^T \mathbf{x}_1 = 0$$

or

$$\alpha_1 = \frac{\mathbf{x}_1^T \mathbf{A}\mathbf{x}^{(o)}}{\mathbf{x}_1^T \mathbf{x}_1}$$

Now let

$$k_2^{(2)} z^{(2)} = \mathbf{y}^{(2)} = \mathbf{A}\mathbf{z}^{(1)} - \alpha_2 \mathbf{x}_1$$

$$\alpha_2 = \frac{\mathbf{x}_1^T \mathbf{A}\mathbf{z}^{(1)}}{\mathbf{x}_1^T \mathbf{x}_1}$$

and $\mathbf{y}^{(2)}$ is orthogonal to $\mathbf{x}_1$. Continuing in this way

$$k_2^{(p)}\mathbf{z}^{(p)} = \mathbf{y}^{(p)} = \mathbf{A}\mathbf{z}^{(p-1)} - \alpha_p \mathbf{x}_1$$

$$\alpha_p = \frac{\mathbf{x}_1^T \mathbf{A}\mathbf{z}^{(p-1)}}{\mathbf{x}_1^T \mathbf{x}_1}$$

and clearly

$$k_2^{(p)} \longrightarrow \lambda_2$$

as p increases and

$$\mathbf{y}^{(p)} = k_2^{(p)}\mathbf{z}^{(p)} \longrightarrow \mathbf{x}_2$$

This is obvious from the analysis for λ_1 and $\mathbf{x}_1$ since the mode of formation of $\mathbf{y}^{(p)}$ guarantees that the vectors $\mathbf{y}^{(p)}$ have no $\mathbf{x}_1$ component.

5.17 Eigenvalues of the Jacobi Matrix

In many of the applications in chemical engineering where matrices arise, such as distillation, absorption, continuous stirred tank reactors, and others, the matrices are of the form

$$\mathbf{A} = \begin{bmatrix} a_1 & b_1 & 0 & & \cdots & 0 \\ c_1 & a_2 & b_2 & & & \vdots \\ 0 & c_2 & a_3 & b_3 & & 0 \\ \vdots & \cdots & & & & b_{n-1} \\ 0 & & & \cdots & c_{n-1} \quad \cdots & a_n \end{bmatrix}$$

where the elements not shown are zero. This is a tridiagonal matrix and occurs frequently enough that it should be considered in some detail. Its presence will be seen in a number of applications in this book. In the transient absorber problem in Chapter 6,

$$\begin{aligned} a_1 &= a_2 = a_3 = \ldots = a_n \\ b_1 &= b_2 \qquad \ldots = b_{n-1} \\ c_1 &= \qquad\quad \ldots = c_{n-1} \end{aligned}$$

and in this case the eigenvalues and eigenvectors may be found analytically. The preceding matrix $\mathbf{A}$ is called a *Jacobi matrix*. Write

$$\det[\mathbf{A} - \lambda\mathbf{I}] = \begin{vmatrix} a_1 - \lambda & b_1 & 0 & 0 & \cdots & 0 \\ c_1 & a_2 - \lambda & b_2 & 0 & \cdots & 0 \\ 0 & c_2 & a_3 - \lambda & b_3 & \cdots & 0 \\ \vdots & & & & \ddots & b_{n-1} \\ 0 & 0 & \cdots & 0 \quad \cdots & c_{n-1} \quad \cdots & a_n - \lambda \end{vmatrix}$$

This determinant may be computed by expanding by elements of the last column. Let $P_n(\lambda)$ be the value of the determinant. Then if the last row and last column are omitted, we get a determinant which we can call $P_{n-1}(\lambda)$, and if the two last rows and columns are omitted, there will be a determinant of the same form, $P_{n-2}(\lambda)$. Expansion gives

$$\begin{aligned} P_n(\lambda) &= (a_n - \lambda)P_{n-1}(\lambda) - b_{n-1}c_{n-1}P_{n-2}(\lambda) \\ P_{n-1}(\lambda) &= (a_{n-1} - \lambda)P_{n-2}(\lambda) - b_{n-2}c_{n-2}P_{n-3}(\lambda) \\ &\vdots \\ P_r(\lambda) &= (a_r - \lambda)P_{r-1}(\lambda) - b_{r-1}c_{r-1}P_{r-2}(\lambda) \\ &\vdots \\ P_3(\lambda) &= (a_3 - \lambda)P_2(\lambda) - b_2c_2P_1(\lambda) \\ P_2(\lambda) &= (a_2 - \lambda)P_1(\lambda) - b_1c_1P_o(\lambda) \\ P_1(\lambda) &= (a_1 - \lambda)P_o(\lambda) \end{aligned} \tag{5.17.1}$$

where $P_0(\lambda)$ is one. These recurrence formulae are convenient to use to compute real eigenvalues since one can assume a value of λ' and calculate easily the sequence

$$P_1(\lambda'), P_2(\lambda'), P_3(\lambda'), \ldots, P_{n-1}(\lambda'), P_n(\lambda')$$

If $P_n(\lambda') = 0$, then λ' is an eigenvalue, and the eigenvalue can be found without obtaining the characteristic equation in detail. The computation of the eigenvectors is simple, for they are obtained as the solution of

$$(\mathbf{A} - \lambda\mathbf{I})\,\mathbf{x} = \mathbf{o}$$

or

$$\begin{aligned}
(a_1 - \lambda)x_1 + b_1x_2 &= 0\\
c_1x_1 + (a_2 - \lambda)x_2 + b_2x_3 &= 0\\
c_2x_2 + (a_3 - \lambda)x_3 + b_3x_4 &= 0\\
c_rx_r + (a_{r+1} - \lambda)x_{r+1} + b_{r+1}x_{r+2} &= 0\\
&\vdots\\
c_{n-2}x_{n-2} + (a_{n-1} - \lambda)x_{n-1} + b_{n-1}x_n &= 0\\
c_{n-1}x_{n-1} + (a_n - \lambda)x_n &= 0
\end{aligned}$$

These may be solved to give

$$\begin{aligned}
x_2 &= -\frac{1}{b_1}(a_1 - \lambda)x_1\\
x_3 &= -\frac{1}{b_2}[(a_2 - \lambda)x_2 + c_1x_1]\\
x_4 &= -\frac{1}{b_3}[(a_3 - \lambda)x_3 + c_2x_2]\\
&\vdots\\
x_n &= -\frac{1}{b_{n-1}}[(a_{n-1} - \lambda)x_{n-1} + c_{n-2}x_{n-2}]
\end{aligned}$$

and the last equation can be omitted or, alternatively, used as a consistency check, that is,

$$c_{n-1}x_n + (a_n - \lambda)x_n = 0$$

Hence one can assume a value for x_1 and the remaining x_j may be computed corresponding to the eigenvalue λ.

5.18 Sturmian Functions

One further comment on the calculation of eigenvalues from Eq. (5.17.1) is in order. One of the ways in which zeroes of a polynomial may be located if they are real is by an examination of Sturm's functions related to that polynomial. Sturm's theorem may be stated for our purposes as follows: Suppose a sequence of polynomials, with real coefficients, $f(x)$, $f_1(x)$, $f_2(x)$, $f_3(x)$, . . . , $f_m(x)$, f_{m+1} where, f_{m+1} is a non-zero constant, is given and suppose the polynomials are related to each other by the formulae

$$c_o f(x) = q_1(x) f_1(x) - f_2(x)$$
$$c_1 f_1(x) = q_2(x) f_2(x) - f_3(x)$$
$$c_2 f_2(x) = q_3(x) f_3(x) - f_4(x)$$
$$\vdots$$
$$c_{m-2} f_{m-2}(x) = q_{m-1}(x) f_{m-1}(x) - f_m(x)$$
$$c_{m-1} f_{m-1}(x) = q_m(x) f_m(x) - f_{m+1}$$

where $\{q_i(x)\}$ is a set of polynomials and where the constants $\{c_i\}$ are introduced only so that numerical fractions can be avoided. Then

$$\frac{c_j f_j(x)}{f_{j+1}(x)} = q_{j+1}(x) - \frac{f_{j+2}(x)}{f_{j+1}(x)}$$

so that the polynomials are related to each other by a division process, $q_{j+1}(x)$ being the quotient and $f_{j+2}(x)$ the remainder. Now consider the sequence in order

$$f(x), f_1(x), f_2(x), \; . \; . \; . \; , f_m(x), f_{m+1}$$

and consider their values at $x = a$ and $x = b$ and let $V(a)$ and $V(b)$ be the number of sign changes in the sequence at $x = a$ and $x = b$ respectively, $a < x < b$. The number of real roots of $f(x)$, N, between $x = a$ and $x = b$ is

$$N = V(a) - V(b)$$

The proof of this theorem is not presented here but may be found in Dickson.*

The fact to be noted here is that the recursion formulae for the characteristic polynomials given in Eq. (5.17.1) form a Sturm series of functions and can therefore be used to locate the real eigenvalues of the characteristic polynomial. To be precise, if one assumes two values of λ, say λ' and λ'' and computes the sequence

$$P_n, P_{n-1}, P_{n-2}, \; . \; . \; . \; , P_2, P_1, P_o$$

then from

$$V(\lambda') - V(\lambda'')$$

one can deduce the number of real eigenvalues in the interval $\lambda' < \lambda < \lambda''$.

In a large number of cases in chemical engineering problems, the $\{b_j\}$

*L. E. Dickson, *New First Course in the Theory of Equations*, New York: John Wiley & Sons, 1939, p. 83.

and $\{c_j\}$ are positive real numbers whereas the $\{a_j\}$ are negative real numbers. Now for λ large and positive

$$P_o(\lambda) = 1 > 0$$
$$P_1(\lambda) = (a_1 - \lambda) < 0$$
$$P_2(\lambda) = (a_2 - \lambda)P_1(\lambda) - b_1c_1 > 0$$
$$P_3(\lambda) = (a_3 - \lambda)P_2(\lambda) - b_2c_2P_1(\lambda) < 0$$
$$\vdots$$
$$P_{n-1}(\lambda) = (a_{n-1} - \lambda)P_{n-2}(\lambda) - b_{n-2}c_{n-2}P_{n-3}(\lambda) \begin{cases} < 0 & n \text{ even} \\ > 0 & n \text{ odd} \end{cases}$$
$$P_n(\lambda) = (a_n - \lambda)P_{n-1}(\lambda) - b_{n-1}c_{n-1}P_{n-2}(\lambda) \qquad \begin{cases} > 0 & n \text{ even} \\ < 0 & n \text{ odd} \end{cases}$$

On the other hand, for λ large and negative

$$P_o(\lambda) = 1 > 0$$
$$P_1(\lambda) > 0$$
$$P_2(\lambda) > 0$$
$$P_3(\lambda) > 0$$
$$\vdots$$
$$P_{n-1}(\lambda) > 0$$
$$P_n(\lambda) > 0$$

and it is clear that the net number of sign changes is n. Thus all eigenvalues for such a matrix will be *real*. It will be shown later that, in fact, the relationship among a_j, b_j, c_j is such that the eigenvalues will be nonpositive.

EXERCISES

Consider the matrix

$$\mathbf{A} = \begin{bmatrix} 2 & -2 & 3 \\ 1 & 1 & 1 \\ 1 & 3 & -1 \end{bmatrix}$$

1. Compute the eigenvalues.
2. Compute a set of eigenvectors and a set of eigenrows and show by computation that they form two biorthogonal sets.
3. Show by direct calculation that the matrix satisfies its own characteristic equation.

4. Compute $\mathbf{A}^4$ using the Hamilton-Cayley theorem.

5. Compute $\mathbf{A}^{-1}$ and check.

6. Show by direct computation that the set of eigenvectors and eigenrows are linearly independent.

7. Calculate the eigenvalues of the matrix

$$3\mathbf{A}^6 + 4\mathbf{A}^{-4} + 5\mathbf{A}^2 - 6\mathbf{A}^{-1} - 7\mathbf{I}$$

8. Suppose m is an arbitrary integer greater than 3. Compute a formula for $\mathbf{A}^m$ using Sylvester's theorem.

9. Suppose m is very large, compute an *approximate* value of $\mathbf{A}^{100}$.

10. Given the system of differential equations which arise in a kinetics problem

$$\frac{dx_1}{d\theta} = 2x_1 - 2x_2 + 3x_3$$

$$\frac{dx_2}{d\theta} = x_1 + x_2 + x_3$$

$$\frac{dx_3}{d\theta} = x_1 + 3x_2 - x_3$$

with initial concentrations

$$x_1 = 1, \quad \theta = 0$$
$$x_2 = 0, \quad \theta = 0$$
$$x_3 = \tfrac{1}{2}, \quad \theta = 0$$

Compute the vector of solutions and exhibit the three solutions.

11. Prove the assertion in Section 5.2 that, if the eigenvalues of a matrix are simple, the rank of the characteristic matrix is $(n - 1)$.

Some Problems in Staged Operations

6

6.1 Introduction

In this chapter we discuss some of the problems which may be conveniently handled by the methods of the previous chapters. These problems may not be important since they are grossly simplified idealizations, but, hopefully, the solution of such models should give some insight into the solution and behavior of more complicated situations. We shall consider problems in multicomponent rectification, transient absorber problems, and transient stirred pot reactors. For completeness, and to show how more realistic problems may be phrased in the language of this book, a steady state distillation problem will be considered for more realistic physical models.

6.2 Multicomponent Rectification

The problem to be considered here is not the general problem of a distillation column consisting of a rectification section and reboiler section with feed at an intermediate stage, but rather the problem of a rectification section with a total condenser at the top, a part of the total condensate returned to the top stage as reflux. We shall then consider the problem to be solved, from our point of view, if, knowing the composition of the product, we can compute the composition at any stage, without computing the compositions at any intermediate stage. A similar problem can be solved for the reboiler section of a full distillation column and a trial-and-error method could then be developed to match the compositions at the feed stage if this were desired. This will not be done here.

As is usual in chemical engineering calculations let the mole fraction of the ith species in the nth stage be x_{ni} for the liquid and y_{ni} for the vapor. We shall consider N stages. Figure 6.2.1 shows a schematic diagram. We shall treat the case of constant molal overflow so that the liquid flow from stage to stage is L, the upward vapor flow is V, and the reflux ratio R is the ratio of the moles of product L returned to the top of the column from the reflux drum to the amount withdrawn from the drum as product D. Under these assumptions, it is well known from mass conservation that

$$y_{n+1,i} = \frac{R}{R+1} x_{ni} + \frac{x_{oi}}{R+1}, \quad i = 1, 2, 3, \ldots, m \tag{6.2.1}$$

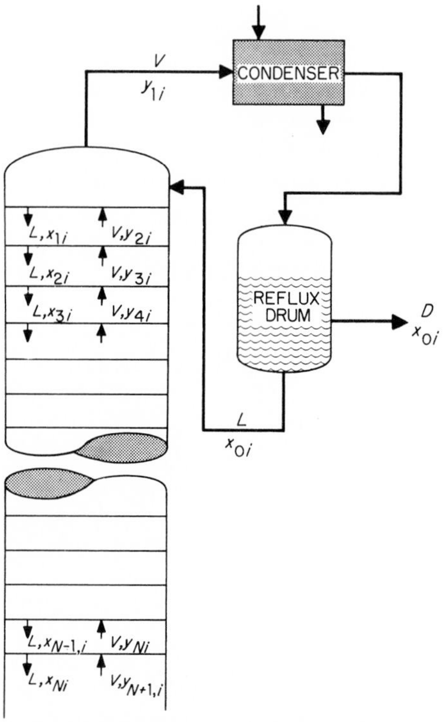

Figure 6.2.1 Schematic diagram of the rectification section of a distillation column with a total condenser.

where x_{oi} is the composition of the product D which, because of the total condenser, is the same as y_{1i}. Because of the simplifying assumptions, this is a set of linear equations connecting liquid composition on one stage with vapor composition coming to that stage. It is assumed that $\{x_{oi}\}$ is known. It will be assumed that there is a relation between liquid and vapor composition on each stage of the form

$$y_{ni} = K_{ni} x_{ni}$$

where K_{ni} is related to vapor pressures, fugacities, temperature, pressure, etc. It is known that this equation may be written in the form,

$$y_{ni} = \frac{\alpha_{ni} x_{ni}}{\sum \alpha_{ni} x_{ni}} \qquad i = 1, 2, 3, \ldots, m \tag{6.2.2}$$

where α_{ni} is the ratio K_{ni}/K_{nj} for two components, and, if j is fixed, there are m such ratios with one of the ratios being one. α_{ni} is called the *relative volatility*, that is, the volatility of the ith species relative to the volatility of the jth species. In order to preserve the symmetry of the formulae, α_{nj} will be preserved and not written as one. Also, for the same reason, although we know that

$$\sum x_{ni} = 1 \quad \text{and} \quad \sum y_{ni} = 1$$

all the preceding m equations will be written instead of substituting for one in terms of the other $m - 1$. Whenever a summation occurs from here on, if no mention is made of the summation variable, it will be over the components from one to m. An additional assumption will be made that the relative volatility is a constant, independent of n but not i. Thus we will write $\alpha_{ni} = \alpha_i$

so

$$y_{ni} = \frac{\alpha_i x_{ni}}{\sum \alpha_i x_{ni}} \tag{6.2.3}$$

This equation may be inverted by dividing by α_i and summing

$$\sum \frac{y_{ni}}{\alpha_i} = \sum x_{ni} (\sum x_{ni} \alpha_i)^{-1} = (\sum x_{ni} \alpha_i)^{-1}$$

and therefore

$$x_{ni} = \frac{y_{ni}/\alpha_i}{\sum y_{ni}/\alpha_i}$$

If we define $p_i = \alpha_i^{-1}$, the reciprocal relative volatility, then

$$x_{ni} = \frac{p_i y_{ni}}{\sum p_i y_{ni}} \tag{6.2.4}$$

Our problem now becomes clear. Starting at the top of the column, $\{y_{1i}\}$ is known; hence, from Eq. (6.2.4), $\{x_{1i}\}$ may be calculated. From Eq. (6.2.1), $\{y_{2i}\}$ may be calculated. $\{x_{2i}\}$ may then be obtained from Eq. (6.2.4) and the whole process repeated. This calculation is somewhat tedious but it is straightforward. We shall now show that it can be systematized and put into a form so that $\{x_{Ni}\}$ may be obtained as a function of $\{x_{oi}\}$, R, $\{p_i\}$, and N, the number of stages under consideration.

$\{x_{1i}\}$ may be computed from

$$x_{1i} = \frac{p_i x_{oi}}{\sum p_i x_{oi}}$$

since $x_{oi} = y_{1i}$. In order to determine the structure of the formulae which obtain let p_i, in this formula, be denoted by X_{1i}, then

$$x_{1i} = \frac{X_{1i} x_{oi}}{\sum X_{1i} x_{oi}}, \qquad X_{1i} = p_i, \quad i = 1, 2, \ldots, m$$

and

$$y_{2i} = \frac{R}{R+1} \frac{X_{1i} x_{oi}}{\sum X_{1i} x_{oi}} + \frac{x_{oi}}{R+1}$$

When this is substituted into Eq. (6.2.4), one obtains

$$x_{2i} = \frac{(Rp_i X_{1i} + p_i \sum X_{1i} x_{oi}) x_{oi}}{\sum (Rp_i X_{1i} + p_i \sum X_{1i} x_{oi}) x_{oi}}$$

Now define

$$X_{2i} = Rp_i X_{1i} + p_i \sum X_{1i} x_{oi}$$

then,

$$x_{2i} = \frac{X_{2i} x_{oi}}{\sum X_{2i} x_{oi}}$$

If x_{3i} is computed, one will obtain

$$x_{3i} = \frac{X_{3i} x_{oi}}{\sum X_{3i} x_{oi}}$$

where

$$X_{3i} = Rp_i X_{2i} + p_i \sum X_{2i} x_{oi}$$

If one defines

$$X_{ni} = Rp_i X_{n-1,i} + p_i \sum X_{n-1,i} x_{oi}$$

then,

$$x_{ni} = \frac{X_{ni} x_{oi}}{\sum X_{ni} x_{oi}} \tag{6.2.5}$$

and this may be easily established. The interesting point here is that, if we can solve the system of equations,

$$\left.\begin{aligned} X_{ni} &= Rp_i X_{n-1,i} + p_i \sum X_{n-1,i}\, x_{oi} \\ X_{1i} &= p_i \end{aligned}\right\} \quad i = 1, 2, \ldots, m \tag{6.2.6}$$

then the composition on the nth stage may be computed from Eq. (6.2.5). Our aim now is to consider in detail the system of Eq. (6.2.6) and the solution will be obtained by two different methods.

6.3 The Solution as a Difference Equation

We note that we seek a solution of Eq. (6.2.6) in the form,

$$X_{ni} = X_{ni}(R, \{p_i\}, \{x_{oi}\}, n)$$

and we shall write the set of Eq. (6.2.6) as a set of linear relations between X_{ni} and $X_{n-1,i}$ for each i to give

$$X_{n1} = (Rp_1 + p_1x_{o1})X_{n-1,1} + p_1x_{o2}X_{n-1,2} + p_1x_{o3}X_{n-1,3} + \cdots + p_1x_{om}X_{n-1,m}$$
$$X_{n2} = p_2x_{o1}X_{n-1,1} + (Rp_2 + p_2x_{o2})X_{n-1,2} + p_2x_{o3}X_{n-1,3} + \cdots + p_2x_{om}X_{n-1,m}$$
$$\vdots$$
$$X_{nm} = p_mx_{o1}X_{n-1,1} + p_mx_{o2}X_{n-1,2} + \cdots + p_mx_{o,m-1}X_{n-1,m-1} + (Rp_m + p_mx_{om})X_{n-1,m}$$

We define vectors

$$\mathbf{X}_n = \begin{bmatrix} X_{n1} \\ X_{n2} \\ \cdot \\ \cdot \\ X_{nm} \end{bmatrix}; \qquad \mathbf{X}_1 = \begin{bmatrix} p_1 \\ p_2 \\ \cdot \\ \cdot \\ \cdot \\ p_m \end{bmatrix} = \mathbf{p}$$

and a matrix

$$\mathbf{A} = \begin{bmatrix} Rp_1 + p_1x_{o1} & p_1x_{o2} & p_1x_{o3} & \cdots & p_1x_{om} \\ p_2x_{o1} & Rp_2 + p_2x_{o2} & p_2x_{o3} & \cdots & p_2x_{om} \\ \cdot & \cdot & \cdot & & \cdot \\ \cdot & \cdot & \cdot & & \cdot \\ \cdot & \cdot & \cdot & & \cdot \\ p_mx_{o1} & p_mx_{o2} & p_mx_{o3} & \cdots & Rp_m + p_mx_{om} \end{bmatrix}$$

then the set of linear relations, Eq. (6.2.6), may be written

$$\mathbf{X}_n = \mathbf{A}\mathbf{X}_{n-1}; \qquad \mathbf{X}_1 = \mathbf{p} \tag{6.3.7}$$

Equation (6.3.7) is a matrix difference equation of the first order in the variable n and may be solved by inspection, since

$$\mathbf{X}_2 = \mathbf{A}\mathbf{X}_1 = \mathbf{A}\mathbf{p}$$
$$\mathbf{X}_3 = \mathbf{A}\mathbf{X}_2 = \mathbf{A}^2\mathbf{p}$$

and, in general,

$$\mathbf{X}_n = \mathbf{A}\mathbf{X}_{n-1} = \mathbf{A}^{n-1}\mathbf{p}$$

We could stop at this point, since $\mathbf{A}$ is a matrix of constants and a simple machine program would give $\mathbf{A}^{n-1}$ and $\mathbf{A}^{n-1}\,\mathbf{p}$; hence $\{x_{ni}\}$. The analysis may, however, be carried further. From Sylvester's formula, Section 5.14, we have

$$\mathbf{A}^{n-1} = \sum_{j=1}^{m} \lambda_j^{n-1} \frac{\operatorname{adj}(\lambda_j \mathbf{I} - \mathbf{A})}{\prod\limits_{i=1,j}^{m} (\lambda_j - \lambda_i)}$$

hence we must compute the eigenvalues of $\mathbf{A}$ and the adj $(\lambda_j\mathbf{I} - \mathbf{A})$. Consider $\mathbf{A} - \lambda\mathbf{I}$ and let

$$\frac{Rp_j - \lambda}{p_j x_{oj}} = a_j$$

then,

$$G = |\mathbf{A} - \lambda\mathbf{I}| = \prod p_i \prod x_{oi} \begin{vmatrix} 1+a_1 & 1 & 1 & \cdots & 1 \\ 1 & 1+a_2 & 1 & \cdots & 1 \\ \cdot & \cdot & & & \\ \cdot & \cdot & & & \\ \cdot & \cdot & & & \\ 1 & 1 & 1 & \cdots & 1+a_m \end{vmatrix}$$

Before expanding the determinant subtract the first row from each of the other rows, then expand by elements of the first column, giving

$$G = \prod_i p_i \prod_i x_{oi}[a_1 a_2 a_3 \ldots a_m + a_2 a_3 a_4 \ldots a_m + a_1 a_3 a_4 \ldots a_m + a_1 a_2 a_4 \ldots a_m + \cdots + a_1 a_2 a_3 \ldots a_{m-1}]$$

The quantity in the square bracket is a polynomial of degree m in $(-\lambda)$ and is not zero for $\lambda = Rp_j$, since there are m terms, and each a_j is missing from one of them. Therefore,

$$G = \prod_i p_i \prod_i x_{oi} \prod_i a_i \left[1 + \frac{1}{a_1} + \frac{1}{a_2} + \frac{1}{a_3} + \cdots \frac{1}{a_m}\right]$$

and the eigenvalues are determined by

$$\phi(\lambda) = \frac{p_1 x_{01}}{\lambda - Rp_1} + \frac{p_2 x_{02}}{\lambda - Rp_2} + \frac{p_3 x_{03}}{\lambda - Rp_3} + \cdots + \frac{p_m x_{om}}{\lambda - Rp_m} = 1$$

If we suppose that $p_1 < p_2 < p_3 < \ldots < p_m$, there are asymptotes at $\{Rp_j\}$ and there must be axis crossings between the values of Rp_{j-1} and Rp_j so that $\phi(\lambda)$ must intersect $\phi(\lambda) = 1$ between these same values. The additional eigenvalue must occur for $\lambda > Rp_m$, since $\phi(\lambda) \to 0$ as λ increases. As the schematic diagram, (Fig. 6.3.1) shows, there will be m real separated eigenvalues so long as the reciprocal relative volatilities are distinct. If there are two equal relative volatilities, then these two species may be treated as one component.

We must obtain $\mathbf{A}^{n-1}\mathbf{p}$ which is given by

$$\mathbf{A}^{n-1}\mathbf{p} = \sum_j (\lambda_j)^{n-1} \frac{\operatorname{adj}(\lambda_j \mathbf{I} - \mathbf{A})\mathbf{p}}{\prod\limits_{i=1,j}^{m} (\lambda_j - \lambda_i)}$$

It seems obvious that we need calculate only the first row of adj $(\lambda_j\mathbf{I} - \mathbf{A})$ since this will be enough to give the first element in the vector

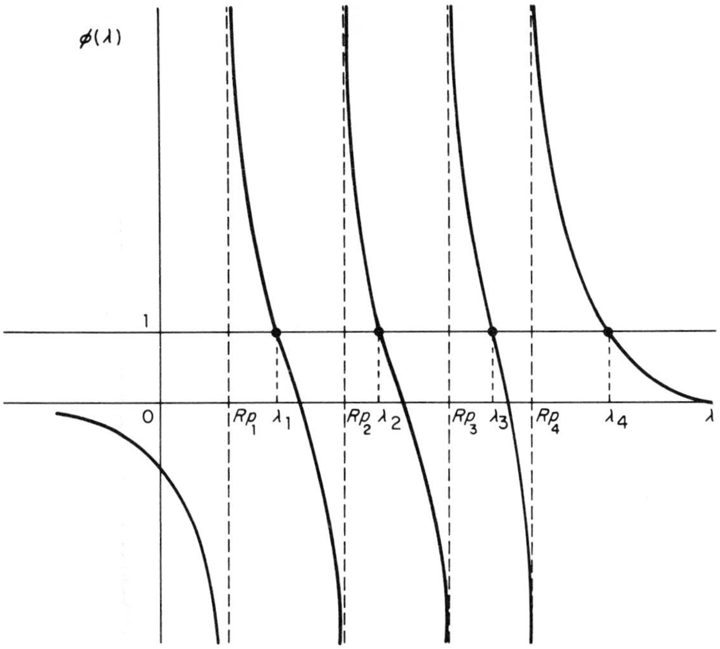

Figure 6.3.1 Schematic diagram of $\phi(\lambda)$ versus λ showing four real values of λ for a four-component system.

$$\operatorname{adj}(\lambda_j \mathbf{I} - \mathbf{A}) \cdot \mathbf{p}$$

Because one component must enter the formulae in the same way that every other component enters, the remaining terms of the vector may be written by analogy. Let the first row of adj $(\mathbf{A} - \lambda_j \mathbf{I})$ be denoted by

$$A_{11} A_{21} A_{31} \ldots A_{m1}$$

[Note that adj $(\mathbf{A} - \lambda_j \mathbf{I}) = (-1)^{m-1}$ adj $(\lambda_j \mathbf{I} - \mathbf{A})$.] Now

$$A_{11} = \prod_{i=2}^{m} p_i \prod_{i=2}^{m} x_{oi} \begin{vmatrix} 1 + a_2 & 1 & 1 & \ldots & 1 \\ 1 & 1 + a_3 & 1 & \ldots & 1 \\ \cdot & & & & \\ \cdot & & & & \\ \cdot & & & & \\ 1 & 1 & 1 & \ldots & 1 + a_m \end{vmatrix}$$

$$= \prod_{i=2}^{m} p_i \prod_{i=2}^{m} x_{oi} \prod_{i=2}^{m} a_i \left[1 + \frac{1}{a_2} + \frac{1}{a_3} + \cdots + \frac{1}{a_m} \right]$$

Also

$$A_{21} = -\prod_{i=1,2}^{m} p_i \prod_{i=2}^{m} x_{oi} \begin{vmatrix} 1 & 1 & 1 & \ldots & 1 \\ 1 & 1 + a_3 & 1 & \ldots & 1 \\ 1 & 1 & 1 + a_4 & \ldots & 1 \\ \cdot & & & & \\ \cdot & & & & \\ \cdot & & & & \\ 1 & 1 & \ldots \; \ldots & \ldots & 1 + a_m \end{vmatrix}$$

Subtract the first row from each of the other rows to obtain zeroes in the first column and expand by elements of the first column, then

$$A_{21} = -\prod_{i=1,2}^{m} p_i \prod_{i=2}^{m} x_{oi} \prod_{i=3}^{m} a_i$$

Also

$$A_{31} = \prod_{i=1,3}^{m} p_i \prod_{i=2}^{m} x_{oi} \begin{vmatrix} 1 & 1 & 1 & \cdots & 1 \\ 1+a_2 & 1 & 1 & \cdots & \\ 1 & 1 & 1+a_4 & \cdots & 1 \\ \cdot & & & & \\ \cdot & & & & \\ \cdot & & & & \\ 1 & 1 & 1 & \cdots & 1+a_m \end{vmatrix}$$

Subtract the first row from each of the other rows and expand the determinant by elements of the first column to give

$$A_{31} = -\prod_{i=1,3}^{m} p_i \prod_{i=2}^{m} x_{oi} \prod_{i=2,3}^{m} a_i$$

Then

$$A_{41} = -\prod_{i=1,4}^{m} p_i \prod_{i=2}^{m} x_{oi} \prod_{i=2,4}^{m} a_i$$

$$\vdots$$

$$A_{m1} = -\prod_{i=1}^{m-1} p_i \prod_{i=2}^{m} x_{oi} \prod_{i=2}^{m-1} a_i$$

The first row of adj $(\lambda_j \mathbf{I} - \mathbf{A})$ multiplied by $\mathbf{p}$ is

$$(-1)^{m-1} \prod_i p_i \left[\prod_{i=2}^{m} a_i \left(1 + \frac{1}{a_2} + \frac{1}{a_3} + \cdots + \frac{1}{a_m}\right) - \prod_{i=3}^{m} a_i - \prod_{i=2,3}^{m} a_i - \prod_{i=2,4}^{m} a_i - \cdots - \prod_{i=2}^{m-1} a_i \right] \prod_{i=2}^{m} x_{oi}$$

$$= (-1)^{m-1} \prod_i p_i \prod_{i=2}^{m} x_{oi} \prod_{i=2}^{m} \left(\frac{Rp_i - \lambda_j}{p_i x_{oi}}\right)$$

$$= (-1)^{m-1} p_1 \prod_{i=2}^{m} (Rp_i - \lambda_j)$$

Since this is the first element of adj $(\lambda_j \mathbf{I} - \mathbf{A})\, \mathbf{p}$, the vector is

$$\operatorname{adj}(\lambda_j \mathbf{I} - \mathbf{A})\mathbf{p} = (-1)^{m-1} \begin{bmatrix} p_1 \prod_{i=2}^{m} (Rp_i - \lambda_j) \\ p_2 \prod_{i=1,2}^{m} (Rp_i - \lambda_j) \\ \cdot \\ \cdot \\ \cdot \\ p_m \prod_{i=1}^{m-1} (Rp_i - \lambda_j) \end{bmatrix} = \mathbf{P}_j (-1)^{m-1}$$

Thus

$$\mathbf{X}_n = \sum_{j=1}^{m} (\lambda_j)^{n-1}(-1)^{m-1} \frac{1}{\prod_{i=1,j}^{m} (\lambda_j - \lambda_i)} \mathbf{P}_j$$

or

$$X_{ni} = \sum_{j=1}^{m} (\lambda_j)^{n-1}(-1)^{m-1} \frac{p_i \prod_{k=1,i}^{m} (Rp_k - \lambda_j)}{\prod_{i=1,j}^{m} (\lambda_j - \lambda_i)}$$

so that

$$x_{ni} = \frac{x_{oi} \sum_{j=1}^{m} \lambda_j^{n-1} \left\{ \left[p_i \prod_{k=1,i}^{m} (Rp_k - \lambda_j) \right] \Big/ \left[\prod_{k=1,j}^{m} (\lambda_j - \lambda_k) \right] \right\}}{\sum_i x_{oi} \sum_{j=1}^{m} \lambda_j^{n-1} \left\{ \left[p_i \prod_{k=1,i}^{m} (Rp_k - \lambda_j) \right] \Big/ \left[\prod_{k=1,j}^{m} (\lambda_j - \lambda_k) \right] \right\}}$$

The denominator may be simplified by changing the orders of summation to give

$$\sum_{j=1}^{m} \lambda_j^{n-1} \frac{1}{\prod_{k=1,j}^{m} (\lambda_j - \lambda_k)} \sum_i (x_{oi} p_i) \prod_{k=1,i}^{m} (Rp_k - \lambda_j)$$

but

$$\sum_i (x_{oi} p_i) \prod_{k=1,i}^{m} (Rp_k - \lambda_j) = -\prod_{k=1}^{m} (Rp_k - \lambda_j)$$

from the equation which determines the eigenvalues. Thus

$$x_{ni} = \frac{x_{oi} \sum_{j=1}^{m} \lambda_j^{n-1} \left\{ \left[p_i \prod_{k=1,i}^{m} (Rp_k - \lambda_j) \right] \Big/ \left[\prod_{k=1,j}^{m} (\lambda_j - \lambda_k) \right] \right\}}{-\sum_{j=1}^{m} \lambda_j^{n-1} \left\{ \left[\prod_{k=1}^{m} (Rp_k - \lambda_j) \right] \Big/ \left[\prod_{k=1,j}^{m} (\lambda_j - \lambda_k) \right] \right\}}$$

and this is the formula sought in terms of $\{p_i\}$, $\{x_{oi}\}$, R and, of course, the eigenvalues of $\mathbf{A}$. As mentioned earlier, an analogous formula for the reboiler section may be derived.

6.4 The Solution by Separation of Variables

We now supply a somewhat more elegant solution of the same problem. Note in Eq. (6.2.6) that if $\mathbf{X}_{ni}$ is written as $\mathbf{X}(n, i)$ this is a set of equations homogeneous in the dependent variables. A method frequently used in the solution of partial differential equations is the method of separation of variables, and we shall show that the same technique may be used on this problem to produce a set of eigenfunctions (in this case, eigenvectors) with a biorthogonality property. Let us assume that Eq. (6.2.6) has a solution of the form

$$X_{ni} = N(n)v(i)$$

That is, there is a product solution, and let us determine the form of the two functions $N(n)$ and $v(i)$. Substitution gives

$$N(n)v(i) = Rp_i N(n-1)v(i) + p_i \sum_i N(n-1)v(i)x_{oi}$$

or

$$\frac{N(n)}{N(n-1)} = \frac{Rp_i v(i) + p_i \sum v(i)x_{oi}}{v(i)}$$

The left-hand side is a function of n alone, whereas the right-hand side is a function of i alone, and if they are to be equal to each other for all n and i, they must be constant, say, λ. Therefore

$$N(n) = \lambda N(n-1)$$

$$Rp_i v(i) + p_i \sum_i v(i)x_{oi} = \lambda v(i)$$

The first gives a simple solution

$$N(n) = N(1)\lambda^{n-1}$$

The second may be written

$$v(i) = -\frac{p_i \sum_i v(i)x_{oi}}{Rp_i - \lambda}$$

where λ is still undetermined. The quantity $\sum v(i)\ x_{oi}$ is a constant and it may be eliminated by multiplying this expression through by x_{oi} and, summing on i,

$$\sum_i x_{oi}v(i) = -\sum_i \frac{p_i x_{oi}}{Rp_i - \lambda} \sum_i v(i)x_{oi}$$

to give

$$\sum_i \frac{p_i x_{oi}}{\lambda - Rp_i} - 1 = 0 \tag{6.4.1}$$

the same equation as was obtained in the previous section for determination of λ. Thus the quantity

$$\lambda_j^{n-1}\frac{p_i}{Rp_i - \lambda_j}$$

will satisfy Eq. (6.2.6) if λ_j is a zero of Eq. (6.4.1), as can be verified by direct substitution. Now the vectors $\mathbf{v}_j$ have a very interesting property, where the ith component of $\mathbf{v}_j$ is

$$\frac{p_i}{Rp_i - \lambda_j} = v_j(i)$$

which may be written

$$p_i = Rp_i v_j(i) - \lambda_j v_j(i)$$

For another eigenvalue λ_k

$$p_i = Rp_i v_k(i) - \lambda_k v_k(i)$$

If the first is multiplied by $v_k(i)\, x_{oi}/p_i$ and the second by $v_j(i)\, x_{oi}/p_i$ and the two subtracted there results

$$(\lambda_k - \lambda_j)\frac{x_{oi}\, v_j(i)\, v_k(i)}{p_i} = x_{oi}\, v_k(i) - x_{oi}\, v_j(i)$$

and when this is summed on i

$$(\lambda_k - \lambda_j)\sum_i \frac{x_{oi}\, v_j(i)\, v_k(i)}{p_i} = \sum_i \frac{x_{oi} p_i}{Rp_i - \lambda_k} - \sum_i \frac{x_{oi} p_i}{Rp_i - \lambda_j} = 0$$

Since $\lambda_k \neq \lambda_j$ if $k \neq j$, it follows that

$$\sum_i \frac{x_{oi}}{p_i} v_j(i) v_k(i) = 0$$

and the vectors $\{\mathbf{v}_j\}$ are orthogonal with a weight function x_{oi}/p_i. Or in other words, the two vector sets with ith components

$$v_j(i) = \frac{p_i}{Rp_i - \lambda_j}, \qquad w_k(i) = \frac{x_{oi}}{Rp_i - \lambda_k}$$

have a biorthogonality property

$$\mathbf{w}_k^T \mathbf{v}_j = 0, \quad j \neq k$$

Now we must determine what happens when $k = j$. Consider now the function $v(i)$, noting that $v(i)$ as previously defined is determined only up to an arbitrary multiplicative constant,

$$v(i) = \frac{p_i}{Rp_i - \lambda}$$

and the expression for $w_k(i)$ written in the forms

$$p_i = Rp_i v(i) - \lambda v(i)$$

$$x_{oi} = Rp_i w_k(i) - \lambda_k w_k(i)$$

Note that $v(i)$ is the same as $v_j(i)$ but for any λ. Multiply the first by $w_k(i)$ and the second by $v(i)$ and subtract

$$p_i w_k(i) - x_{oi} v(i) = (\lambda_k - \lambda)\, v(i)\, w_k(i)$$

Summed on i this gives

$$-1 - \sum_i x_{oi} v(i) = (\lambda_k - \lambda) \sum_i v(i) w_k(i)$$

or

$$\sum_i v(i) w_k(i) = \frac{1 + \sum_i x_{oi} v(i)}{-\lambda_k + \lambda}$$

In the limit as $\lambda \longrightarrow \lambda_k$ the right-hand side is an indeterminate form. Application of L'Hopital's rule gives

$$\sum_i v_k(i) w_k(i) = \sum_i \frac{x_{oi} p_i}{(Rp_i - \lambda_k)^2}$$

or
$$\mathbf{w}_j^T \mathbf{v}_j = \sum_i \frac{x_{oi} p_i}{(Rp_i - \lambda_j)^2}$$

Now a vector solution to Eq. (6.2.6) is

$$\mathbf{X}_n = \begin{bmatrix} \lambda_j^{n-1} v_j(1) \\ \lambda_j^{n-1} v_j(2) \\ \cdot \\ \cdot \\ \cdot \\ \lambda_j^{n-1} v_j(m) \end{bmatrix} = \lambda_j^{n-1} \mathbf{v}_j$$

but for $n = 1$ it will not satisfy the condition $\mathbf{X}_1 = \mathbf{p}$. Hence we shall assume a linear combination of these solutions of the form

$$\mathbf{X}_n = \sum_j c_j \lambda_j^{n-1} \mathbf{v}_j$$

If the condition at $n = 1$ is to be satisfied, then

$$\mathbf{p} = \sum_j c_j \mathbf{v}_j$$

The coefficients c_j may be determined from the biorthogonality condition, so

$$\mathbf{w}_k^T \mathbf{p} = \sum_j c_j \mathbf{w}_k^T \mathbf{v}_j$$

hence
$$c_j = \frac{\mathbf{w}_j^T \mathbf{p}}{\mathbf{w}_j^T \mathbf{v}_j} = -\frac{1}{\mathbf{w}_j^T \mathbf{v}_j}$$

or
$$\mathbf{X}_n = -\sum_j \lambda_j^{n-1} \frac{\mathbf{v}_j}{\mathbf{w}_j^T \mathbf{v}_j}$$

$$X_{ni} = -\sum_j \lambda_j^{n-1} \frac{p_i}{Rp_i - \lambda_j} \left[\sum_i \frac{x_{oi} p_i}{(Rp_i - \lambda_j)^2} \right]^{-1}$$

and this expresses X_{ni} in a slightly different form than previously obtained. Finally,

$$x_{ni} = \frac{x_{oi} \sum_j \lambda_j^{n-1} \dfrac{p_i}{Rp_i - \lambda_j} \left[\sum_i \dfrac{x_{oi} p_i}{(Rp_i - \lambda_j)^2} \right]^{-1}}{-\sum_j \lambda_j^{n-1} \left[\sum_i \dfrac{x_{oi} p_i}{(Rp_i - \lambda_j)^2} \right]^{-1}}$$

and the reader should show that the two solutions are identical.

6.5 The Transient Analysis of a Staged Absorber

We shall consider a staged absorber containing N stages. The liquid and gaseous phases pass countercurrent through the column being contacted in each stage and brought to equilibrium. It will be supposed that the inert carrier gas carries a solute in concentration of y pounds of solute per pound of inert gas while the liquid carries the same solute in concentration x pounds of solute per pound of inert liquid. It is assumed that the inert liquid is non-volatile and the inert gas is insoluble in the liquid phase in each stage. Since each stage is an equilibrium stage, it will be assumed that in each stage $y = Kx$, where K is an absolute constant. Figure 6.5.1 shows a schematic diagram of a plate absorber. L is the liquid inert rate and G is the gaseous inert rate. The conservation equation on each stage is then, with the assumption that the only capacity term which needs consideration is that for the liquid,

$$h\frac{dx_j}{d\theta} = Lx_{j-1} + Gy_{j+1} - Lx_j - Gy_j,$$

$$j = 1, 2, 3, \ldots, N$$

where h is the constant holdup liquid inert in each stage and x_j and y_j are the liquid and gaseous concentrations in stage j. Now one would usually specify y_{N+1} and x_o, the gaseous feed to stage N and the liquid feed to stage one, respectively, and these may be considered as known. The initial values of the liquid composition must also be specified at time zero. The foregoing system may then be written

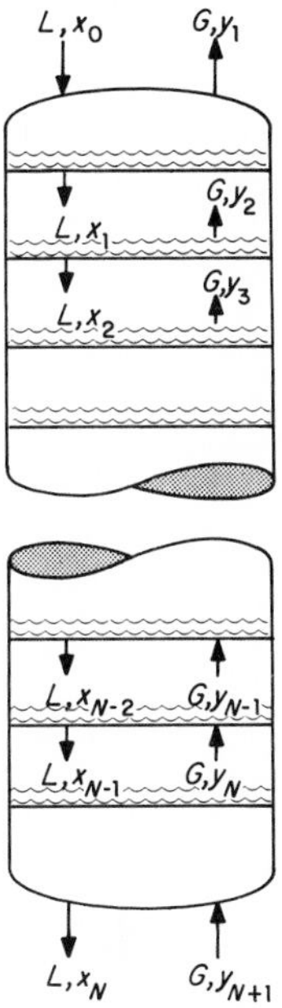

Figure 6.5.1 Schematic diagram of an absorption system.

$$\frac{d\mathbf{x}}{d\tau} = \mathbf{A}\mathbf{x} + \mathbf{b} \qquad \tau > 0 \tag{6.5.1}$$

where $\tau h = \theta$ and

$$\mathbf{x} = \begin{bmatrix} x_1 \\ x_2 \\ \cdot \\ \cdot \\ \cdot \\ x_N \end{bmatrix}, \quad \mathbf{b} = \begin{bmatrix} Lx_o \\ 0 \\ 0 \\ \cdot \\ \cdot \\ \cdot \\ 0 \\ Gy_{N+1} \end{bmatrix}$$

and

$$\mathbf{A} = \begin{bmatrix} -(L+GK) & GK & 0 & 0 & \ldots & 0 \\ L & -(L+GK) & GK & 0 & \ldots & 0 \\ 0 & L & -(L+GK) & GK\ 0 & \ldots & 0 \\ \cdot & & & & & \\ \cdot & & & & & GK \\ \cdot & & & & & \\ 0\ \ldots & \ldots & \ldots & \ldots & L & -(L+GK) \end{bmatrix}$$

and

$$\mathbf{x} = \mathbf{x}^{(i)}, \quad \tau = 0 \tag{6.5.2}$$

We shall assume in this problem that the $\mathbf{x}^{(i)}$ is arbitrary and x_0 and y_{N+1} are constants.

We must obtain the eigenvalues of $\mathbf{A}$ and the corresponding eigenvectors. Now $\mathbf{A}$ is of the form,

$$\mathbf{A} = \begin{bmatrix} -(\alpha+\beta) & \alpha & 0 & 0 & 0 & \cdots & 0 \\ \beta & -(\alpha+\beta) & \alpha & 0 & 0 & \cdots & 0 \\ 0 & \beta & -(\alpha+\beta) & \alpha & 0 & \cdots & 0 \\ \cdot & & & & & & \\ \cdot & & & & & & \alpha \\ \cdot & & & & & & \\ 0 & \cdots & \cdots & \cdots & \cdots & \beta & -(\alpha+\beta) \end{bmatrix}$$

This is a matrix of Jacobi type and a special case of the one considered at the end of Section 5.17. There it was shown that

$$P_n(\lambda) = (-\alpha - \beta - \lambda)P_{n-1}(\lambda) - \alpha\beta P_{n-2}(\lambda) \tag{6.5.3}$$

where

$$P_n(\lambda) = \det\,[\mathbf{A}_n - \lambda\mathbf{I}]$$

where $\mathbf{A}_n$ is the tridiagonal matrix of nth order and $\mathbf{A}_N = \mathbf{A}$ and

$$P_0(\lambda) = 1, \quad P_1(\lambda) = -(\alpha + \beta + \lambda)$$

Equation (6.5.3) is a second-order difference equation which may be solved by standard techniques. Consider the equation written in the form

$$P_n(\lambda) + (\alpha + \beta + \lambda)P_{n-1}(\lambda) + \alpha\beta P_{n-2}(\lambda) = 0$$

Assume a solution of the form

$$P_n(\lambda) = \gamma^n$$

then

$$\gamma^n + (\alpha + \beta + \lambda)\gamma^{n-1} + \alpha\beta\gamma^{n-2} = 0$$

and

$$\gamma^2 + (\alpha + \beta + \lambda)\gamma + \alpha\beta = 0$$

so

$$\gamma = \frac{-(\alpha+\beta+\lambda) \pm \sqrt{(\alpha+\beta+\lambda)^2 - 4\alpha\beta}}{2}$$

and, calling the respective roots γ_+ and γ_-, a general solution to Eq. (6.5.3) is

$$P_n(\lambda) = c_1\gamma_+^n + c_2\gamma_-^n$$

For $N = 0$ and $N = 1$,

$$1 = c_1 + c_2$$

$$-(\alpha + \beta + \lambda) = c_1\gamma_+ + c_2\gamma_-$$

Solving

$$c_1 = \frac{\gamma_+}{\sqrt{(\alpha+\beta+\lambda)^2 - 4\alpha\beta}}$$

$$c_2 = -\frac{\gamma_-}{\sqrt{(\alpha+\beta+\lambda)^2 - 4\alpha\beta}}$$

one obtains

$$P_n(\lambda) = \frac{\gamma_+^{n+1} - \gamma_-^{n+1}}{\gamma_+ - \gamma_-}$$

and

$$P_N(\lambda) = \frac{1}{\sqrt{(\alpha + \beta + \lambda)^2 - 4\alpha\beta}}[\gamma_+^{N+1} - \gamma_-^{N+1}] = 0$$

implies

$$\left(\frac{\gamma_+}{\gamma_-}\right)^{N+1} = +1$$

The roots of

$$z^{N+1} - 1 = 0$$

are

$$z_k = e^{2\pi ki/(N+1)}, \quad k = 0, 1, 2, 3, \ldots, N$$

and hence

$$\frac{-(\alpha + \beta + \lambda) + \sqrt{(\alpha + \beta + \lambda)^2 - 4\alpha\beta}}{-(\alpha + \beta + \lambda) - \sqrt{(\alpha + \beta + \lambda)^2 - 4\alpha\beta}} = e^{2\pi ki/(N+1)}$$

Solution for λ gives after considerable manipulation,

$$\lambda_k = -\alpha - \beta \pm \sqrt{4\alpha\beta} \cos\frac{\pi k}{N+1}, \quad k = 0, 1, 2, \ldots, N$$

There are extraneous roots, however, and after eliminating these by substitution into the previous formula, the eigenvalues are

$$\lambda_k = -\alpha - \beta - \sqrt{4\alpha\beta} \cos\frac{\pi k}{N+1}, \quad k = 1, 2, 3, \ldots, N$$

(Note that these λ_k are always negative.)

Now in order to generate the solution we refer to Section 5.7 and particularly to Eq. (5.7.7). In this problem we need the eigenvectors and the row vectors of **A** as well as the inverse of **A**. In order to compute the eigenvectors and row vectors, we need only one column and one row of

$$adj\,(\mathbf{A} - \lambda_k\mathbf{I})$$

and because the vector **b** has only two non-zero elements, the top and bottom, only the first column and last column of the inverse of **A** need be obtained. Using the preceding notation, direct computation will show that the first column of the adjoint of $(\mathbf{A} - \lambda_k\mathbf{I})$ is

$$\mathbf{z}_k = \begin{bmatrix} P_{N-1}(\lambda_k) \\ (-\beta)P_{N-2}(\lambda_k) \\ (-\beta)^2 P_{N-3}(\lambda_k) \\ (-\beta)^3 P_{N-4}(\lambda_k) \\ \cdot \\ \cdot \\ \cdot \\ (-\beta)^{N-1} P_o(\lambda_k) \end{bmatrix}$$

whereas the first row of adj $(\mathbf{A} - \lambda_k \mathbf{I})$ is

$$\mathbf{w}_k = \begin{bmatrix} P_{N-1}(\lambda_k) \\ (-\alpha)P_{N-2}(\lambda_k) \\ (-\alpha)^2 P_{N-3}(\lambda_k) \\ (-\alpha)^3 P_{N-4}(\lambda_k) \\ \cdot \\ \cdot \\ \cdot \\ (-\alpha)^{N-1} P_o(\lambda_k) \end{bmatrix}$$

Now

$$\begin{aligned} P_n(\lambda_k) &= \frac{1}{\sqrt{(\alpha + \beta + \lambda)^2 - 4\alpha\beta}} [\gamma_+^{n+1} - \gamma_-^{n+1}] \\ &= (\alpha\beta)^{n/2} \frac{\sin [\pi k(n + 1)/(N + 1)]}{\sin \pi k/(N + 1)} \end{aligned}$$

so that the elements of $\mathbf{z}_k$ and $\mathbf{w}_k$ may be calculated directly to give

$$\mathbf{z}_k = \begin{bmatrix} (\alpha\beta)^{(N-1)/2} \dfrac{\sin [\pi kN/(N + 1)]}{\sin [\pi k/(N + 1)]} \\ -\beta(\alpha\beta)^{(N-2)/2} \dfrac{\sin [\pi k(N - 1)/(N + 1)]}{\sin [\pi k/(N + 1)]} \\ (-\beta)^2(\alpha\beta)^{(N-3)/2} \dfrac{\sin [\pi k(N - 2)[(N + 1)]}{\sin \pi k/(N + 1)} \\ \cdot \\ \cdot \\ \cdot \\ (-\beta)^{N-1} \end{bmatrix} \tag{6.5.4}$$

$$\mathbf{w}_k = \begin{bmatrix} (\alpha\beta)^{(N-1)/2} \dfrac{\sin [\pi kN/(N + 1)]}{\sin \pi k/(N + 1)} \\ -\alpha(\alpha\beta)^{(N-2)/2} \dfrac{\sin [\pi k(N - 1)/(N + 1)]}{\sin [\pi k/(N + 1)]} \\ (-\alpha)^2(\alpha\beta)^{(N-3)/2} \dfrac{\sin [\pi k(N - 2)/(N + 1)]}{\sin \pi k/(N + 1)} \\ \cdot \\ \cdot \\ \cdot \\ (-\alpha)^{N-1} \end{bmatrix} \tag{6.5.5}$$

Therefore in Eq. (5.7.7) on may write

$$\mathbf{w}_j^T \mathbf{z}_j = (\alpha\beta)^{N-1} \sum_{n=1}^{N} \left(\frac{\sin [\pi\, jn/(N + 1)]}{\sin [\pi\, j(N + 1)]} \right)^2 \tag{6.5.6}$$

The value of the determinant of **A**, det **A**, may be found in a way analogous to the foregoing. If

$$D_N = \det |\mathbf{A}|$$

then

$$D_n = (-\alpha - \beta)D_{n-1} - \alpha\beta\, D_{n-2}$$
$$D_1 = -(\alpha + \beta)$$
$$D_0 = 1$$

The solution of this difference equation may be shown to be

$$D_n = \frac{(-1)^{n+1}}{\alpha - \beta}(\beta^{n+1} - \alpha^{n+1})$$

Then the first and last column of $\mathbf{A}^{-1}$ may be shown to be, from the definition of the inverse of **A**,

$$\begin{bmatrix} -\dfrac{\beta^N - \alpha^N}{\beta^{N+1} - \alpha^{N+1}} \\ -\beta\dfrac{\beta^{N-1} - \alpha^{N-1}}{\beta^{N+1} - \alpha^{N+1}} \\ -\beta^2\dfrac{\beta^{N-2} - \alpha^{N-2}}{\beta^{N+1} - \alpha^{N+1}} \\ -\beta^3\dfrac{\beta^{N-3} - \alpha^{N-3}}{\beta^{N+1} - \alpha^{N+1}} \\ \vdots \\ -\beta^{N-1}\dfrac{\beta - \alpha}{\beta^{N+1} - \alpha^{N+1}} \end{bmatrix}, \quad \begin{bmatrix} -\alpha^{N-1}\dfrac{\beta - \alpha}{\beta^{N+1} - \alpha^{N+1}} \\ -\alpha^{N-2}\dfrac{\beta^2 - \alpha^2}{\beta^{N+1} - \alpha^{N+1}} \\ -\alpha^{N-3}\dfrac{\beta^3 - \alpha^3}{\beta^{N+1} - \alpha^{N+1}} \\ \vdots \\ -\alpha\dfrac{\beta^{N-1} - \alpha^{N-1}}{\beta^{N+1} - \alpha^{N+1}} \\ -\dfrac{(\beta^N - \alpha^N)}{\beta^{N+1} - \alpha^{N+1}} \end{bmatrix}$$

and it follows then that

$$\mathbf{A}^{-1}\mathbf{b} = -\begin{bmatrix} Lx_0\dfrac{\beta^N - \alpha^N}{\beta^{N+1} - \alpha^{N+1}} + Gy_{N+1}\alpha^{N-1}\dfrac{\beta - \alpha}{\beta^{N+1} - \alpha^{N+1}} \\ Lx_0\beta\dfrac{\beta^{N-1} - \alpha^{N-1}}{\beta^{N+1} - \alpha^{N+1}} + Gy_{N+1}\alpha^{N-2}\dfrac{\beta^2 - \alpha^2}{\beta^{N+1} - \alpha^{N+1}} \\ Lx_0\beta^2\dfrac{\beta^{N-2} - \alpha^{N-2}}{\beta^{N+1} - \alpha^{N+1}} + Gy_{N+1}\alpha^{N-3}\dfrac{\beta^3 - \alpha^3}{\beta^{N+1} - \alpha^{N+1}} \\ \vdots \\ Lx_0\beta^{N-1}\dfrac{\beta - \alpha}{\beta^{N+1} - \alpha^{N+1}} + Gy_{N+1}\dfrac{\beta^N - \alpha^N}{\beta^{N+1} - \alpha^{N+1}} \end{bmatrix}$$

and this may be written

$$\mathbf{A}^{-1}\mathbf{b} = -\frac{1}{1-(\alpha/\beta)^{N+1}}\begin{bmatrix} x_0\left[1-\left(\frac{\alpha}{\beta}\right)^{N}\right] + \frac{y_{N+1}}{K}\left[\left(\frac{\alpha}{\beta}\right)^{N} - \left(\frac{\alpha}{\beta}\right)^{N+1}\right] \\ x_0\left[1-\left(\frac{\alpha}{\beta}\right)^{N-1}\right] + \frac{y_{N+1}}{K}\left[\left(\frac{\alpha}{\beta}\right)^{N-1} - \left(\frac{\alpha}{\beta}\right)^{N+1}\right] \\ \vdots \\ x_0\left[1-\left(\frac{\alpha}{\beta}\right)^{N-n+1}\right] + \frac{y_{N+1}}{K}\left[\left(\frac{\alpha}{\beta}\right)^{N-n+1} - \left(\frac{\alpha}{\beta}\right)^{N+1}\right] \\ \vdots \\ x_0\left[\left(1-\frac{\alpha}{\beta}\right)\right] + \frac{y_{N+1}}{K}\left[\left(\frac{\alpha}{\beta}\right) - \left(\frac{\alpha}{\beta}\right)^{N+1}\right] \end{bmatrix} \tag{6.5.7}$$

Let the nth element be

$$-\frac{1}{1-(\alpha/\beta)^{N+1}}\left\{x_0\left[1-\left(\frac{\alpha}{\beta}\right)^{N-n+1}\right] + \frac{y_{N+1}}{K}\left[\left(\frac{\alpha}{\beta}\right)^{N-n+1} - \left(\frac{\alpha}{\beta}\right)^{N+1}\right]\right\}$$

then the nth element of $\mathbf{x}^{(i)} + \mathbf{A}^{-1}\mathbf{b}$ is

$$c_n = x_n^{(i)} - \frac{1}{1-(\alpha/\beta)^{N+1}}\left\{x_0\left[1-\left(\frac{\alpha}{\beta}\right)^{N-n+1}\right] + x_{N+1}\left[\left(\frac{\alpha}{\beta}\right)^{N-n+1} - \left(\frac{\alpha}{\beta}\right)^{N+1}\right]\right\} \tag{6.5.8}$$

The solution to Eq. (6.5.1) subject to the initial condition, Eq. (6.5.2), is

$$\mathbf{x} = \sum_{j=1}^{N} \frac{\mathbf{w}_j^T(\mathbf{x}^i + \mathbf{A}^{-1}\mathbf{b})}{\mathbf{w}_j^T\mathbf{z}_j}\mathbf{z}_j e^{\lambda_j\theta} - \mathbf{A}^{-1}\mathbf{b}$$

where $\mathbf{z}_j$, $\mathbf{w}_j^T$, $\mathbf{w}_j^T\mathbf{z}_j$, $\mathbf{A}^{-1}\mathbf{b}$, $\mathbf{x}^i + \mathbf{A}^{-1}\mathbf{b}$ are given by Eq. (6.5.4), (6.5.5), (6.5.6), (6.5.7), and (6.5.8). Two things need mentioning: First, all the eigenvalues are negative, so that the transient state approaches a unique steady state; second, the steady state is given by the elements of $-\mathbf{A}^{-1}b$, or

$$x_n = \frac{1}{1-(\alpha/\beta)^{N+1}}\left\{x_0\left[1-\left(\frac{\alpha}{\beta}\right)^{N-n+1}\right] + x_{N+1}\left[\left(\frac{\alpha}{\beta}\right)^{N-n+1} - \left(\frac{\alpha}{\beta}\right)^{n+1}\right]\right\}$$

In this section the matrix $\mathbf{A}$ had a rather special form. If the matrix had been of the form

$$\begin{bmatrix} -\gamma & \alpha & 0 & 0 & 0 & \dots & 0 \\ \beta & -\gamma & \alpha & 0 & 0 & \dots & 0 \\ 0 & \beta & -\gamma & \alpha & 0 & \dots & 0 \\ \vdots & & & & & & \vdots \\ 0 & 0 & \dots & \dots & \beta & \dots & -\gamma \end{bmatrix}$$

then the eigenvalues of $\mathbf{B}$ would be

$$\lambda_k = -\gamma - 2\sqrt{\alpha\beta}\cos\frac{\pi k}{N+1}, \quad k = 1, 2, \ldots, N$$

6.6 Reactions in Stirred Tank Reactors and Batch Reactors

We consider here a sequence of reactions

$$A_1 \underset{k_1'}{\overset{k_1}{\rightleftarrows}} A_2 \underset{k_2'}{\overset{k_2}{\rightleftarrows}} A_3 \underset{k_3'}{\overset{k_3}{\rightleftarrows}} A_4 \underset{k_4'}{\overset{k_4}{\rightleftarrows}} A_5 \rightleftarrows \cdots \underset{k_m'}{\overset{k_m}{\rightleftarrows}} A_{m+1} \tag{6.6.1}$$

where all reactions are first order, and the velocity constants k have the units of reciprocal time, and where the reactions are carried out either in a sequence of stirred tank reactors or in a single batch reactor. The problems will not be completely solved but only set up for solution. Suppose that the batch reactor is isothermal and of constant volume and let A_j also stand for the concentration of species A_j. The reactor is started with some initial concentration of reactants $\{A_j^i\}$. It is clear that the system of differential equations which describes the system is

$$\frac{d\mathbf{a}}{d\theta} = = \mathbf{A}\mathbf{a} \tag{6.6.2}$$

$$\mathbf{a} = \mathbf{a}^{(i)}, \quad \theta = 0$$

where

$$\mathbf{a}^T = [A_1 A_2 A_3 \ldots A_{m+1}]$$

and

$$\mathbf{A} = \begin{bmatrix} -k_1 & k_1' & 0 & 0 & 0 & \cdots & 0 \\ k_1 & -(k_2 + k_1') & k_2' & 0 & 0 & \cdots & 0 \\ 0 & k_2 & -(k_3 + k_2') & k_3' & 0 & \cdots & 0 \\ 0 & 0 & k_3 & -(k_4 + k_3') & k_4' & \cdots & 0 \\ \vdots & & & \ddots & \ddots & \ddots & \vdots \\ & & & & & -(k_m + k_{m-1}') & k_m' \\ 0 & \cdots & \cdots & & & k_m & -k_m' \end{bmatrix}$$

We know what the form of the solution is for this case, for it is given as a linear combination of exponentials as in Eq. (5.7.7) with $\mathbf{b} = 0$. We are, however, immediately struck here by this fact: if all the eigenvalues are negative, then $\mathbf{a} \longrightarrow \mathbf{o}$ as $\theta \longrightarrow \infty$ or if they are not negative, then one or more of the concentrations must become arbitrarily large. Both these cases are impossible, since the former implies that all mass disappears and the latter implies that the mass becomes either positively or negatively arbitrarily large. The difficulty lies in that if one examines the matrix $\mathbf{A}$ and adds all the rows, that is, adds each row to the last one, for example, a row of zeroes is produced; hence **A** is a singular matrix. This means that the characteristic equation will have no constant term in it and $\lambda = 0$ is an eigenvalue. Let the eigenvalues be $0, \lambda_1, \lambda_2, \ldots, \lambda_m$ and let the corresponding eigenvectors and eigenrows be as before $\{\mathbf{z}_j\}$ and $\{\mathbf{w}_j\}$, respectively, with $\mathbf{z}_o$ and $\mathbf{w}_o$ corresponding to the zero eigenvalue. The solution will be

$$\mathbf{a} = \sum_{j=1}^{m} \frac{\mathbf{w}_j^T \mathbf{a}^{(i)}}{\mathbf{w}_j^T \mathbf{z}_j} \mathbf{z}_j e^{\lambda_j \theta} + \frac{\mathbf{w}_0^T \mathbf{a}^i}{\mathbf{w}_o^T \mathbf{z}_o} \mathbf{z}_o \tag{6.6.3}$$

so that the steady state is given by

$$\mathbf{a}_{ss} = \frac{\mathbf{w}_o^T \mathbf{a}^i}{\mathbf{w}_o^T \mathbf{z}_o} \mathbf{z}_o$$

provided that all the other eigenvalues are negative. Let us now show that this is the case by considering the Sturmian sequence of polynomials. We showed in Section 5.17 that the sequence

$$P_o, P_1, P_2, \ldots, P_{m+1}$$

had *no* sign changes for large negative λ. Consider now the same Sturmian sequence but for $\lambda = 0$. For $\lambda = 0$ the sequence of polynomials is a sequence of determinants formed along the main diagonal since $P_2(\lambda)$ is $P_n(\lambda)$ for $n = 2$. These determinants are called the *principal minors* of the main determinant. For example, $P_o(0) = 1$ and $P_1(0) = -k_1$. Now consider $P_2(0)$. Direct calculation shows it to be positive, $P_2(0) = -k_2 P_1(0) > 0$. In order to compute $P_3(0)$ add the first and the second rows to the third row and expand by elements of the third column obtaining

$$P_3(0) = -k_3 P_2(0) < 0$$

Consider $P_4(0)$ and do likewise. Add the first, second, and third rows to the fourth and expand by elements of the fourth column, obtaining

$$P_4 = -k_4 P_3(0) > 0$$

This may be repeated to show that $P_0, P_1, P_2, \ldots, P_m$ alternate in sign so that there are m sign changes. Since there were no sign changes for λ large and negative, there must be m negative eigenvalues.

It is somewhat more convenient to use the system of Eq. (6.6.2) than to eliminate one equation since preserving the form of the solution allows us to calculate the eigenvalues from the Sturmian sequence of polynomials. If one equation is to be eliminated from the set, it can be done as follows: Since **A** is singular and since the sum of the rows is zero

$$\sum_{j=1}^{m+1} \frac{dA_j}{d\theta} = 0$$

or

$$\sum_{j=1}^{m+1} A_j = \sum_{j=1}^{m+1} A_j^{(i)}$$

and
$$A_{m+1} = \sum_{j=1}^{m+1} A_j^{(i)} - \sum_{j=1}^{m} A_j$$

Thus the set of equations may be written

$$\frac{d\mathbf{a}'}{d\theta} = \mathbf{B}\mathbf{a}' + \mathbf{c}$$

$$\mathbf{A}' = \mathbf{A}^{(i)},$$

where $\mathbf{a}'$ is $\mathbf{a}$ with the last element omitted and

$$\mathbf{B} = \begin{bmatrix} -k_1 & k_1' & 0 & 0 & 0 & 0 & 0 \\ -k_1 & -(k_2+k_1') & k_2' & & & \cdot & \vdots \\ 0 & -k_2 & -(k_3+k_2') & k_3' & & \cdot & 0 \\ 0 & 0 & \cdot & \cdot & \cdot & 0 & 0 \\ \vdots & \vdots & & \ddots & & k_{m-2}' & 0 \\ 0 & 0 & 0 & \cdots & -k_{m-2} & -(k_{m-1}+k_{m-2}') & k_{m-1}' \\ -k_m' & -k_m' & \cdots & \cdots & \cdots\ -k_m' & -(k_m'-k_{m-1}) & -(k_m'+k_m-k_{m-1}') \end{bmatrix}$$

$$\mathbf{c} = \begin{bmatrix} 0 \\ 0 \\ \cdot \\ \cdot \\ \cdot \\ 0 \\ k_m' \sum_{j=1}^{m+1} A_j^{(i)} \end{bmatrix}$$

From a computational point of view, this is clearly more difficult to handle.

For a series of stirred pot reactors in which the effluent from one becomes the influent to the next and where q is the volumetric flow rate and V is the volume of each reactor with A_{jn} as the concentration of A_j in the nth reactor, under the usual assumptions

$$V\frac{d\mathbf{a}_n}{d\theta} = q\mathbf{a}_{n-1} - q\mathbf{a}_n + V\mathbf{A}\mathbf{a}_n$$

We must also specify the initial condition of each of the species in each reactor

$$\mathbf{a}_n = \mathbf{a}_n^i, \quad \theta = 0$$

and of course $\mathbf{a}_0$ is the influent concentration to the first reactor. This is a set of differential difference equations whose solution is best obtained by the Laplace transformation. Note that in this set of equations, in contradistinction to the batch reactor, there is not a dependent equation which can be eliminated. These equations may be written

$$\theta_h\frac{d\mathbf{a}_n}{d\theta} = \mathbf{a}_{n-1} - \mathbf{a}_n + \theta_h\mathbf{A}\mathbf{a}_n$$

or
$$\frac{d\mathbf{a}_n}{d\tau} - \mathbf{M}\mathbf{a}_n = \mathbf{a}_{n-1}$$

where
$$\theta_h^{-1}\theta = \tau$$
$$\mathbf{M} = \theta\mathbf{A} - \mathbf{I}$$

and $\mathbf{M}$ is clearly not singular unless θ_h^{-1} is an eigenvalue of $\mathbf{A}$. The formal solution of these equations will be developed in a later chapter.

6.7 Multicomponent Distillation

Although the development of this section makes no formal use of the deeper properties of vectors and matrices, the full power of matrix notation, when brought to bear on a problem, frequently rather simplifies the statement of that problem. The distillation column shown in Fig 6.7.1 is used

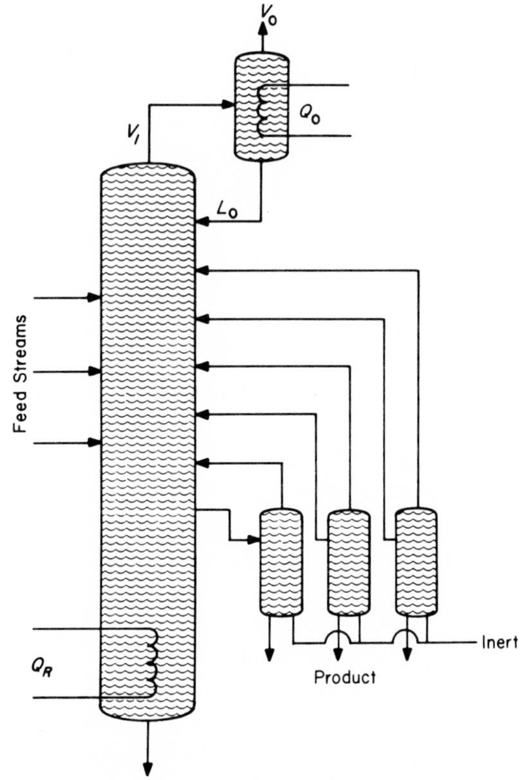

Figure 6.7.1 Schematic diagram of a distillation column with side-stream strippers and multiple-feed streams and partial condenser.

as a model. We assume that each stage is an ideal stage with equilibrium established. The use of an inert material is also allowed for in the vapor phase and it is assumed that it does not condense out. Inert may be injected into each stage and there may be feed streams to each stage or product drawoffs from each stage. There will be N stages plus a partial condenser, the latter being the zeroth stage and the reboiler the Nth. There may be arbitrary heat additions Q_j to each stage. Let the following quantities be defined in moles/unit time:

V_j = noninert vapor rate from the jth stage
I_j = inert rate from the jth stage
U_j = total vapor rate from the jth stage $= V_j + I_j$
S_j = side stream drawoff from jth stage
F_j = feed rate to the jth stage
L_j = liquid rate from jth stage $= Y_j + V_{j+1}$
s_j = inert injection rate to the jth stage
Y_j = sum of the feeds minus sum of the drawoffs from top to jth stage inclusive $= \sum_{k=0}^{i} (F_k - S_k)$

Composition variables in mole fractions will be

x_{ji} = mole fraction of ith component in the jth stage liquid
y_{ji} = mole fraction of ith component in jth stage vapor
x_{Fji} = mole fraction of ith component in the feed to the jth stage

Enthalpies of various streams in BTU per pound mole are defined as

H_j = enthalpy of vapor from the jth stage
H_{ji} = partial molar enthalpy of the ith component in the vapor from the jth stage
h_j = liquid enthalpy from the jth stage
h_{ji} = partial molar enthalpy of the ith component in the jth stage liquid
H_{sj} = partial molar enthalpy of inert in jth stage vapor
$\tilde{H}_{sj}$ = enthalpy of inert injected into jth stage.

A mass balance on the ith component around the jth stage gives

$$L_{j-1}x_{j-1,i} + U_{j+1}y_{j+1,i} + F_jx_{Fji} - L_jx_{ji} - U_jy_{ji} - S_jx_{ji} = 0 \quad (6.7.1)$$

This equation holds for all stages if

$$x_{ji} = 0, \quad j < 0, \quad j > N$$
$$U_j = 0, \quad j > N$$
$$Y_0 = -V_0$$
$$Y_{-1} = -V_0$$

If one assumes an equilibrium relation of the form

$$y_{ji} = K_{ji}x_{ji}$$

Equation (6.7.1) becomes

$$(Y_{j-1} + V_j)x_{j-1,i} - (Y_{j-1} + F_j + V_{j+1} + U_jK_{ji})x_{ji} + U_{j+1}K_{j+1,i}x_{j+1,i} = F_jx_{Fji} \qquad (6.7.2)$$

Equation (6.7.2) may be written as a matrix equation

$$\mathbf{M}_i\mathbf{X}_i = -\mathbf{F}_i, \quad i = 1, 2, 3, \ldots, m$$

where if the elements of $\mathbf{M}_i$ are denoted by a_{pq}, $\mathbf{M}_i$ is a Jacobi matrix, with

$$\begin{aligned} a_{jj} &= -(Y_{j-1} + F_j + V_{j+1} + U_jK_{ji}), && 0 \leqslant j \leqslant N \\ a_{j,j+1} &= U_{j+1}K_{j+1,i}, && 0 \leqslant j \leqslant N-1 \\ a_{j,j-1} &= V_j + Y_{j-1}, && 1 \leqslant j \leqslant N \end{aligned}$$

and where all other elements are zero. Also

$$\mathbf{X}_i = \begin{bmatrix} x_{oi} \\ x_{1i} \\ x_{2i} \\ x_{3i} \\ \cdot \\ \cdot \\ \cdot \\ x_{Ni} \end{bmatrix}, \quad \mathbf{F}_i = \begin{bmatrix} F_ox_{Foi} \\ F_1x_{F1i} \\ F_2x_{F2i} \\ \cdot \\ \cdot \\ \cdot \\ F_Nx_{FNi} \end{bmatrix}$$

The heat balance around the jth stage gives

$$Q_j + (L_{j-1}h_{j-1} + V_{j+1}H_{j+1} + I_{j+1}H_{s,j+1}) - (L_jh_j + V_jH_j + I_jH_{sj}) + F_jH_{Fj} - S_jh_j + s_j\tilde{H}_{sj} = 0$$

From

$$I_j - I_{j+1} = s_j$$

and

$$Y_j - Y_{j-1} = F_j - S_j$$

it follows that

$$\begin{aligned} V_j(h_{j-1} - H_j) + V_{j+1}(H_{j+1} - h_j) = &-Q_j + Y_{j-1}(h_j - h_{j-1}) + F_j(h_j - H_{Fj}) \\ &+ I_j(H_{sj} - H_{s,j+1}) \\ &+ s_j(H_{s,j+1} - \tilde{H}_{sj}) \end{aligned}$$

or

$$\begin{aligned} V_j(h_{j-1} - H_j) + V_{j+1}(H_{j+1} - h_j) = &-Q_j + Y_{j-1}(h_j - h_{j-1}) \\ &+ F_j(h_j - H_{Fj}) \\ &+ I_{j+1}(H_{sj} - H_{s,j+1}) \\ &+ s_j(H_{sj} - \tilde{H}_{sj}) \end{aligned}$$

This set of equations may be written in matrix form

$$\mathbf{HV} = \mathbf{G}$$

where the elements of $\mathbf{V}$ are V_j, $j = 1$ to N and the elements of $\mathbf{G}$ are c_j

$$c_j = -Q_j + Y_{j-1}(h_j - h_{j-1}) + F_j(h_j - H_{Fj}) + I_{j+1}(H_{sj} - H_{s,j+1}) + s_j(H_{sj} - \tilde{H}_{sj})$$

$$j = 1 \quad \text{to} \quad N$$

and the elements b_{pq} of $\mathbf{H}$ are

$$b_{jj} = h_{j-1} - H_j \quad 1 \leqslant j \leqslant N$$
$$b_{j,j+1} = H_{j+1} - h_j \quad 1 \leqslant j \leqslant N - 1$$

all other elements being zero. Note that $\mathbf{H}$ is in triangular form with nonzero elements on the main diagonal and upper adjacent diagonal only.

A complete heat balance around the column but excluding the partial condenser (that is, an envelope which excludes the partial condenser) is

$$Q_N = V_1 H_1 + I_1 H_{s1} - \sum_{j-1}^{N} F_j H_{Fj} + \sum_{j=1}^{N} S_j h_j - \sum_{j=1}^{N} s_j \tilde{H}_{sj} - L_o h_o - \sum_{j=1}^{N-1} Q_j$$
$$= (R+1) V_o H_1 - R V_o h_o + I_1 H_{s1} - \sum_{j=1}^{N} F_j H_{Fj} + \sum_{j=1}^{N} S_j h_j - \sum_{j=1}^{N} s_j \tilde{H}_{sj} - \sum_{j=1}^{N-1} Q_j$$

so that Q_N, the heat to the reboiler, may be eliminated from the enthalpy matrix. Ordinarily it is assumed that Q_j, $j = 1$ to $N - 1$, is zero. The heat balance around the partial condenser is

$$Q_o = V_o[H_o - H_1 + R(h_o - H_1)]$$

Now consider the equations

$$\mathbf{X}_i = -\mathbf{M}_i^{-1}\mathbf{F}_i$$
$$\mathbf{V} = \mathbf{H}^{-1}\mathbf{G}$$

and these contain the m unknown mole fractions in $N + 1$ stages and the N unknown V_j's. In addition to this the temperature is unknown. Although the equations for the column are written in matrix-vector notation, in reality these equations are highly nonlinear in the temperature and $\mathbf{M}_i$ has elements which are determined by the thermal characteristics of the column and compositions. The equilibrium constant K_{ji} is generally far from a constant; in fact, it can be shown that, for multicomponent systems, to assume that it is a constant leads to gross inconsistencies. In fact K_{ji} usually depends upon the local composition, temperature, and total pressure. For the present, we assume that K_{ji} is a function of pressure, which is fixed a priori, and the temperature, and the functional form usually assumed is polynomial, as

$$K_{ji} = \alpha_i + \beta_i T_j + \gamma_i T_j^2 + \Delta_i T_j^3$$

where α_i, β_i, γ_i, and Δ_i are fixed by the total pressure and are specific for a given component.

Not all the compositions in each stage are independent since, for mole fractions,

$$\sum_{i=1}^{m} x_{ji} = 1, \quad j = 0, 1, 2, 3, \ldots, N$$

Now in the vapor phase of each stage, the mole fraction of inert to total vapor is I_j/U_j and therefore the sum of the mole fractions in the vapor phase corresponding to liquid phase mole fractions is

$$\sum_{i=1}^{m} K_{ji}x_{ji} = \sum_{i=1}^{m} y_{ji} = \frac{V_j}{U_g} = g_j$$

where g_j will be a number less than one, but usually near one. (In the case of strippers, to be discussed later, g_j may be small, say 0.1.)

The problem to be solved is generally one which may be reduced to the following form: Specify in advance x_{Fji}, F_j, V_o, S_j, s_j, H_{Fj}, the total pressure and the reflux ratio. This is equivalent to specifying the feed, its composition, the amount of the drawoffs from each stage, the thermal condition of the feed, and the inert injection to each stage and its enthalpy. As a first approximation, one can assume constant molal overflow, in which case the V_j are constant in various sections of the column and may be computed in advance once and for all. One usually has a fairly good idea about the top and bottom temperature so that if these are assumed, the other stage temperatures may be estimated by linear interpolation. Now with the temperatures all fixed and the total pressure assumed to be constant, K_{ji} for each component may be calculated from the polynomial expression (or picked from a graph). The m equations

$$\mathbf{X}_i = -\mathbf{M}_i^{-1}\mathbf{F}_i$$

may then be solved, since all the elements of $\mathbf{M}_i$ and $\mathbf{F}_i$ are fixed, to give $\mathbf{X}_i$, the $m(N+1)$ mole fractions. In general from this calculation $\sum_{i=1}^{m} x_{ji} \neq 1$. Note that in $\sum_{i=1}^{m} x_{ji}$ each of the m terms in the sum comes from a different vector $\mathbf{X}_i$. Also $\sum_{i=1}^{m} K_{ji}x_{ji} \neq g_i$, since it would have been fortuitous had the correct stage temperatures been assumed. We must therefore devise a scheme for correcting the temperatures. Before this is begun, however, it should be recognized that in the sum $\sum_{i=1}^{m} K_{ji}x_{ji}$, the x_{ji} are not correct either. In order to allow for this fact we will assume that in the temperature correction stage the mole fractions on each stage have been normalized. Suppose now we take the Taylor series expansion of $\sum_{i=1}^{m} K_{ji}x_{ji}$ as

$$\left[\sum_{i=1}^{m} K_{ji}x_{ji}\right]_{T_j} = \left[\sum_{i=1}^{m} K_{ji}x_{ji}\right]_{T'_j} + \left[\frac{\partial \sum_{i=1}^{m} K_{ji}x_{ji}}{\partial T_j}\right]_{T'_j}(T - T'_j)$$

If T'_j was the assumed temperature in the jth stage and if T_j is to be the corrected temperature, then

$$\sum_{i=1}^{m} K_{ji}x_{ji} = g_j$$

with x_{ji} normalized, and making use of the normalization factor in the other terms, it follows that

$$g_j \sum_{i=1}^{m} x_{ji} = \sum_{i=1}^{m} K_{ji}x_{ji} + \left[\frac{\partial \sum_{i=1}^{m} K_{ji}x_{ji}}{\partial T_j}\right]_{T_j} (T_j - T'_j)$$

where the x_{ji} here are unnormalized. Therefore,

$$T_j - T'_j = \left[g_j \sum_{i=1}^{m} x_{ji} - \sum_{i=1}^{m} K_{ji}x_{ji}\right]\left[\frac{\partial \sum_{i=1}^{m} K_{ji}x_{ji}}{\partial T_j}\right]^{-1}_{T'_j} \qquad (6.7.2)$$

and so a temperature correction is available in terms of previously calculated quantities. We shall assume in making the temperature correction that

$$\frac{\partial \sum_{i=1}^{m} K_{ji}x_{ji}}{\partial T_j} = \sum_{i=1}^{m} x_{ji}\frac{\partial K_{ji}}{\partial T_j}$$

and therefore

$$\frac{\partial K_{ji}}{\partial T_j} = \beta_i + 2\gamma_i T_j + 3\Delta_i T_j^2$$

Hence a new temperature may be estimated and the foregoing calculations repeated in an iterative way until convergence results, convergence being indicated here by

$$\sum_{i=1}^{m} x_{ji} = 1$$

$$\sum_{i=1}^{m} K_{ji}x_{ji} = g_j$$

This solves only the "constant molal overflow problem." The elements of the matrix **H** and **G** are simple to calculate, however, since

$$H_j = \sum_{i=1}^{m} y_{ji}H_{ji}$$

$$h_j = \sum_{i=1}^{m} x_{ji}h_{ji}$$

and from

$$\mathbf{V} = \mathbf{H}^{-1}\mathbf{G}$$

using the temperatures and compositions just obtained the vapor rates **V** may be obtained. Alteration of the vapor rates changes the corresponding quantities in $\mathbf{M}_i$, hence new compositions must be computed from $\mathbf{X}_i = -\mathbf{M}^{-1}\mathbf{F}_i$. Thus an iterative procedure on the two sets of equations is instituted, calculating compositions, vapor rates, compositions, new temperatures, new compositions, etc., until convergence by the preceding criteria results. No numerical results are presented here, for they can be found in the original literature and our main purpose is to show the formulation of the problem in matrix terms.

In many hydrocarbon distillations, the side streams which are withdrawn are introduced into the top of a stripper with a few stages and into the bottom of which is injected inert, usually steam. Analogous equations to those previously given, only simpler, may be derived for the stripper stages and the calculations performed in the same way. These will not be described here. But we do need a word of caution about such iterative calculations for although the whole procedure seems straightforward, all is generally not what it seems, Frequently, such processes diverge, oscillate, and otherwise do not give reasonable answers, and devious schemes must be resorted to in order to make the steps converge. For example, although intuitively it seems that the stripper calculations should be without difficulty, the process previously described for the distillation column will not converge for the stripper unless, in the calculations, the results of the last two iterations for the vapor rates are averaged when used in the next iteration. Obviously, other schemes might improve the convergence. In the distillation calculations, many schemes which seemed elegant and certain to work in the temperature correction procedure diverged in practice.

Further Properties of Matrices 7

7.1 Introduction

This chapter considers certain additional topics, including the important ones of reduction of a matrix to diagonal form, the reduction of a quadratic form to canonical form, and the reduction of a system of differential equations to normal form. In connection with differential systems it is natural to consider the exponential function defined on matrices and this leads to the question of the definition of other functions defined on matrices and infinite series of square matrices.

7.2 Reduction of a Matrix to Diagonal Form

In Section 2.6 it was shown that a matrix could be reduced in general to diagonal form by a finite number of pre- and postmultiplicative nonsingular matrices. Hence there exist, in general, two matrices **P** and **Q** such that

$$\mathbf{PAQ} = \mathbf{D}$$

where **D** contains only diagonal elements. The matrices **P** and **Q** are certainly not unique and the matrix **D** will certainly change with different choices of **P** and **Q**. **D** may of course be transformed to the unit matrix by appropriate pre- or postmultiplications.

Two matrices **S** and **T** related by the following transformation are said to be *similar*

$$\mathbf{P}^{-1}\mathbf{S}\mathbf{P} = \mathbf{T}$$

and the transformation from **S** to **T** is called a *similarity transformation* (also called a *collineatory* transformation).

An important characteristic of the similarity transformation is that the eigenvalues of the transformed matrix are the same as the eigenvalues of the matrix itself; that is, the λ's which satisfy

$$\det(\mathbf{A} - \lambda\mathbf{I}) = 0$$

are the same as those which satisfy

$$\det(\mathbf{P}^{-1}\mathbf{A}\mathbf{P} - \lambda\mathbf{I}) = 0$$

This follows directly from

$$\begin{aligned}\det[\mathbf{P}^{-1}(\mathbf{A} - \lambda\mathbf{I})\mathbf{P}] &= \det[\mathbf{P}^{-1}\mathbf{A}\mathbf{P} - \lambda\mathbf{I}] \\ &= \det\mathbf{P}^{-1}\det(\mathbf{A} - \lambda\mathbf{I})\det\mathbf{P} \\ &= \det(\mathbf{A} - \lambda\mathbf{I})\end{aligned}$$

since the determinant of a product of square matrices is equal to the product of the determinants of the matrices.

An important question is whether it is possible to reduce a given matrix **A** by a transformation, where **S** is nonsingular,

$$\mathbf{D} = \mathbf{S}^{-1}\mathbf{A}\mathbf{S}$$

to diagonal form. Frequent use is made of this idea in the applications of matrices.

We will suppose that **A** is a real matrix. Consider now the eigenvalue problem

$$\mathbf{A}\mathbf{x}_j = \lambda_j\mathbf{x}_j$$

and consider also the matrix made up of the eigenvectors of **A** as

$$\mathbf{X} = \begin{bmatrix} x_{11} & x_{12} & \cdots & x_{1n} \\ x_{21} & x_{22} & \cdots & x_{2n} \\ x_{31} & & & \\ \cdot & \cdot & & \cdot \\ \cdot & \cdot & & \cdot \\ \cdot & \cdot & & \cdot \\ x_{n1} & x_{n2} & \cdots & x_{nn} \end{bmatrix}$$

where the jth column is $\mathbf{x}_j$, the eigenvector which belongs to λ_j. Then

$$\mathbf{AX} = \begin{bmatrix} \lambda_1 x_{11} & \lambda_2 x_{12} & \dots & \lambda_n x_{1n} \\ \lambda_1 x_{21} & \lambda_2 x_{22} & \dots & \lambda_n x_{2n} \\ \cdot & & & \\ \cdot & & & \\ \cdot & & & \\ \lambda_1 x_{n1} & \lambda_2 x_{n2} & \dots & \lambda_n x_{nn} \end{bmatrix} = \begin{bmatrix} x_{11} & x_{12} & \dots & x_{1n} \\ x_{21} & x_{22} & \dots & x_{2n} \\ \cdot & \cdot & & \cdot \\ \cdot & \cdot & & \cdot \\ \cdot & \cdot & & \cdot \\ x_{n1} & x_{n2} & \dots & x_{nn} \end{bmatrix} \begin{bmatrix} \lambda_1 & 0 & \dots & 0 \\ 0 & \lambda_2 & \dots & 0 \\ \cdot & & \cdot & \cdot \\ \cdot & & \cdot & \cdot \\ \cdot & & \cdot & \cdot \\ 0 & \dots & \dots & \lambda_n \end{bmatrix}$$

$$= \mathbf{X\Lambda}$$

where $\mathbf{\Lambda}$ is a diagonal matrix with the eigenvalues as non-zero elements. Therefore

$$\mathbf{X}^{-1}\mathbf{AX} = \mathbf{\Lambda}$$

Now we recall that if $\mathbf{y}_j$ are the eigenvectors of

$$\mathbf{A}^T\mathbf{y}_j = \lambda_j\mathbf{y}_j$$

then

$$\mathbf{y}_j^T\mathbf{x}_i = 0, \quad i \neq j$$

and if $\mathbf{Y}^T$ is a matrix whose rows are the eigenrows $\mathbf{y}_j^T$ then

$$\mathbf{A}^T\mathbf{Y} = \mathbf{Y\Lambda} \quad \text{and} \quad \mathbf{Y}^{-1}\mathbf{A}^T\mathbf{Y} = \mathbf{\Lambda}$$

An important special case which we shall return to is the one in which $\mathbf{A}$ is symmetric. If $\mathbf{A}$ is symmetric and if the eigenvectors have been normalized, then

$$\mathbf{X}^T\mathbf{X} = \mathbf{I}$$

since $\mathbf{Y}^T = \mathbf{X}^T$, so that

$$\mathbf{X}^T = \mathbf{X}^{-1}$$

hence

$$\mathbf{X}^T\mathbf{AX} = \mathbf{\Lambda}$$

Such a transformation is said to be by an orthogonal matrix, as was defined in Section 2.12. The term *orthogonal* has particular significance and will be returned to later. Consider the following examples:

(a) Symmetric matrix of Example a, Section 5.6:

$$\mathbf{A} = \begin{bmatrix} 1 & -1 & 2 \\ -1 & 2 & 0 \\ 2 & 0 & -1 \end{bmatrix}; \qquad \mathbf{X} = \begin{bmatrix} 2 & -5 + \sqrt{13} & -5 - \sqrt{13} \\ -2 & -2 & -2 \\ 1 & 6 - 2\sqrt{13} & 6 + 2\sqrt{13} \end{bmatrix}$$

where $\mathbf{X}$ has not been normalized. Let the normalization factor for each column be a, b, and c respectively, then

$$a = 3$$

$$b = \sqrt{130 - 34\sqrt{13}}$$

$$c = \sqrt{130 + 34\sqrt{13}}$$

then

$$\begin{bmatrix} \frac{2}{a} & -\frac{2}{a} & \frac{1}{a} \\ \frac{-5+\sqrt{13}}{b} & -\frac{2}{b} & \frac{6-2\sqrt{13}}{b} \\ \frac{-5-\sqrt{13}}{c} & -\frac{2}{c} & \frac{6+2\sqrt{13}}{c} \end{bmatrix} \begin{bmatrix} 1 & -1 & 2 \\ -1 & 2 & 0 \\ 2 & 0 & -1 \end{bmatrix} \begin{bmatrix} \frac{2}{a} & \frac{-5+\sqrt{13}}{b} & \frac{-5-\sqrt{13}}{c} \\ -\frac{2}{a} & -\frac{2}{b} & -\frac{2}{c} \\ \frac{1}{a} & \frac{6-2\sqrt{13}}{b} & \frac{6+2\sqrt{13}}{c} \end{bmatrix}$$

$$= \begin{bmatrix} \frac{2}{a} & -\frac{6}{a} & \frac{3}{a} \\ \frac{9-3\sqrt{13}}{b} & \frac{1-\sqrt{13}}{b} & \frac{-16+4\sqrt{13}}{b} \\ \frac{9+3\sqrt{13}}{c} & \frac{1+\sqrt{13}}{c} & \frac{-16-4\sqrt{13}}{c} \end{bmatrix} \begin{bmatrix} \frac{2}{a} & \frac{-5+\sqrt{13}}{b} & \frac{-5-\sqrt{13}}{c} \\ -\frac{2}{a} & -\frac{2}{b} & -\frac{2}{c} \\ \frac{1}{a} & \frac{6-2\sqrt{13}}{b} & \frac{6+2\sqrt{13}}{c} \end{bmatrix}$$

$$= \begin{bmatrix} 3 & 0 & 0 \\ 0 & -\frac{1}{2}+\frac{1}{2}\sqrt{13} & 0 \\ 0 & 0 & -\frac{1}{2}-\frac{1}{2}\sqrt{13} \end{bmatrix}$$

(b) Unsymmetric matrix

$$\mathbf{A} = \begin{bmatrix} -1 & 2 \\ -1 & 1 \end{bmatrix}$$

$$\det[\mathbf{A} - \lambda\mathbf{I}] = \lambda^2 + 1 = 0; \qquad \lambda_1 = i, \lambda_2 = -i$$

then

$$\mathbf{X} = \begin{bmatrix} -1+i & -2 \\ 1 & 1+i \end{bmatrix}; \qquad \mathbf{X}^{-1} = \begin{bmatrix} \frac{i}{2} & \frac{1}{2}+\frac{i}{2} \\ -\frac{i}{2} & \frac{1}{2}-\frac{i}{2} \end{bmatrix}$$

$$\mathbf{X}^{-1}\mathbf{A}\mathbf{X} = \begin{bmatrix} i & 0 \\ 0 & -i \end{bmatrix}$$

7.3 Quadratic Forms

Closely related to matrices are second-order expressions of the following kind

$$\begin{aligned} G = {} & a_{11}x_1^2 + a_{12}x_1x_2 + \cdots + a_{1n}x_1x_n \\ & + a_{21}x_2x_1 + a_{22}x_2^2 + \cdots + a_{2n}x_2x_n \\ & \quad \vdots \\ & + a_{n1}x_nx_1 + a_{n2}x_nx_2 + \cdots + a_{nn}x_n^2 \\ & + b_1x_1 + b_2x_2 + \cdots + b_nx_n \\ & + c \end{aligned}$$

Suppose one makes a change of variable, which is a translation of axes,

$$x_j = y_j + c_j, \quad j = 1, 2, \dots, n$$

where the $\{c_j\}$ at this point are still undetermined, then

$$\begin{aligned} G &= \sum_i \sum_j a_{ij} x_i x_j + \sum_i b_i x_j + c \\ &= \sum_i \sum_j a_{ij}(y_i + c_i)(y_j + c_j) + \sum b_i(y_i + c_i) + c \\ &= \sum_i \sum_j a_{ij} y_i y_j + \sum_i \sum_j a_{ij} y_i c_j + \sum_i \sum_j a_{ij} y_j c_i + \sum_i \sum_j a_{ij} c_i c_j \\ &\qquad + \sum_i b_i y_i + \sum_i b_i c_i + c \end{aligned}$$

or

$$\begin{aligned} G = \sum_i \sum_j a_{ij} y_i y_j + 2 \sum_i \sum_j a_{ij} y_i c_j + \sum_i b_i y_i \\ + \sum_i \sum_j a_{ij} c_i c_j + \sum b_i c_i + c \end{aligned}$$

where it has been assumed that $a_{ij} = a_{ji}$. Now,

$$\sum_i \sum_j a_{ij} c_i c_j + \sum_i b_i c_i + c = a$$

a constant, and we ask whether it is possible to choose the $\{c_j\}$ such that the linear terms in G will vanish; that is,

$$2 \sum_i \sum_j a_{ij} y_i c_j + \sum_i b_i y_i = 0$$

This can be done, provided that c_j may be determined in such a way that

$$\sum_j a_{ij} c_j = -\frac{b_i}{2}, \quad i = 1, 2, 3, \dots, n$$

or that

$$\begin{aligned} a_{11}c_1 + a_{12}c_2 + \cdots + a_{1n}c_n &= -\frac{b_1}{2} \\ a_{21}c_1 + a_{22}c_2 + \cdots + a_{2n}c_n &= -\frac{b_2}{2} \\ &\vdots \\ a_{n1}c_1 + a_{n2}c_2 + \cdots + a_{nn}c_n &= -\frac{b_n}{2} \end{aligned}$$

If the rank of **A**, the matrix of the coefficients, is the same as the rank of aug **A** then such a reduction will be possible. To determine what this means, it is instructive to examine the second-degree form

$$G = ax^2 + 2bxy + cy^2 + dx + ey + f$$

The linear terms may be removed by the substitution,

$$\begin{aligned} x &= x_1 + c_1 \\ y &= y_1 + c_2 \end{aligned}$$

provided that there is a c_1 and c_2 which satisfy the equations:

$$ac_1 + bc_2 = -\frac{d}{2}$$

$$bc_1 + cc_2 = -\frac{e}{2}$$

The condition for solution is that

$$b^2 - ac \neq 0$$

which is precisely the condition that G be a conic of other than parabolic type. It must therefore be a central conic, an ellipse or hyperbola, if the linear terms are to be removable. In the future we shall deal only with central conics and if we replace $G - a$ with F, then

$$\begin{aligned} F = a_{11}x_1^2 + a_{12}x_1x_2 + \cdots + a_{1n}x_1x_n \\ a_{21}x_1x_2 + a_{22}x_2^2 + \cdots + a_{2n}x_2x_n \\ \cdot \\ \cdot \\ \cdot \\ a_{n1}x_1x_n + a_{n2}x_2x_n + \cdots + a_{nn}x_n^2 \end{aligned}$$

The quantity F will be referred to as a *quadratic form* and its geometrical significance will be that of a quadric surface, although such geometric visualization loses some of its value in spaces of higher dimensions.

If one considers the linear transformation

$$\mathbf{y} = \mathbf{A}\mathbf{x}$$

then

$$F = \mathbf{x}^T\mathbf{y} = \mathbf{x}^T\mathbf{A}\mathbf{x}$$

as may be directly verified by direct calculation where

$$\mathbf{A} = \begin{bmatrix} a_{11} & a_{12} & \cdots & a_{1n} \\ a_{21} & a_{22} & \cdots & a_{2n} \\ \cdot & & & \\ \cdot & & & \\ \cdot & & & \\ a_{n1} & a_{n2} & \cdots & a_{nn} \end{bmatrix}$$

The coefficient of x_ix_j in F is $a_{ij} + a_{ji}$; thus, there is no loss in generality in the treatment of quadratic forms to assume that the matrix $\mathbf{A}$ is symmetric. Given a quadratic form one may obtain the matrix with ease once it is realized that

$$a_{ij} = \frac{1}{2}\frac{\partial^2 F}{\partial x_i\, \partial x_j}$$

7.4 The Reduction of a Quadratic Form to the Canonical Form

As shown previously, the quadratic form may be written in terms of a matrix as

$$F = \mathbf{x}^T \mathbf{A} \mathbf{x}$$

In applications it is frequently convenient to transform the quadratic form to a form which contains only a sum of squares as

$$F = \mathbf{A}_1 x_1^2 + \mathbf{A}_2 x_2^2 + \cdots + \mathbf{A}_n x_n^2$$

and this is called the *canonical* form of the quadratic form. If one considers the change of variable characterized by

$$\mathbf{x} = \mathbf{C}\mathbf{x}'$$

where $\mathbf{C}$ is a nonsingular matrix and $\mathbf{x}'$ is a vector in the primed axes, then the quadratic form may be written as

$$\begin{aligned} F &= (\mathbf{C}\mathbf{x}')^T \mathbf{A} \mathbf{C} \mathbf{x}' \\ &= \mathbf{x}'^T \mathbf{C}^T \mathbf{A} \mathbf{C} \mathbf{x}' \end{aligned}$$

and the matrix of the form is now $\mathbf{C}^T\mathbf{A}\mathbf{C}$ in the primed axes. $\mathbf{A}$ is a symmetric matrix, and as was shown in Section 7.2, there is a matrix $\mathbf{X}$, the matrix of normalized eigenvectors belonging to the eigenvalues of $\mathbf{A}$, which will have the property

$$\mathbf{X}^T \mathbf{A} \mathbf{X} = \mathbf{\Lambda}$$

where $\mathbf{\Lambda}$ is the diagonal matrix with the eigenvalues of $\mathbf{A}$ on the main diagonal. Therefore, if $\mathbf{C}$ is chosen as $\mathbf{X}$,

$$\begin{aligned} F &= \mathbf{x}'^T \mathbf{\Lambda} \mathbf{x}' \\ &= \lambda_1 x_1'^2 + \lambda_2 x_2'^2 + \cdots + \lambda_n x_n'^2 \end{aligned}$$

and the transformation of variables which will perform the reduction to canonical form is

$$\mathbf{x} = \mathbf{X}\mathbf{x}'$$

where $\mathbf{X}$ is an orthogonal matrix. The term *orthogonal* has particular significance here because the original unprimed coordinate system was a rectangular coordinate system and the new axes, in which the principal axes of the quadric surface lie along the coordinate axes, also form a rectangular system. Therefore the problem which has been solved is the familiar one in two-dimensional analytic geometry of finding the new axes which will remove the cross product term in an expression of the form

$$ax^2 + 2bxy + cy^2$$

This problem may be approached from another point of view. If one considers the quadratic form

$$F = \mathbf{x}^T\mathbf{A}\mathbf{x}$$

with F a constant, then as $\mathbf{x}$ varies, a surface will be swept out by the end of the vector emanating from the origin, and in the two-dimensional case, an ellipse or hyperbola will result. The length of the vector from the origin to the surface will be

$$(\mathbf{x}^T\mathbf{x})^{1/2}$$

If one thinks of an ellipse, since the vector has its terminus on the surface, the length of the vector will go through a maximum value and a minimum value, a maximum when the vector is the length of the semimajor axis and a minimum when the vector coincides with the semiminor axis. This leads naturally to a method for finding the lengths of these axes by determining the extreme values of the length of the vector subject to the condition, the vector terminus lies on the quadric surface. If the square root of the function is to be a maximum, the function is also a maximum so that we can maximize the length squared as well. We are thus led to the problem to find the maximum of

$$\sum_j x_j^2 = \mathbf{x}^T\mathbf{x}$$

subject to the restriction that

$$\mathbf{x}^T\mathbf{A}\mathbf{x} = F$$

where F is a constant. If γ is a Lagrange multiplier, then we can maximize the function,

$$\begin{aligned} f(\mathbf{x}) &= \mathbf{x}^T\mathbf{x} - \gamma(\mathbf{x}^T\mathbf{A}\mathbf{x} - F) \\ &= \sum_j x_j^2 - \gamma\left[\sum_i \sum_j a_{ij}x_ix_j - F\right] \end{aligned}$$

The partial derivatives with respect to x_j are

$$\frac{\partial f}{\partial x_j} = 2x_j - \gamma\left[2\sum_i a_{ij}x_i\right] = 0 \quad \text{for each} \quad j \tag{7.4.1}$$

or

$$\left(\mathbf{A} - \frac{1}{\gamma}\mathbf{I}\right)\mathbf{x} = 0 \tag{7.4.2}$$

Hence the values of γ which will make $\mathbf{x}^T\mathbf{x}$ an extremum are such that γ^{-1} is an eigenvalue of $\mathbf{A}$. If Eq. (7.4.1) is multiplied through by x_j and summed on j, then

$$\sum_j x_j^2 = \gamma_i \sum_j \sum_i a_{ij}x_ix_j = \gamma_i F$$

where F was the constant in $\mathbf{x}^T\mathbf{A}\mathbf{x} = F$; hence the extremal lengths are $\sqrt{\gamma_i F}$, the lengths of the n principal axes. The directions of these axes can also be obtained, since the directions are given by the directions of the eigenvectors belonging to the set of eigenvalues $\{\gamma^{-1}\}$, the eigenvectors being the normalized solutions of Eq. (7.4.2) for each eigenvalue γ_i^{-1}. In fact, in this case the direction cosines of the principal axes are the elements of the eigenvectors themselves since the vectors have been normalized. Hence each eigenvalue (actually the square root of the eigenvalue) and eigenvector gives, if F is taken as unity, the length of a principal axis and its direction. Because the matrix **A** was symmetric, the principal axes are mutually orthogonal and the eigenvalues are real, although, of course, some may be negative. In that case, the length of the vector will be imaginary, as it would for a hyperbola in that direction (in two dimensions).

As was shown earlier, if the principal axes are taken as the coordinate axes for representation of the quadric surface, then

$$\begin{aligned} F &= \mathbf{x}'^T\mathbf{A}\mathbf{x}' \\ &= \frac{1}{\gamma_1}x_1'^2 + \frac{1}{\gamma_2}x_2'^2 + \cdots + \frac{1}{\gamma_n}x_n'^2 \end{aligned}$$

When there are multiple eigenvalues or if **A** is singular, then the situation may be treated, but as stated earlier in this work these pathological cases are not dealt with.

7.5 An Associated Problem

In connection with an example to be treated later, a problem associated with that given at the end of Section 7.4 is worth considering. Instead of finding the extreme values of the length of a vector subject to the restraint that the vector lie on the quadric surface, we shall consider the problem of finding the maximum value of $F = \mathbf{x}^T\mathbf{A}\mathbf{x}$ subject to the condition that $\mathbf{x}^T\mathbf{x} = R$. The maximum problem with λ as a Lagrange multiplier is then to find the extrema of

$$g(\mathbf{x}) = \mathbf{x}^T\mathbf{A}\mathbf{x} - \lambda(\mathbf{x}^T\mathbf{x} - R)$$

the partial derivatives are

$$\frac{\partial g}{\partial x_j} = 2\sum_i a_{ij}x_i - 2\lambda x_j = 0$$

or

$$(\mathbf{A} - \lambda\mathbf{I})\mathbf{x} = 0$$

so that the extreme values occur for the eigenvalues of **A** and the extreme values themselves are

$$F_{\text{ext}} = \mathbf{x}_j^T\mathbf{A}\mathbf{x}_j = \mathbf{x}_j^T\lambda_j\mathbf{x}_j = \lambda_j R$$

Hence there are n extreme values, and they occur in the directions given by the eigenvectors belonging to the eigenvalues of **A**. In this problem, **x** does not trace out a quadric surface, but from the transformation

$$\mathbf{y} = \mathbf{A}\mathbf{x}; \qquad \mathbf{x} = \mathbf{A}^{-1}\mathbf{y}$$

the quadratic form is

$$F = \mathbf{x}^T\mathbf{y} = \mathbf{y}^T\mathbf{A}^{-1}\mathbf{y}$$

and the sphere $\mathbf{x}^T\mathbf{x} = R$ becomes

$$R = (\mathbf{A}^{-1}\mathbf{y})^T\mathbf{A}^{-1}\mathbf{y}$$

or

$$R = \mathbf{y}^T\mathbf{A}^{-1^T}\mathbf{A}^{-1}\mathbf{y} = \mathbf{y}^T\mathbf{A}^{-2}\mathbf{y}$$

and therefore the vector $\mathbf{y}$ does trace out a quadric surface. The square of the lengths of the semi-axes of this quadric will be proportional to the reciprocal of the eigenvalues of $\mathbf{A}^{-2}$, or the length of the semi-axes will be proportional to $\{\lambda_i\}$. The direction and length of the semi-axes may be easily computed as in the foregoing, but the interesting point is that when $\mathbf{x}$ is confined to a sphere, $\mathbf{y}$ traces out a quadric surface. Note that the quadric surface traced out by $\mathbf{y}$ is always ellipsoidal in character.

7.6 Definite Quadratic Forms

Given a general quadratic form

$$Q = \mathbf{x}^T\mathbf{A}\mathbf{x} \tag{7.6.1}$$

it may take on plus or minus signs depending upon the vector $\mathbf{x}$. If for all vectors $\mathbf{x}$, Q has only one sign, say positive, and is zero only if $\mathbf{x} = \mathbf{0}$ then Q is said to be *positive definite*. An important theorem relative to the positive definiteness of Q was proved by Sylvester. Consider the matrix $\mathbf{A}$ and let it be of order n and suppose det $\mathbf{A} = A_n$. The principal minors of $\mathbf{A}$ are those minors $A_{n-1}, A_{n-2}, \ldots, A_2, A_1$ which are obtained successively by striking out the last row and last column, the last two rows and two columns, etc., respectively, until finally A_1 is the determinant which consists of the element a_{11} of $\mathbf{A}$. Sylvester's theorem states that the necessary and sufficient condition for the quadratic form (7.6.1) to be positive definite is that each of the principal minors of the matrix of the quadratic form be greater than zero. That is, if

$$\mathbf{A} = \begin{bmatrix} a_{11} & a_{12} & \cdots & a_{1n} \\ a_{21} & a_{22} & \cdots & a_{2n} \\ \cdot & & & \\ \cdot & & & \\ \cdot & & & \\ a_{n1} & a_{n2} & \cdots & a_{nn} \end{bmatrix}$$

with $a_{ij} = a_{ji}$, then the necessary and sufficient condition that $Q > 0$ is each of the determinants

$$\mathbf{A}_1 = a_{11}, \quad \mathbf{A}_2 = \begin{vmatrix} a_{11} & a_{12} \\ a_{21} & a_{22} \end{vmatrix}, \quad \mathbf{A}_3 = \begin{vmatrix} a_{11} & a_{12} & a_{13} \\ a_{21} & a_{22} & a_{23} \\ a_{31} & a_{32} & a_{33} \end{vmatrix}, \ldots, \mathbf{A}_n$$

be greater than zero. We whall not prove this theorem. A negative definite quadratic form is one which is always negative for all vectors $\mathbf{x}$ save for $\mathbf{x} = 0$. It is clear that the preceding theorem may be applied in this case by considering the form minus Q.

Since a quadratic form may be reduced to canonical form, its sign will depend upon the signs of the eigenvalues of its matrix. It seems obvious that a necessary and sufficient condition for a quadratic form to be positive definite is that the eigenvalues of its matrix be non-negative.

7.7 Maxima and Minima of Functions of Several Variables

Given a function of a single variable $y = f(x)$, the necessary condition that it have a relative minimum or maximum at $x = a$ is $f'(a) = 0$ and from a consideration of the Taylor series about the point $x = a$, where $x - a = h$,

$$f(a + h) = f(a) + f'(a)h + \frac{f''(a)}{2!}h^2 + \frac{f'''(a)}{3!}h^3 + \cdots$$

it follows that if

$$f(a + h) - f(a) = \frac{f''(a)}{2!}h^2 + \frac{f'''(a)}{3!}h^3 + \cdots$$

is to be of constant sign for small but arbitrary h, then

$$f''(a) < 0, \quad \text{for } x = a \quad \text{a maximum}$$

$$f''(a) > 0, \quad \text{for } x = a \quad \text{a minimum}$$

$f''(a) = 0$ is a doubtful case and further analysis is required.

For a function of n variables, $y = f(x_1, x_2, \ldots, x_n)$, the corresponding necessary condition for a maximum, minimum, or other extremum is that all the first partial derivatives be zero; that is,

$$\frac{\partial f}{\partial x_i} = 0, \quad i = 1, 2, \ldots, n$$

at a point $\mathbf{x} = \mathbf{a}$. The Taylor series expansion is, for $\mathbf{x} = \mathbf{a} + \mathbf{h}$,

$$f(\mathbf{a} + \mathbf{h})$$
$$= f(\mathbf{a}) + \sum_{n=1}^{\infty} \frac{1}{n!}\left[\left(h_1 \frac{\partial}{\partial x_1} + h_2 \frac{\partial}{\partial x_2} + \cdots + h_n \frac{\partial}{\partial x_n}\right)^n f(x_1, x_2, \ldots, x_n)\right]_{\mathbf{x} = \mathbf{a}}$$

where the quantity in parentheses is an operator.* If the quantity

$$\mathbf{f}(\mathbf{a}+\mathbf{h}) - f(\mathbf{a}) = \sum_{n=2}^{\infty} \frac{1}{n!}\left[\left(h_1 \frac{\partial}{\partial x_1} + h_2 \frac{\partial}{\partial x_2} + \cdots + h_n \frac{\partial}{\partial x_n}\right)^n f(x_1, x_2, \ldots, x_n)\right]_{\mathbf{x}=\mathbf{a}}$$

is to be of one sign, either negative for a relative maximum or positive for a relative minimum, the term

$$\left[\left(h_1 \frac{\partial}{\partial x_1} + h_2 \frac{\partial}{\partial x_2} + \cdots + \frac{\partial}{\partial x_n}\right)^2 f(x_1, x_2, \ldots, x_n)\right]_{\mathbf{x}=\mathbf{a}}$$

for $\mathbf{h}$ small but arbitrary must be of one sign and zero for $\mathbf{h} = 0$. This may be written out, where all partial derivatives are evaluated at $\mathbf{x} = \mathbf{a}$ with

$$\frac{\partial^2 f}{\partial x_i \, \partial x_j} = f_{ij}$$

as

$$\begin{aligned} & f_{11}h_1^2 + f_{12}h_1h_2 + \cdots + f_{1n}h_1h_n \\ & + f_{21}h_2h_1 + f_{22}h_2^2 + \cdots + f_{1n}h_1h_n \\ & \quad \vdots \\ & + f_{n1}h_nh_1 + f_{n2}h_nh_2 + \cdots + f_{nn}h_n^2 \end{aligned}$$

Since, under appropriate hypotheses, $f_{ij} = f_{ji}$, the foregoing may be written as

$$\mathbf{S} = \mathbf{h}^T\mathbf{F}\mathbf{h}$$

where $\mathbf{F}$ is the symmetric matrix $\{f_{ij}\}$. Thus, whether the function $f(x_1, x_2, \ldots, x_n)$ has a maximum or a minimum at $\mathbf{x} = \mathbf{a}$ depends upon whether the quadratic form S is negative definite or positive definite. In short, a sufficient condition that $f(x_1, x_2, \ldots, x_n)$ have a maximum (minimum) at $\mathbf{x} = \mathbf{a}$ is that $\mathbf{F}$ be negative (positive). If all the second partial derivatives are zero, then further analysis is necessary.

7.8 The Minimal Properties of the Eigenvalues

In many applications it is not necessary to obtain the eigenvectors but only the eigenvalues. For example, in Section 5.7, it was shown that the eigenvalues of the matrix $\mathbf{A}$ were the reciprocal of the time constants in the solutions of the differential equations, and these are frequently the most

*For example,

$$\left(h_1 \frac{\partial}{\partial x_1} + h_2 \frac{\partial}{\partial x_2}\right)^2 f = \left(h_1^2 \frac{\partial^2}{\partial x_1^2} + 2h_1h_2 \frac{\partial^2}{\partial x_1 \, \partial x_2} + h_2^2 \frac{\partial^2}{\partial x_2^2}\right) f$$

important quantities in many applications, particularly the largest and the smallest. Unfortunately, the theory is presentable only for symmetric matrices; hence, we shall be content with that here. In the previous section, certain aspects of maximization and minimization of quadratic forms led to eigenvalues of the matrix of the quadratic form. In this section, we consider this problem further, not in depth, but only in survey form.

Suppose we are given a symmetric matrix $\mathbf{A}$ and its eigenvalues and normalized eigenvectors, $\{\lambda_j\}$ and $\{\mathbf{x}_j\}$, respectively, then *any* vector $\mathbf{x}$ can be expanded

$$\mathbf{x} = \sum_j c_j \mathbf{x}_j$$

where we assume, as always, that the eigenvalues are simple. Further,

$$\mathbf{x}^T\mathbf{x} = \sum_j c_j^2$$

$$\mathbf{A}\mathbf{x} = \sum_j c_j \lambda_j \mathbf{x}_j$$

$$\mathbf{x}^T\mathbf{A}\mathbf{x} = \sum_j \lambda_j c_j^2$$

hence

$$\frac{\mathbf{x}^T\mathbf{A}\mathbf{x}}{\mathbf{x}^T\mathbf{x}} = \frac{\sum_j \lambda_j c_j^2}{\sum_j c_j^2}$$

Now suppose the eigenvalues are arranged in decreasing algebraic order so that

$$\lambda_1 > \lambda_2 > \lambda_3 \cdots > \lambda_n$$

Let λ_i be the one which is largest in absolute value, then

$$\left|\sum_j \lambda_j c_j^2\right| \le \sum_j |\lambda_j| c_j^2 \le |\lambda_i|_{\max} \sum c_j^2$$

so that

$$\left|\frac{\mathbf{x}^T\mathbf{A}\mathbf{x}}{\mathbf{x}^T\mathbf{x}}\right| \le |\lambda_i|_{\max}$$

which holds for *any* vector $\mathbf{x}$. The equality sign is necessary, since if $\mathbf{x}$ is the eigenvector $\mathbf{x}_i$, the equality sign holds. Also

$$\lambda_1 - \frac{\mathbf{x}^T\mathbf{A}\mathbf{x}}{\mathbf{x}^T\mathbf{x}} = \frac{\sum (\lambda_1 - \lambda_j) c_j^2}{\sum_j c_j^2} \ge 0 \tag{7.8.1}$$

and since $\lambda_1 \ge \lambda_j$ for all j

$$\lambda_1 \ge \frac{\mathbf{x}^T\mathbf{A}\mathbf{x}}{\mathbf{x}^T\mathbf{x}}$$

and the equality sign holds for $\mathbf{x} = \mathbf{x}_1$ the eigenvector belonging to λ_1. On the other hand,

$$\frac{\mathbf{x}^T\mathbf{A}\mathbf{x}}{\mathbf{x}^T\mathbf{x}} - \lambda_n = \frac{\sum_j (\lambda_j - \lambda_n) c_j^2}{\sum_j c_j^2} \ge 0$$

or
$$\frac{\mathbf{x}^T\mathbf{A}\mathbf{x}}{\mathbf{x}^T\mathbf{x}} \geq \lambda_n$$

and thus
$$\lambda_n \leq \frac{\mathbf{x}^T\mathbf{A}\mathbf{x}}{\mathbf{x}^T\mathbf{x}} \leq \lambda_1$$

This is a remarkable result, since it says that the ratio

$$\frac{\mathbf{x}^T\mathbf{A}\mathbf{x}}{\mathbf{x}^T\mathbf{x}}$$

called the *Rayleigh quotient*, is bounded between the largest and smallest eigenvalues for *any vector* $\mathbf{x}$ and where the equality signs hold for $\mathbf{x}_1$ and $\mathbf{x}_n$. λ_1 is the largest algebraic eigenvalue and we can find λ_2 in the following manner: Suppose that we have found λ_1 and $\mathbf{x}_1$. This may be done by computing with Eq. (7.8.1) using random vectors $\mathbf{x}$ until one is reasonably satisfied that the maximum value of the Rayleigh quotient has been obtained. Then the set of equations $\mathbf{A}\mathbf{x} = \lambda_1\mathbf{x}$ may be solved for $\mathbf{x}_1$. Suppose now we compute the Rayleigh quotient but only with vectors that are orthogonal to $\mathbf{x}_1$

$$\mathbf{x} = \sum_{j=2}^{n} d_j\mathbf{x}_j$$

then
$$\mathbf{A}\mathbf{x} = \sum_{j=2}^{n} d_j\lambda_j\mathbf{x}_j$$

$$\mathbf{x}^T\mathbf{A}\mathbf{x} = \sum_{j=2}^{n} d_j^2\lambda_j$$

$$\mathbf{x}^T\mathbf{x} = \sum_{j=2}^{n} d_j^2$$

and
$$\lambda_2 - \frac{\mathbf{x}^T\mathbf{A}\mathbf{x}}{\mathbf{x}^T\mathbf{x}} = \frac{\sum_{j=2}^{n} (\lambda_2 - \lambda_j)d_j^2}{\sum_{j=2}^{n} d_j^2} \geq 0$$

where the equality sign holds for $\mathbf{x} = \mathbf{x}_2$. Then

$$\lambda_2 \geq \frac{\mathbf{x}^T\mathbf{A}\mathbf{x}}{\mathbf{x}^T\mathbf{x}}$$

for any vector $\mathbf{x}$ orthogonal to $\mathbf{x}_1$. Similarly, it is obvious that

$$\lambda_3 \geq \frac{\mathbf{x}^T\mathbf{A}\mathbf{x}}{\mathbf{x}^T\mathbf{x}}$$

for any vector $\mathbf{x}$ orthogonal to both $\mathbf{x}_1$ and $\mathbf{x}_2$. In general,

$$\lambda_r \geq \frac{\mathbf{x}^T\mathbf{A}\mathbf{x}}{\mathbf{x}^T\mathbf{x}} \tag{7.8.2}$$

for any vector $\mathbf{x}$ orthogonal to each of the vectors $\mathbf{x}_1, \mathbf{x}_2, \ldots, \mathbf{x}_{r-1}$. This, in theory, gives a technique for calculating each of the eigenvalues. In practice, the calculation of the vectors is determined to a lesser degree of approximation than the eigenvalues so that the approximation of succeeding eigenvalues becomes poorer and poorer. The important thing, however, is that Eq. (7.8.2) gives a lower bound on λ_r. The problem of avoiding the computation of eigenvectors altogether was solved by Courant-Fischer; the computations required are quite extensive, however, and the reader is referred to other works.

7.9 Functions Defined on Matrices. Infinite Series

In earlier sections, we showed that polynomials on matrices could be defined, and by applying the Hamilton-Cayley theorem, it was shown that any polynomial in $\mathbf{A}$ and $\mathbf{A}^{-1}$ could be reduced to a polynomial in $\mathbf{A}$ of maximum degree $n - 1$. It is natural then to consider the problem of infinite series of square matrices, and we would now like to investigate infinite power series of matrices of the form

$$a_0\mathbf{I} + a_1\mathbf{A} + a_2\mathbf{A}^2 + a_3\mathbf{A}^3 + \cdots + a_m\mathbf{A}^m + \cdots \tag{7.9.1}$$

where the $\{a_j\}$ are scalar coefficients. Let

$$\mathbf{S}_m = a_0\mathbf{I} + a_1\mathbf{A} + a_2\mathbf{A}^2 + \cdots + a_m\mathbf{A}^m \tag{7.9.2}$$

then $\mathbf{S}_m$ is a square matrix of the same order as $\mathbf{A}$ and if

$$\mathbf{A} = \begin{bmatrix} a_{11} & a_{12} & \cdots & a_{1n} \\ a_{21} & a_{22} & \cdots & a_{2n} \\ \cdot & \cdot & & \cdot \\ \cdot & \cdot & & \cdot \\ \cdot & \cdot & & \cdot \\ a_{n1} & a_{n2} & \cdots & a_{nn} \end{bmatrix}$$

then $\mathbf{S}_m$ may be written

$$\mathbf{S}_m = \begin{bmatrix} a_{11/m} & a_{12/m} & \cdots & a_{1n/m} \\ a_{21/m} & a_{22/m} & \cdots & a_{2n/m} \\ \cdot & & & \\ \cdot & & & \\ \cdot & & & \\ a_{n1/m} & a_{n2/m} & \cdots & a_{nn/m} \end{bmatrix} \tag{7.9.3}$$

If

$$\mathbf{A}^k = \begin{bmatrix} b_{11/k} & b_{12/k} & \cdots & b_{1n/k} \\ b_{21/k} & b_{22/k} & \cdots & b_{2n/k} \\ \cdot & & & \\ \cdot & & & \\ \cdot & & & \\ b_{n1/k} & b_{n2/k} & \cdots & b_{nn/k} \end{bmatrix}$$

then

$$a_{ij/m} = \sum_{k=0}^{m} a_k b_{ij/k}$$

and if

$$\lim_{m\to\infty} a_{ij/m} = \lim_{m\to\infty} \left[\sum_{k=0}^{m} a_k b_{ij/k} \right]$$

exists and equals S_{ij} for all i and j, then the series is said to *converge* and have the sum $\mathbf{S}$ where

$$\mathbf{S} = \begin{bmatrix} S_{11} & S_{12} & \dots & S_{1n} \\ S_{21} & S_{22} & \dots & S_{2n} \\ \cdot & & & \\ \cdot & & & \\ \cdot & & & \\ S_{n1} & S_{n2} & \dots & S_{nn} \end{bmatrix}$$

that is,

$$\lim_{m\to\infty} \mathbf{S}_m = \mathbf{S}$$

Now this definition, like many definitions in mathematics, is difficult to use as a test for convergence so that another path must be followed.

Suppose for the moment that $\mathbf{T}$ is any square nonsingular matrix, then $\mathbf{A}$ and $\mathbf{C}$ are said to be *transforms* of each other if a relation of the form

$$\mathbf{C} = \mathbf{TAT}^{-1}$$

exists, and indeed, this is a similarity transformation.

Now

$$\mathbf{C}^2 = \mathbf{TAT}^{-1}\mathbf{TAT}^{-1} = \mathbf{TA}^2\mathbf{T}^{-1}$$

and in general

$$\mathbf{C}^k = \mathbf{TA}^k\mathbf{T}^{-1}$$

where k is an integer. If a_k is a scalar

$$a_k\mathbf{C}^k = \mathbf{T}a_k\mathbf{A}^k\mathbf{T}^{-1}$$

and

$$P_m(\mathbf{C}) = \sum_{k=0}^{m} a_k\mathbf{C}^k = \mathbf{T}\sum_{k=0}^{m} a_k\mathbf{A}^k\mathbf{T}^{-1}$$

so that

$$P_m(\mathbf{C}) = \mathbf{T}P_m(\mathbf{A})\mathbf{T}^{-1}$$

and thus if $\mathbf{A}$ and $\mathbf{C}$ are transforms and $P_m(x)$ is any polynomial, then $P_m(\mathbf{A})$ and $P_m(\mathbf{C})$ are also transforms. It is apparent also that

$$P_m(\mathbf{A}) = \mathbf{T}^{-1}P_m(\mathbf{C})\mathbf{T}$$

Now consider the partial sum in Eq. (7.9.2) for the infinite series given in Eq. (7.9.1) and where $\mathbf{S}_m$ has the explicit form given in Eq. (7.9.3) and suppose $\mathbf{T}$ is a matrix such that

$$\mathbf{C} = \mathbf{TAT}^{-1}$$

and therefore

$$\mathbf{S}_m(\mathbf{C}) = \mathbf{TS}_m(\mathbf{A})\mathbf{T}^{-1}$$

where **T** and **C** are yet to be defined. Now if

$$\mathbf{TS}_m(\mathbf{A})\mathbf{T}^{-1} = \begin{bmatrix} t_{11} & t_{12} & \dots & t_{1n} \\ t_{21} & t_{22} & \dots & t_{2n} \\ \cdot \\ \cdot \\ \cdot \\ t_{n1} & t_{n2} & \dots & t_{nn} \end{bmatrix} \begin{bmatrix} a_{11/m} & a_{12/m} & \dots & a_{1n/m} \\ a_{21/m} & a_{22/m} & \dots & a_{2n/m} \\ \cdot \\ \cdot \\ \cdot \\ a_{n1/m} & a_{n2/m} & \dots & a_{nn/m} \end{bmatrix} \begin{bmatrix} s_{11} & s_{12} & \dots & s_{1n} \\ s_{21} & s_{22} & \dots & s_{2n} \\ \cdot \\ \cdot \\ \cdot \\ s_{n1} & s_{n2} & \dots & s_{nn} \end{bmatrix}$$

where t_{ij} and s_{ij} are typical elements of **T** and $\mathbf{T}^{-1}$, respectively, then

$$\mathbf{TS}(\mathbf{A})\mathbf{T}^{-1} = \left[\sum_{l=1}^{n} \sum_{k=1}^{n} t_{ik} a_{kl/m} s_{lj}\right]$$

where the quantity inside the square brackets is the element in the ith row and jth column of $\mathbf{S}_m(\mathbf{C})$. Now,

$$\lim_{m\to\infty} \left(\sum_{l=1}^{n} \sum_{k=1}^{n} t_{ik} a_{kl/m} s_{lj}\right)$$

may be taken term by term since it is a finite double sum and therefore if

$$\lim_{m\to\infty} a_{kl/m}$$

exists for all k and l, then

$$\lim_{m\to\infty} \mathbf{S}_m(\mathbf{C})$$

exists; hence, if the series in Eq. (7.9.1) converges, the transform also converges and conversely; that is, if **A** and **C** are transforms, the series

$$a_0\mathbf{I} + a_1\mathbf{A} + a_2\mathbf{A}^2 + a_3\mathbf{A}^3 + \cdots$$
$$a_0\mathbf{I} + a_1\mathbf{C} + a_2\mathbf{C}^2 + a_3\mathbf{C}^3 + \cdots$$

converge or diverge together.

Up to this point no particular matrix **T** has been specified. Let $\mathbf{T} = \mathbf{X}^{-1}$ where **X** is the matrix of normalized eigenvectors of **A**, then

$$\mathbf{X}^{-1}\mathbf{AX} = \mathbf{\Lambda}$$

where $\mathbf{\Lambda}$ is the diagonal matrix of eigenvalues and **A** and $\mathbf{\Lambda}$ are transforms. Hence instead of considering the convergence of the series defined on **A**, we can consider the same series but defined on $\mathbf{\Lambda}$,

$$\mathbf{S}_m(\mathbf{\Lambda}) = a_0\mathbf{I} + a_1\mathbf{\Lambda} + a_2\mathbf{\Lambda}^2 + \cdots + a_m\mathbf{\Lambda}^m$$

and since

$$\mathbf{\Lambda}^k = \begin{bmatrix} \lambda_1^k & 0 & 0 & \dots & 0 \\ 0 & \lambda_2^k & 0 & \dots & 0 \\ \cdot \\ \cdot \\ \cdot \\ 0 & & & \dots & \lambda_n^k \end{bmatrix}$$

then

$$\mathbf{S}_m(\mathbf{\Lambda}) = \begin{bmatrix} \sum_0^m a_k\lambda_1^k & 0 & \cdots & \cdots & 0 \\ 0 & \sum_0^m a_k\lambda_2^k & 0 & \cdots & 0 \\ \cdot & & & & \\ \cdot & & & & \\ \cdot & & & & \\ 0 & 0 & \cdots & & \sum_0^m a_k\lambda_n^k \end{bmatrix}$$

and

$$\lim_{m\to\infty} \mathbf{S}_m(\mathbf{\Lambda})$$

exists if each of the limits

$$\lim_{m\to\infty}\left(\sum_{k=0}^m a_k\lambda_j^k\right)$$

exists. Thus one is led to consideration of the convergence of the scalar power series

$$S = \lim_{m\to\infty} S_m(x) = \sum_0^\infty a_k x^k$$

If each of the eigenvalues lies within the circle of convergence of this power series, then the matrix power series Eq. (7.9.1) will converge.

Example: Let

$$\mathbf{S} = \mathbf{I} + \mathbf{A} + \mathbf{A}^2 + \mathbf{A}^3 + \mathbf{A}^4 + \cdots$$

for

$$\mathbf{A} = \begin{bmatrix} 0.15 & -0.01 \\ -0.25 & 0.15 \end{bmatrix}$$

What is $\mathbf{S}$ the sum of the series?

First it must be determined whether the series converges. Consider

$$\begin{vmatrix} 0.15 - \lambda & -0.01 \\ -0.25 & 0.15 - \lambda \end{vmatrix} = \lambda^2 - 0.30\lambda + 0.020 = (\lambda - 0.2)(\lambda - 0.1) = 0$$

$$\lambda_1 = 0.10, \quad \lambda_2 = 0.20$$

Since the series

$$1 + x + x^2 + x^3 + x^4 + \cdots$$

converges for $|x| < 1$ it follows that the series defined on $\mathbf{A}$ converges.

Thus

$$\lim_{m\to\infty} \mathbf{S}_m = \lim_{m\to\infty} \sum_{k=0}^m \mathbf{A}^k$$

exists. Write Sylvester's formula as

$$\mathbf{S}_m = \sum_{j=1}^2 S_m(\lambda_j) \frac{\operatorname{adj}(\lambda_j I - \mathbf{A})}{\prod_{i=1,j}^2 (\lambda_j - \lambda_i)}$$

and for

$$\lambda_1 = 0.10, \quad \text{adj}\,(\lambda_1 \mathbf{I} - \mathbf{A}) = \begin{bmatrix} -0.05 & -0.01 \\ -0.25 & -0.05 \end{bmatrix}$$

$$\lambda_2 = 0.20, \quad \text{adj}\,(\lambda_2 \mathbf{I} - \mathbf{A}) = \begin{bmatrix} 0.05 & -0.01 \\ -0.25 & 0.05 \end{bmatrix}$$

Also
$$S_m(x) = 1 + x + x^2 + \cdots + x^m = \frac{1 - x^{m+1}}{1 - x}$$

So
$$\mathbf{S}_m(\mathbf{A}) = \frac{1 - (0.10)^{m+1}}{0.90}(0.10 - 0.20)^{-1}\begin{bmatrix} -0.05 & -0.01 \\ -0.25 & -0.05 \end{bmatrix}$$
$$+ \frac{1 - (0.20)^{m+1}}{0.80}(0.20 - 0.10)^{-1}\begin{bmatrix} 0.05 & -0.01 \\ -0.25 & 0.05 \end{bmatrix}$$

$$\mathbf{S} = -\tfrac{100}{9}\begin{bmatrix} -0.05 & -0.01 \\ -0.25 & -0.05 \end{bmatrix} + \tfrac{100}{8}\begin{bmatrix} 0.05 & -0.01 \\ -0.25 & 0.05 \end{bmatrix}$$
$$= \tfrac{1}{72}\begin{bmatrix} 85 & -1 \\ -25 & 85 \end{bmatrix}$$

The foregoing analysis now allows us to define those functions of matrices defined by power series. Although many could be considered we will treat the exponential function only and the others such as sin **A**, tan **A**, etc., can be handled in a similar fashion. The series

$$\mathbf{I} + \mathbf{A} + \frac{\mathbf{A}^2}{2!} + \frac{\mathbf{A}^3}{3!} + \cdots + \frac{\mathbf{A}^m}{m!} + \cdots$$

converges for any matrix **A**, since the corresponding scalar series in x is the series for e^x which converges for all x. Also if

$$\mathbf{S}_m(\mathbf{A}) = \mathbf{I} + \mathbf{A} + \frac{\mathbf{A}^2}{2!} + \frac{\mathbf{A}^3}{3!} + \cdots + \frac{\mathbf{A}^m}{m!}$$

then
$$\mathbf{S}_m(\mathbf{A}) = \sum_{j=1}^{n} S_m(\lambda_j)\frac{\text{adj}\,(\lambda_j \mathbf{I} - \mathbf{A})}{\prod_{i=1,j}^{n}(\lambda_j - \lambda_i)}$$

and
$$\lim_{m\to\infty} \mathbf{S}_m(\mathbf{A}) = \sum_{j=1}^{n} e^{\lambda_j}\frac{\text{adj}\,(\lambda_j \mathbf{I} - \mathbf{A})}{\prod_{i=1}^{n}(\lambda_j - \lambda_i)}$$

and so we call the sum of the series $\mathbf{S}(\mathbf{A})$

$$\mathbf{S}(\mathbf{A}) = \lim_{m\to\infty} \mathbf{S}_m(\mathbf{A}) = e^{\mathbf{A}}$$

With the sum

$$\mathbf{S}_m(\mathbf{A}t) = \mathbf{I} + t\mathbf{A} + \frac{t^2\mathbf{A}^2}{2!} + \frac{t^3\mathbf{A}^3}{3!} + \cdots + \frac{t^m\mathbf{A}^m}{m!}$$

with t an arbitrary but fixed number, the corresponding scalar series will converge for all t and the matrix series will converge for all t and all matrices $\mathbf{A}$. The series

$$\lim_{m\to\infty} \mathbf{S}_m(\mathbf{A}t) = e^{\mathbf{A}t}$$

$$\mathbf{S}'_m(\mathbf{A}t) = \frac{d\mathbf{S}_m(\mathbf{A}t)}{dt} = \mathbf{A} + \frac{t\mathbf{A}^2}{1} + \frac{t^2\mathbf{A}^3}{2!} + \frac{t^3\mathbf{A}^4}{3!} + \cdots + \frac{t^{m-1}\mathbf{A}^m}{(m-1)!}$$
$$= \mathbf{A}\left(\mathbf{I} + t\mathbf{A} + \frac{t^2\mathbf{A}^2}{2!} + \frac{t^3\mathbf{A}^3}{3!} + \cdots + \frac{t^{m-1}\mathbf{A}^{m-1}}{(m-1)!}\right)$$
$$= \left(\mathbf{I} + t\mathbf{A} + \frac{t^2\mathbf{A}^2}{2!} + \frac{t^3\mathbf{A}^3}{3!} + \cdots + \frac{t^{m-1}\mathbf{A}^{m-1}}{(m-1)!}\right)\mathbf{A}$$

and

$$\lim_{m\to\infty} \mathbf{S}'_m(\mathbf{A}t)$$

exists and is equal to

$$\mathbf{A}e^{\mathbf{A}t} = e^{\mathbf{A}t}\mathbf{A}$$

Hence the first derivative exists and the exponential function exp $(\mathbf{A}t)$ commutes with $\mathbf{A}$. A similar statement holds for the higher derivatives. From Sylvester's formula,

$$\mathbf{A}e^{\mathbf{A}t} = \sum_{j=1}^{n} \lambda_j e^{\lambda_j t} \frac{\operatorname{adj}(\lambda_j\mathbf{I} - \mathbf{A})}{\prod_{i=1,j}^{n} (\lambda_j - \lambda_i)} = e^{\mathbf{A}t}\mathbf{A}$$

With functions defined on matrices, some care must be used in their manipulation for

$$e^{\mathbf{A}}e^{\mathbf{B}} \neq e^{\mathbf{B}}e^{\mathbf{A}}$$

It may be shown that the series for exp $\mathbf{A}$ and exp $\mathbf{B}$ may be multiplied together to get the product but because $\mathbf{A}$ and $\mathbf{B}$, in general, do not commute when multiplied ($\mathbf{AB} \neq \mathbf{BA}$) it will follow that the two series representations for the foregoing will not be the same. The series will not agree from the second-degree terms on since, in general,

$$\frac{\mathbf{A}^2}{2!} + \mathbf{AB} + \frac{\mathbf{B}^2}{2!} \neq \frac{\mathbf{A}^2}{2!} + \mathbf{BA} + \frac{\mathbf{B}^2}{2!}$$

7.10 Systems of First-order Differential Equations

A system of first-order differential equations was treated in Section 5.7 of the form

$$\frac{dx_1}{dt} = a_{11}x_1 + a_{12}x_2 + \cdots + a_{1n}x_n + b_1$$

$$\frac{dx_2}{dt} = a_{21}x_1 + a_{22}x_2 + \cdots + a_{2n}x_n + b_2$$

$$\vdots$$

$$\frac{dx_n}{dt} = a_{n1}x_1 + a_{n2}x_2 + \cdots + a_{nn}x_n + b_n$$

and where a set of conditions

$$\left.\begin{aligned} x_1 &= x_1^i \\ x_2 &= x_2^i \\ &\vdots \\ x_n &= x_n^i \end{aligned}\right\} t = 0$$

must also be specified in order to provide a unique solution. The complete system may be written

$$\frac{d\mathbf{x}}{dt} = \mathbf{Ax} + \mathbf{b}$$

$$\mathbf{x} = \mathbf{x}^i, \quad t = 0$$

Now

$$\mathbf{x} = e^{\mathbf{A}t}\mathbf{c}$$

has the property

$$\frac{d\mathbf{x}}{dt} = e^{\mathbf{A}t}\mathbf{Ac} = \mathbf{A}e^{\mathbf{A}t}\mathbf{c} = \mathbf{Ax}$$

so that it is a solution of the homogeneous equation with n arbitrary constants. Further the expression

$$-\mathbf{A}^{-1}\mathbf{b}$$

is a particular solution of the nonhomogeneous equation and therefore the general solution of the nonhomogeneous equation is

$$\mathbf{x} = e^{\mathbf{A}t}\mathbf{c} - \mathbf{A}^{-1}\mathbf{b}$$

To satisfy the auxiliary condition

$$\mathbf{x}^i = \mathbf{c} - \mathbf{A}^{-1}\mathbf{b}$$

so

$$\mathbf{x} = e^{\mathbf{A}t}(\mathbf{x}^i + \mathbf{A}^{-1}\mathbf{b}) - \mathbf{A}^{-1}\mathbf{b}$$

The solution may also be obtained by writing the equation in the form

$$\frac{d\mathbf{x}}{dt} - \mathbf{Ax} = \mathbf{b}$$

and
$$\frac{d}{dt}(e^{-\mathbf{A}t}\mathbf{x}) = e^{-\mathbf{A}t}\mathbf{b}$$

Integrating

$$e^{-\mathbf{A}t}\mathbf{x}\Big]_0^t = \int_0^t e^{-\mathbf{A}t}\mathbf{b}\,dt$$

or
$$\mathbf{e}^{-\mathbf{A}t}\mathbf{x} - \mathbf{x}^i = -\mathbf{A}^{-1}(\mathbf{e}^{-\mathbf{A}t} - \mathbf{I})\mathbf{b}$$

Then
$$\begin{aligned}\mathbf{x} &= e^{\mathbf{A}t}\mathbf{x}^i - e^{\mathbf{A}t}\mathbf{A}^{-1}(\mathbf{e}^{\mathbf{A}t} - \mathbf{I})\mathbf{b} \\ &= e^{\mathbf{A}t}(\mathbf{A}^{-1}\mathbf{b} + \mathbf{x}^i) - \mathbf{A}^{-1}\mathbf{b}\end{aligned}$$

The reader should justify the many steps in this reduction, particularly that $\mathbf{A}^{-1}$ commutes with $\exp(\mathbf{A}t)$. The solution obtained here seems to be dissimilar to that obtained in Eq. (5.7.7), and the reader should convince himself that they are really the same. Note in the foregoing if $\mathbf{b}$ is a vector with elements a function of the time, the solution is

$$\begin{aligned}\mathbf{x} &= e^{\mathbf{A}t}\mathbf{x}_i + e^{\mathbf{A}t}\int_0^t \mathbf{e}^{-\mathbf{A}\tau}\mathbf{b}(\tau)\,d\tau \\ &= e^{\mathbf{A}t}\mathbf{x}_i + \int_0^t \mathbf{e}^{-\mathbf{A}(\tau-t)}\mathbf{b}(\tau)\,d\tau\end{aligned}$$

In Section 6.6 the integration of the equations describing the transient behavior of the stirred tank reactors was not performed. These were

$$\frac{d\mathbf{A}_n}{d\tau} - \mathbf{M}\mathbf{A}_n = \mathbf{A}_{n-1}$$

$$\mathbf{A}_n = \mathbf{A}_n^i, \quad \tau = 0$$

and they may be integrated by a successive substitution scheme to give

$$\mathbf{A}_1 = e^{\mathbf{M}\tau}\mathbf{A}_1^i + e^{\mathbf{M}\tau}\int_0^\tau e^{-\mathbf{M}s}\mathbf{A}_0\,ds$$

If $\mathbf{A}_0$ is a constant this reduces to

$$\mathbf{A}_1 = e^{\mathbf{M}\tau}\mathbf{A}_1^i + \mathbf{M}^{-1}e^{\mathbf{M}\tau}\mathbf{A}_0 - \mathbf{M}^{-1}\mathbf{A}_0$$

Also

$$\begin{aligned}\mathbf{A}_2 &= e^{\mathbf{M}\tau}(\mathbf{A}_2^i + \tau\mathbf{A}_1^i) + e^{\mathbf{M}\tau}\int_0^\tau\int_0^s e^{-\mathbf{M}s_1}\mathbf{A}_0\,ds_1\,ds \\ &= e^{\mathbf{M}\tau}(\mathbf{A}_2^i + \tau\mathbf{A}_1^i) + e^{\mathbf{M}\tau}\int_0^\tau (\tau - s)e^{-\mathbf{M}s}\mathbf{A}_0\,ds\end{aligned}$$

and if $\mathbf{A}_0$ is a constant vector

$$\mathbf{A}_2 = e^{\mathbf{M}\tau}(\mathbf{A}_2^i + \tau\mathbf{A}_1^i) + e^{\mathbf{M}\tau}\mathbf{M}^{-2}[\mathbf{M}\tau - \mathbf{I}]\mathbf{A}_0 + \mathbf{M}^{-2}\mathbf{A}_0$$

Further substituting, again one obtains

$$\mathbf{A}_3 = e^{\mathbf{M}\tau}\left(\mathbf{A}_3^i + \tau\mathbf{A}_2^i + \frac{\tau^2}{2}\mathbf{A}_1^i\right) + e^{\mathbf{M}\tau}\int_0^\tau \frac{(\tau - s)^2}{2!}e^{-\mathbf{M}s}\mathbf{A}_0\,ds$$

and if $\mathbf{A}_0$ is a constant vector

$$\mathbf{A}_3 = e^{\mathbf{M}\tau}\left(\mathbf{A}_3^i + \tau\mathbf{A}_2^i + \frac{\tau^2}{2!}\mathbf{A}_1^i\right) + e^{\mathbf{M}\tau}\mathbf{M}^{-3}\left[\frac{(\mathbf{M}\tau)^2}{2!} - \mathbf{M}\tau + \mathbf{I}\right]\mathbf{A}_0 - \mathbf{M}^{-3}\mathbf{A}_0$$

For $n = 4$, one can show

$$\mathbf{A}_4 = e^{\mathbf{M}\tau}\left(\mathbf{A}_4^i + \tau\mathbf{A}_3^i + \frac{\tau^2}{2}\mathbf{A}_2^i + \frac{\tau^3}{3!}\mathbf{A}_1^i\right) + e^{\mathbf{M}\tau}\int_0^\tau \frac{(\tau - s)^3}{3!} e^{-\mathbf{M}s}\mathbf{A}_0\, ds$$

and if $\mathbf{A}_0$ is a constant vector

$$\mathbf{A}_4 = e^{\mathbf{M}\tau}\left(\mathbf{A}_4^i + \tau\mathbf{A}_3^i + \frac{\tau^2}{2}\mathbf{A}_2^i + \frac{\tau^3}{3!}\mathbf{A}_1^i\right) + e^{\mathbf{M}\tau}\mathbf{M}^{-4}\left[\frac{(\mathbf{M}\tau)^3}{3!} - \frac{(\mathbf{M}\tau)^2}{2!} + \mathbf{M}\tau - I\right] + \mathbf{M}^{-4}\mathbf{A}_0$$

By induction one can show in general

$$\mathbf{A}_n = e^{\mathbf{M}\tau}\sum_{j=1}^{n}\mathbf{A}_{n-j+1}\frac{\tau^{j-1}}{(j-1)!} + e^{\mathbf{M}\tau}\int_0^\tau \frac{(\tau - s)^{n-1}}{(n-1)!} e^{-\mathbf{M}s}\mathbf{A}_0\, ds$$

and for a constant vector $\mathbf{A}_0$

$$\mathbf{A}_n = e^{\mathbf{M}\tau}\sum_{j=1}^{n}\mathbf{A}_{n-j+1}\frac{\tau^{j-1}}{(j-1)!} - e^{\mathbf{M}\tau}\mathbf{M}^{-n}\sum_{j=1}^{n}\frac{(-\mathbf{M}\tau)^{j-1}}{(j-1)!}\mathbf{A}_0 + (-\mathbf{M})^{-n}\mathbf{A}_0$$

The transient part of the solution is the first and second summation, and the steady state portion is

$$\mathbf{A}_{nss} = (-\mathbf{M})^{-n}\mathbf{A}_0$$

For small n, the computations can be made directly with this formula; for large n, the quantity $\mathbf{M}^{-n}$ may be computed using Sylvester's formula.

7.11 The Matrizant

Up to this time, although we have expended considerable space on differential equations of the first order we have considered only the case of constant coefficients. We now consider the case of variable coefficients, and the solution will be represented in terms of the *matrizant*, a function which is a generalization of the exponential function. We therefore write the system of equations in the form

$$\frac{d\mathbf{y}}{d\theta} = \mathbf{A}\mathbf{y} \tag{7.11.1}$$

$$\mathbf{y} = \mathbf{y}_i, \quad \theta = 0$$

where the elements of $\mathbf{A}$ are functions of θ, $\mathbf{A} = \mathbf{A}(\theta)$.

Our method will be a generalization of the method of successive substitutions used for a single scalar equation, so we write

$$\mathbf{y} = \mathbf{y}_i + \int_0^\theta \mathbf{A}(\theta_1)\mathbf{y}(\theta_1)\,d\theta_1 \tag{7.11.2}$$

Substituting a known value for $\mathbf{y}$ under the integral, say $\mathbf{y}_i$, a new $\mathbf{y}$ will be generated, call it $\mathbf{y}^{(1)}$ which, hopefully, is an improvement on the initial guess y_i, thus

$$\mathbf{y}^{(1)} = \mathbf{y}_i + \int_0^\theta \mathbf{A}(\theta_1)\,\mathbf{y}_i\,d\theta_1$$

Repeating this process

$$\begin{aligned}\mathbf{y}^{(2)} &= \mathbf{y}_i + \int_0^\theta \mathbf{A}(\theta_1)\left[\mathbf{y}_i + \int_0^{\theta_1} \mathbf{A}(\theta_2)\mathbf{y}_i\,d\theta_2\right]d\theta_1\\ &= \mathbf{y}_i + \int_0^\theta \mathbf{A}(\theta_1)\mathbf{y}_i\,d\theta_1 + \int_0^\theta \mathbf{A}(\theta_1)\,d\theta_1 \int_0^{\theta_1} \mathbf{A}(\theta_2)\mathbf{y}_i\,d\theta_2\end{aligned}$$

and again a second substitution into Eq. (7.9.2) gives

$$\begin{aligned}\mathbf{y}^{(3)} = \mathbf{y}_i &+ \int_0^\theta \mathbf{A}(\theta_1)\mathbf{y}_i\,d\theta_1 + \int_0^\theta \mathbf{A}(\theta_1)\,d\theta_1 \int_0^{\theta_1} \mathbf{A}(\theta_2)\mathbf{y}_i\,d\theta_2\\ &+ \int_0^\theta \mathbf{A}(\theta_1)\,d\theta_1 \int_0^{\theta_1} \mathbf{A}(\theta_2)\,d\theta_2 \int_0^{\theta_2} \mathbf{A}(\theta_3)\mathbf{y}_i\,d\theta_3\end{aligned}$$

Repeated substitution gives

$$\mathbf{y}^{(m)} = \Omega^{(m)}(\theta)\,\mathbf{y}_i$$

where

$$\begin{aligned}\Omega^{(m)}(\theta) = \mathbf{I} &+ \int_0^\theta \mathbf{A}(\theta_1)\,d\theta_1 + \int_0^\theta \mathbf{A}(\theta_1)\,d\theta_1 \int_0^{\theta_1} \mathbf{A}(\theta_2)\,d\theta_2\\ &+ \int_0^\theta \mathbf{A}(\theta_1)\,d\theta_1 \int_0^{\theta_1} \mathbf{A}(\theta_2)\,d\theta_2 \int_0^{\theta_2} \mathbf{A}(\theta_3)\,d\theta_3 + \cdots +\\ &+ \underbrace{\int_0^\theta \mathbf{A}(\theta_1)\,d\theta_1 \cdots \int_0^{\theta_m} \mathbf{A}(\theta_m)\,d\theta_m}_{\text{m-fold iterated integral}}\end{aligned}$$

We then shall try to prove that given an $\epsilon > 0$ there exists an M such that

$$|\mathbf{y} - \mathbf{y}^{(m)}| < \epsilon \quad \text{for } m \geqslant M$$

or we desire that

$$\lim_{m\to\infty} \Omega^{(m)}(\theta) = \Omega(\theta)$$

is such that

$$\mathbf{y} = \Omega(\theta)\,\mathbf{y}_i$$

is a solution to Eq. (7.11.1). We now show that the series defining $\Omega(\theta)$ is absolutely and uniformly convergent. Now suppose that in an interval $0 \leq \theta \leq \tau$, the matrix $\mathbf{A}(\theta)$ is bounded so that

$$|\mathbf{A}(\theta)| \leqslant U\mathbf{J}$$

where $|a_{ij}(\theta)| < U$ and $\mathbf{J}$ is an nth-order matrix, each of whose elements is unity. Then

$$\left|\int_0^\theta \mathbf{A}(\theta_1)\, d\theta_1\right| \leqslant \left|\int_0^\theta |\mathbf{A}(\theta_1)|\, d\theta_1\right| \leqslant \theta U \mathbf{J}$$

Similarly

$$\left|\int_0^\theta \mathbf{A}(\theta_1)\, d\theta_1 \int_0^{\theta_1} \mathbf{A}(\theta_2)\, d\theta_2\right| \leqslant \left|\int_0^\theta \int_0^{\theta_1} |\mathbf{A}(\theta_1)|\, |\mathbf{A}(\theta_2)|\, d\theta_1\, d\theta_2\right|$$

$$\leqslant \left|\int_0^\theta U\mathbf{J}\theta_1\, U\mathbf{J}\, d\theta_1\right|$$

$$\leqslant \frac{nU^2\theta^2\mathbf{J}}{2}$$

Also

$$\left|\int_0^\theta \mathbf{A}(\theta_1)\, d\theta_1 \int_0^{\theta_1} \mathbf{A}(\theta_2)\, d\theta_2 \int_0^{\theta_2} \mathbf{A}(\theta_3)\, d\theta_3\right| \leqslant \frac{n^2U^3\theta^3\mathbf{J}}{3!}$$

and in general

$$\underbrace{\int_0^\theta A(\theta_1)\, d\theta_1 \cdots \int_0^{\theta_{m-1}} A(\theta_m)\, d\theta_m \leqslant \frac{n^{m-1}U^m\theta^m\mathbf{J}}{m!}}_{\text{m-fold iterated integral}}$$

where

$$\mathbf{J}\mathbf{J} = n\mathbf{J}$$

$$\mathbf{J}^m = n^{m-1}\mathbf{J}$$

Hence

$$|\Omega_m(\theta)| \leqslant \mathbf{I} + U\theta\mathbf{J} + \frac{nU^2\theta^2}{2!}\mathbf{J} + \frac{n^2U^3\theta^3}{3!}\mathbf{J} + \cdots + \frac{n^{m-1}U^m\theta^m}{m!}\mathbf{J} + \cdots$$

$$\leqslant \frac{n - 1 + e^{nU\theta}}{n}\mathbf{J}, \quad \text{for all } m$$

and since each of the series in $\mathbf{\Omega}$ is dominated by the series on the right, it follows that the matrix series Ω converges absolutely and uniformly in any closed subinterval of the original interval since this is true for the dominating series. Now consider the formal term-by-term derivative

$$\frac{d\Omega}{d\theta} = \mathbf{A}(\theta)\mathbf{\Omega}(\theta)$$

where it can be shown by the same analysis as the foregoing that the series $\mathbf{A}(\theta)\ \mathbf{\Omega}(\theta)$ converges absolutely and uniformly. Therefore, the series

$$\mathbf{y} = \mathbf{\Omega}(\theta)\, \mathbf{y}_i$$

satisfies the differential equation and

$$\lim_{\theta \to 0} \mathbf{\Omega}(\theta) = \mathbf{I}$$

where the limit may be taken term by term because of the uniform convergence, so

$$\lim_{\theta \to 0} \mathbf{y} = \mathbf{y}_i$$

The quantity $\boldsymbol{\Omega}(\theta)$ is called the *matrizant* and it should be clear that it reduces to

$$e^{\mathbf{A}\theta}$$

if $\mathbf{A}$ is a constant matrix.

In an application to be considered later, some properties of the matrizant will be required. We shall write $\boldsymbol{\Omega}$ in the form

$$\mathbf{y}(\theta) = \boldsymbol{\Omega}_0^\theta(\mathbf{A})\,\mathbf{y}(0)$$

to denote the interval of integration more explicitly and to denote the dependency on $\mathbf{A}(\theta)$. If the interval of integration is (θ_0, θ) then

$$\mathbf{y}(\theta) = \boldsymbol{\Omega}_{\theta_0}^\theta(\mathbf{A})\mathbf{y}(\theta_0)$$

If θ' is in the interval $\theta_0 < \theta' < \theta$, then

$$\mathbf{y}(\theta) = \boldsymbol{\Omega}_{\theta'}^\theta(\mathbf{A})\mathbf{y}(\theta')$$

$$\mathbf{y}(\theta') = \boldsymbol{\Omega}_{\theta_0}^{\theta'}(\mathbf{A})\mathbf{y}(\theta_0)$$

or

$$\mathbf{y}(\theta) = \boldsymbol{\Omega}_{\theta'}^\theta(\mathbf{A})\boldsymbol{\Omega}_{\theta_0}^{\theta'}(\mathbf{A})\mathbf{y}(\theta_0)$$

and therefore

$$\boldsymbol{\Omega}_{\theta_0}^\theta(\mathbf{A}) = \boldsymbol{\Omega}_{\theta'}^\theta(\mathbf{A})\boldsymbol{\Omega}_{\theta_0}^{\theta'}(\mathbf{A})$$

A second relationship of great utility may be derived in the following way: Let

$$\mathbf{X} = \boldsymbol{\Omega}_{\theta_0}^\theta(\mathbf{A}), \quad \mathbf{Y} = \boldsymbol{\Omega}_{\theta_0}^\theta(\mathbf{A} + \mathbf{B})$$

and from

$$\mathbf{Y} = \mathbf{XC}$$

then

$$(\mathbf{A} + \mathbf{B})\mathbf{XC} = \mathbf{AXC} + \mathbf{X}\frac{d\mathbf{C}}{d\theta}$$

$$\frac{d\mathbf{C}}{d\theta} = \mathbf{X}^{-1}\mathbf{BXC}$$

and thus

$$\mathbf{C} = \Omega_{\theta_0}^\theta(\mathbf{X}^{-1}\mathbf{BX})$$

or

$$\boldsymbol{\Omega}_{\theta_0}^\theta(\mathbf{A} + \mathbf{B}) = \Omega_{\theta_0}^\theta(\mathbf{A})\Omega_{\theta_0}^\theta[\boldsymbol{\Omega}_{\theta_0}^{\theta-1}(\mathbf{A})\mathbf{B}\boldsymbol{\Omega}_{\theta_0}^\theta(\mathbf{A})] \tag{7.11.3}$$

One last thing remains to be accomplished. Consider now the system of equations

$$\mathbf{A}(\theta)\mathbf{y} + \mathbf{b}(\theta) = \frac{d\mathbf{y}}{d\theta}$$

$$\mathbf{y} = \mathbf{y}_i, \quad \theta = 0$$

then the solution is in the form

$$\mathbf{y} = \mathbf{\Omega}_0^\theta(\mathbf{A})\mathbf{z}$$

where $\mathbf{z}$ is to be determined. Differentiating

$$\frac{d\mathbf{y}}{d\theta} = \mathbf{A}\mathbf{\Omega}_0^\theta(\mathbf{A})\mathbf{z} + \mathbf{\Omega}_0^\theta(\mathbf{A})\frac{d\mathbf{z}}{d\theta} = \mathbf{A}\mathbf{\Omega}_0^\theta(\mathbf{A})\mathbf{z} + \mathbf{b}(\theta)$$

or

$$\frac{d\mathbf{z}}{d\theta} = \mathbf{\Omega}_0^{\theta^{-1}}(\mathbf{A})\mathbf{b}(\theta)$$

and

$$\mathbf{z} = \int_0^\theta \mathbf{\Omega}_0^{\tau^{-1}}(\mathbf{A})\mathbf{b}(\tau)\,d\tau + \mathbf{c}$$

where $\mathbf{c}$ is a constant. Then,

$$\mathbf{y} = \mathbf{\Omega}_0^\theta(\mathbf{A})\int_0^\theta \mathbf{\Omega}_0^{\tau^{-1}}(\mathbf{A})\mathbf{b}(t)\,d\tau + \mathbf{\Omega}_0^\theta(\mathbf{A})\mathbf{c}$$

or

$$\mathbf{y} = \mathbf{\Omega}_0^\theta(\mathbf{A})\int_0^\theta \mathbf{\Omega}_0^{\tau^{-1}}(\mathbf{A})\mathbf{b}(\tau)\,d\tau + \mathbf{\Omega}_0^\theta(\mathbf{A})\mathbf{y}_i \tag{7.11.4}$$

This gives the required solution in a form recognizable as a generalization of the case for constant $\mathbf{A}$ and $\exp(\mathbf{A}\theta)$.

The matrizant is also related to the matrix of a fundamental set of solutions. If one considers the set of equations

$$\frac{d\mathbf{y}_j}{d\theta} = \mathbf{A}\mathbf{y}_j$$

$$\mathbf{y}_j = \mathbf{e}_j, \quad \theta = 0$$

where $\mathbf{e}_j$ is a unit vector with one in the jth position and zeroes elsewhere, then n solutions will be generated, one for each value of j. The matrix of these solutions may be written

$$\mathbf{Y} = [\mathbf{y}_1\mathbf{y}_2 \cdots \mathbf{y}_n]$$

where the row vector $\mathbf{Y}$ has for each element a column vector. It is apparent that $\mathbf{Y}$ as a matrix satisfies the system

$$\frac{d\mathbf{Y}}{d\theta} = \mathbf{A}\mathbf{Y}$$

$$\mathbf{Y} = \mathbf{I}, \quad \theta = 0$$

and that $\mathbf{y}$, the solution of Eq. (7.11.1), is

$$\mathbf{y} = \mathbf{Y}\mathbf{y}_i$$

as may be verified by direct substitution. Hence $\mathbf{Y}$ must be the matrizant itself

$$\mathbf{Y} = \mathbf{\Omega}(\mathbf{A})$$

or the matrizant is the matrix of fundamental solutions.

7.12 Reduction of a System of Equations to Normal Form

In this section we consider a system of homogeneous differential equations with the constant coefficients

$$\frac{d\mathbf{x}}{d\theta} = \mathbf{A}\mathbf{x} \tag{7.12.1}$$

The normal form of a set of differential equations is the same set of equations but uncoupled in the sense that the system is changed to n independent equations each one of which may be solved without reference to the others. Suppose we make a change of variables

$$\mathbf{x} = \mathbf{B}\mathbf{y}$$

then

$$\mathbf{B}\frac{d\mathbf{y}}{dt} = \mathbf{A}\mathbf{B}\mathbf{y}$$

If **B** is nonsingular, then

$$\mathbf{B}^{-1}\mathbf{B}\frac{d\mathbf{y}}{dt} = \mathbf{B}^{-1}\mathbf{A}\mathbf{B}\mathbf{y}$$

or

$$\frac{d\mathbf{y}}{dt} = \mathbf{B}^{-1}\mathbf{A}\mathbf{B}\mathbf{y}$$

From Section 7.2, it is known that the matrix **A** is reduced to diagonal form in which the diagonal elements are the eigenvalues of **A** if **B** is chosen as **X**, a matrix with columns as the eigenvectors of **A**, then

$$\mathbf{X}^{-1}\mathbf{A}\mathbf{X} = \mathbf{\Lambda}$$

so that

$$\frac{d\mathbf{y}}{dt} = \mathbf{\Lambda}\mathbf{y} \tag{7.12.2}$$

or

$$\frac{dy_1}{dt} = \lambda_1 y_1$$

$$\frac{dy_2}{dt} = \lambda_2 y_2$$

$$\vdots$$

$$\frac{dy_n}{dt} = \lambda_n y_n$$

and the equations are uncoupled by the change of variables

$$\mathbf{x} = \mathbf{X}\mathbf{y} \quad \text{or} \quad \mathbf{y} = \mathbf{X}^{-1}\mathbf{x}$$

In many problems, one is more interested in the character of the solution than in the solution itself. This character is made evident in these equations, since the solutions are of the form

$$y_j = y_j^{(i)} e^{\lambda_j t}$$

and whether the solution in the $\{y_j\}$ tends to grow with time or to approach zero depends upon whether λ_j is real and positive or real and negative. It may happen that some of the eigenvalues of **A** are complex, and if, indeed, this is the case, such complex λ_j must occur in pairs of complex conjugates. If the matrix **A** is symmetric, the eigenvalues are real and the uncoupling of the equations is accomplished in fact. If there are complex conjugate eigenvalues, then the transformation $\mathbf{y} = \mathbf{X}^{-1}\mathbf{x}$ is not real, in the sense that, although we are speaking of physical quantities $\mathbf{x}$ which are real numbers, the eigenvectors belonging to the complex eigenvalues have elements which are complex and the transformation loses its significance somewhat, at least for the coordinates in the new system corresponding to these complex eigenvalues. Suppose that two of the eigenvalues which are complex conjugates are λ_+ and λ_-, then the corresponding eigenvectors are also complex conjugates $\mathbf{x}_+$ and $\mathbf{x}_-$, respectively. Suppose the matrix **X** is

$$\mathbf{X} = \begin{bmatrix} x_{11} & x_{12} & x_{13} & \cdots & x_{1n} \\ x_{21} & x_{22} & x_{23} & \cdots & x_{2n} \\ x_{31} & x_{32} & x_{33} & \cdots & x_{3n} \\ \cdot & & & & \\ \cdot & & & & \\ \cdot & & & & \\ x_{n1} & x_{n2} & x_{n3} & \cdots & x_{nn} \end{bmatrix}$$

Let $\mathbf{X}^{-1}$ be

$$\mathbf{X}^{-1} = \begin{bmatrix} X_{11} & X_{21} & X_{31} & \cdots & X_{n1} \\ X_{12} & X_{22} & X_{32} & \cdots & X_{n2} \\ X_{13} & X_{23} & X_{33} & \cdots & X_{n3} \\ \cdot & & & & \\ \cdot & & & & \\ \cdot & & & & \\ X_{1n} & X_{2n} & X_{3n} & \cdots & X_{nn} \end{bmatrix}$$

and let the first two columns of **X** be the complex conjugate eigenvectors, that is, so that, if $x_{i1} = \alpha_{i1} + i\beta_{i1}$, $x_{i2} = \alpha_{i1} - i\beta_{i1}$. Now it may be shown that a determinant with two columns as complex conjugates is equal to i times a real determinant. Hence it follows that each element in the inverse $\mathbf{X}^{-1}$ not in the first two rows is real, since it is the ratio of two such determinants. The elements in the same columns in the first two rows of $\mathbf{X}^{-1}$ are complex conjugates, since they are the ratios of a determinant with one complex column and $|\mathbf{X}|$. Thus the first two rows of $\mathbf{X}^{-1}$ are complex conjugates and all the other elements are real, so that

$$\mathbf{y} = \mathbf{X}^{-1}\mathbf{x}$$

since $\mathbf{x}$ is real, y_1 and y_2 are complex conjugates and $y_3, y_4, \ldots, y_n$ are real.

Hence the system of Eq. (7.12.1) has been uncoupled in the real sense to give the system of Eq. (7.12.2), if the eigenvalues of the matrix **A** are all real. If there are complex eigenvalues, say λ_1 and λ_2, then the eigenvectors $\mathbf{x}_1$ and $\mathbf{x}_2$ are complex conjugates and the two equations corresponding in Eq. (7.12.2) are not real, so that in this case, in fact, uncoupling has not occurred to give a real system. As just shown, however, y_1 and y_2 are complex conjugates, and these have been uncoupled from the rest to give, if $y_1 = y_+$ and $y_2 = y_-$,

$$y_+ = y_+^{(i)} e^{\lambda_+ t}$$
$$y_- = y_-^{(i)} e^{\lambda_- t}$$

where $\lambda_+ = \alpha + i\beta$, $\lambda_- = \alpha - i\beta$, and y_+ and y_- are the corresponding coordinates. Since if r and θ are polar coordinates

$$|y_+ y_-| = r^2 = r^{(i)^2} e^{2\mathrm{Re}\lambda_+ t}$$

or

$$r = r^{(i)} e^{\mathrm{Re}\lambda_+ t} = r^{(i)} e^{\alpha t} \tag{7.12.3}$$

Also

$$\begin{aligned} \theta = \arg y_+ &= \tan^{-1} \frac{y_+ - y_-}{i(y_+ + y_-)} \\ &= \tan^{-1}\left[\frac{y_+^{(i)} e^{\lambda_+ t} - y_-^{(i)} e^{\lambda_- t}}{i(y_+^{i} e^{\lambda_+ t} + y_-^{(i)} e^{\lambda_- t})}\right] \\ &= \beta t + \theta^{(i)} \end{aligned} \tag{7.12.4}$$

where $\theta^{(i)} = \arg y_+^{(i)}$. Thus y_1 and y_2 cannot be uncoupled from each other in the rectangular cartesian sense but they can be combined to give the polar coordinates (r, θ) in terms of the eigenvalues, initial vectors, and the time in Eq. (7.12.3) and Eq. (7.12.4). Thus the complete solution of the system in the new set of coordinates is given by

$$y_j = y_j^{(i)} e^{\lambda_j t}, \quad \lambda_j \text{ real}$$

$$\left.\begin{aligned} r &= r^{(i)} e^{\mathrm{Re}\lambda t} \\ \theta &= \theta^{(i)} + Im\lambda t \end{aligned}\right\} \quad \text{for each pair of complex eigenvalues}$$

Hence the behavior of a system, even though the solutions are not obtained explicitly, may be determined qualitatively. If all the eigenvalues are real and negative, the solution tends to approach a steady state asymptotically. If one of the eigenvalues is real and positive, the solution will tend to become arbitrarily large. If there are complex eigenvalues, the character of the system is determined by the sign of the real part of the eigenvalues. If Reλ, is negative the solution will spiral into a steady state and will have the character of a damped oscillation. Reλ positive indicates that the solution must spiral away from a state and oscillate with ever larger amplitude. If Reλ is zero, then there must be a sustained harmonic oscillation.

7.13 A Routh-Hurwitz Criterion for Eigenvalues

It is clear from the preceding discussion that a knowledge of the character of the eigenvalues may determine much of the behavior of a system in a general way. If the eigenvalues have negative real parts or are negative, the system described by the equations will presumably approach some steady state, whereas in all other cases, the system will oscillate or become arbitrarily large with time. It is therefore apparent that some test or criterion which would give information on the eigenvalues would be extremely useful. Such a test was developed by Routh and Hurwitz and reads as follows. Consider the polynomial

$$p_n(z) = a_0 z^n + a_1 z^{n-1} + a_2 z^{n-2} + \cdots + a_{n-1} z + a_n, \quad a_0 > 0$$

and the following determinants

$$\Delta_1 = a_1$$

$$\Delta_2 = \begin{vmatrix} a_1 & a_3 \\ a_0 & a_2 \end{vmatrix}$$

$$\Delta_3 = \begin{vmatrix} a_1 & a_3 & a_5 \\ a_0 & a_2 & a_4 \\ 0 & a_1 & a_3 \end{vmatrix}$$

$$\Delta_4 = \begin{vmatrix} a_1 & a_3 & a_5 & a_7 \\ a_0 & a_2 & a_4 & a_6 \\ 0 & a_1 & a_3 & a_5 \\ 0 & a_0 & a_2 & a_4 \end{vmatrix}$$

$$\vdots$$

$$\Delta_n = \begin{vmatrix} a_1 & a_3 & a_5 & a_7 & \cdots & 0 \\ a_0 & a_2 & a_4 & a_6 & \cdots & 0 \\ 0 & a_1 & a_3 & a_5 & \cdots & \\ 0 & a_0 & a_2 & a_4 & & \\ \vdots & \vdots & \vdots & \vdots & \ddots & \\ 0 & & & & \cdots & a_n \end{vmatrix}$$

where it is to be noted that the determinant for Δ_n is of nth-order and contains $a_1 a_2 a_3 \ldots a_n$ on the main diagonal.

The necessary and sufficient condition that the polynomial $p_n(z)$ have zeroes with negative real parts, is that all determinants, $\Delta_1, \Delta_2, \Delta_3, \ldots, \Delta_n$ be greater than zero.

For example, consider the second-degree polynomial

$$P_2(z) = a_0 z^2 + a_1 z + a_2, \quad a_0 > 0$$

and if

$$\Delta_1 = a_1 > 0 \quad \text{and} \quad \Delta_2 = \begin{vmatrix} a_1 & 0 \\ a_0 & a_2 \end{vmatrix} > 0$$

then

$$a_1 > 0, \quad a_2 > 0$$

and for a quadratic, it is necessary and sufficient that all the coefficients have the same sign in order that the roots have negative real parts, as is well known. For the cubic polynomial,

$$P_3(z) = a_0 z^3 + a_1 z^2 + a_2 z + a_3, \quad a_0 > 0$$

and if

$$\Delta_1 = a_1 > 0, \quad \Delta_2 = \begin{vmatrix} a_1 & a_3 \\ a_0 & a_2 \end{vmatrix} > 0, \quad \Delta_3 = \begin{vmatrix} a_1 & a_3 & 0 \\ a_0 & a_2 & 0 \\ 0 & a_1 & a_3 \end{vmatrix} > 0$$

then the conditions for zeroes with negative real parts are

$$a_1 > 0, \quad a_1 a_2 - a_0 a_3 > 0, \quad a_3 > 0$$

the middle condition is

$$a_1 a_2 > a_0 a_3$$

and hence it is *not* enough that all the coefficients be positive in order that the roots have negative real parts.

It is shown in more advanced books that there is a redundancy in the Routh-Hurwitz conditions and that fewer conditions may be used as a test for the character of the roots. In automatic control theory, the character of the roots determines whether a system is stable, and many methods have been proposed for this purpose.

7.14 The Exploration of Response Surfaces

A problem of interest to chemical engineers is that of determining the optimum value, either maximum or minimum, of some variable which is related to other variables. The engineer usually prefers to arrive at mathematical models by some rational approach to the fundamentals of the problem, but in some cases this is impossible usually because the rational model is too complicated or because the parameters are too difficult to obtain. Not being able to do otherwise, he may then proceed on an empirical basis to determine the response of a given system in terms of what he believes to be the pertinent variables in the problem. The term *response* as used here is the statistician's term rather than that of the control engineer and refers only to the dependent variable as the engineer sees it. It is used here because the techniques to be described have been successfully exploited by G. E. P. Box and others, and Box uses the term which is the title of this section.

In general in determining the optimum value of some variable a preliminary exploration of that variable in terms of the other variables is performed. If the equation of the response surface were available, say,

$$y = \phi(x_1, x_2, \ldots, x_n)$$

then at any point $x_1', x_2', \ldots, x_n'$ the direction in which the function ϕ is increasing the most rapidly is that of the vector

$$\text{grad}\ \phi = \left[\frac{\partial\phi}{\partial x_1}\frac{\partial\phi}{\partial x_2}\cdots\frac{\partial\phi}{\partial x_n}\right]$$

This gives the direction of greatest increase, and the rate of greatest increase is

$$\left|\text{grad}\ \phi\right| = \sqrt{\sum_i \left(\frac{\partial\phi}{\partial x_i}\right)^2}$$

If the response surface is not known, as it certainly is not in most cases, an estimate of the partial derivatives may be obtained by performing experiments to obtain $\phi(x_1, x_2, \ldots, x_n)$ and

$$\phi(x_1, x_2, \ldots, x_{j-1}, x_j + \Delta x_j, x_{j+1}, \ldots, x_n$$

for each j. From these one obtains an estimate of the partial derivative experimentally as the difference quotient

$$\frac{\partial\phi}{\partial x_j} \cong \frac{\phi(x_1, x_2, \ldots, x_{j-1}, x_j + \Delta x_j, x_{j+1}, \ldots, x_n) - \phi(x_1, x_2, \ldots, x_n)}{\Delta x_j}$$

One can then write the equations of the straight line along which changes in the system should be made in order to increase ϕ. These are

$$\frac{x_1' - x_1}{\phi_{x_1'}} = \frac{x_2' - x_2}{\phi_{x_2'}} = \cdots \frac{x_n' - x_n}{\phi_{x_n'}}$$

where

$$\phi_{x_j'} = \left(\frac{\partial\phi}{\partial x_j}\right)_{x_j = x_j'}$$

In order to increase the value of ϕ, the physical or chemical process is shifted so that the independent controllable variables follow this straight line until a point is reached which is prohibited by restrictions or until ϕ no longer increases. The preceding scheme may then be repeated and the process changed as indicated by the new direction. This general procedure in most industrial cases leads to some difficulties since the measurements are subject to error and the estimates of the partial derivatives may therefore be very unreliable. An alternative procedure is to determine a plane

$$y = A_0 + A_1x_1 + A_2x_2 + \cdots + A_nx_n$$

and to determine the constants $\{A_j\}$ by least squares from a suitable experimental design. These constants are

$$\phi_{x'_j} = A_j$$

This procedure may be continued, but in industrial problems an impasse is reached, for it becomes obvious that although the local approximation by a plane may be valid at some distance from an optimum point it is quite likely not valid near the optimum, and finally that the optimum value itself frequently is pathological. This anomalous behavior manifests itself in ridge systems as described by Box. For example, if the hump which is being investigated has contour curves which are elliptical with one axis considerably longer than the other, then there will be essentially a line of values or a region of almost equal optima and this may offer a choice in operating parameters which may be significant from a practical point of view. Box has discussed in some detail the kinds of behavior which may obtain. It becomes clear, however, that some more detailed exploration of the response surface than the linear exploration is in order, and the obvious choice is the surface of next higher order, the second-order one. A second-order surface of the form

$$\begin{aligned} y = b_0 + b_1x_1 + b_2x_2 + \cdots + b_nx_n \\ + \beta_{11}x_1^2 + \beta_{12}x_1x_2 + \cdots + \beta_{1n}x_1x_n \\ + \beta_{22}x_2^2 + \cdots + \beta_{2n}x_2x_n \\ + \beta_{33}x_3^2 + \cdots + \beta_{3n}x_3x_n \\ \vdots \\ + \beta_{nn}x_n^2 \end{aligned} \tag{7.14.1}$$

is then fitted to the system by the data obtained from a suitable experimental design where the data collected are assumed to close on the optimum point. The estimation of the parameters in this equation is obtained by least squares and it is important to estimate their significance in a statistical sense as sketched in Section 2.13. In general, the geometrical interpretation of a quadratic expression in n variables is impossible to visualize. For two variables, however, it is known that there are three general cases in which the contours for constant y are elliptic, hyperbolic, or parabolic. In the elliptic case there will be a true maximum or minimum; the hyperbolic case is a saddle point in which there is a minimax; and in the parabolic case there is a rising or falling ridge for which there is no optimum. From a general expression with many variables as in Eq. (7.14.1), it is apparent that the geometry of the n-dimensional contours may be very complicated and it is not obvious that there will be optimum values of the dependent variables. The quantitative aspects of the contours can, however, be elucidated if the function y is transformed to its canonical form. In order to illustrate the procedure an example quoted by Box* will be presented. The example is one in which tests have been run on a chemical reaction at different temperatures, concentration of reactant, and for different reaction times. A composite experimental design was constructed involving some fifteen different experiments, and the variables—temperature, concentration, and time—were coded linearly into the three variables x_1, x_2, x_3 with the yield y to give an equation for the yield

$$y = 67.711 + 1.944\,x_1 + 0.906\,x_2 + 1.069\,x_3 - 1.539\,x_1^2 - 0.264\,x_2^2 - 0.676\,x_3^2 - 3.088\,x_1x_2 - 2.188\,x_1x_3 - 1.212\,x_2x_3 \tag{7.14.2}$$

We desire to determine the character of the response surface in the neighborhood of the maximum, if there is one. The computation of the partial derivatives in order to establish the necessary condition for an extremum gives

$$\begin{aligned} 3.078x_1 + 3.088x_2 + 2.188x_3 &= 1.944 \\ 3.088x_1 + 0.528x_2 + 1.212x_3 &= 0.906 \\ 2.188x_1 + 1.212x_2 + 1.352x_3 &= 1.069 \end{aligned}$$

which have a solution

$$x_1 = 0.060, \quad x_2 = 0.215, \quad x_3 = 0.501$$

In order to determine the character of this point, it is necessary, according to Sections 7.6 and 7.7, to examine the matrix of second partial derivatives. The partial derivatives of $(-y)$ may be easily calculated for this case, and they are

$$\begin{array}{lll} y_{11} = 3.078 & y_{12} = 3.088 & y_{13} = 2.188 \\ y_{21} = 3.088 & y_{22} = 0.528 & y_{23} = 1.212 \\ y_{31} = 2.188 & y_{32} = 1.212 & y_{33} = 1.352 \end{array}$$

Note that if y has a maximum, $-y$ will have a minimum. We must now consider the determinants

$$y_{11}, \quad \begin{vmatrix} y_{11} & y_{12} \\ y_{21} & y_{22} \end{vmatrix}, \quad \begin{vmatrix} y_{11} & y_{12} & y_{13} \\ y_{21} & y_{22} & y_{23} \\ y_{31} & y_{32} & y_{33} \end{vmatrix}$$

Direct calculation will show that the second- and third-order determinants

*G. E. P. Box, *Design and Analysis of Industrial Experiments*, ed. Owen L. Davies, London: Oliver and Boyd, 1956, pp. 544–49. Example reproduced by permission of the author and publisher.

are negative; hence the corresponding quadratic form is not positive or negative definite and therefore $(-y)$ is not a minimum or a maximum. It is apparent in this problem that the matrix of partial derivatives is the same as the matrix of the quadratic part of the function y. In the general case, the optimum point will be determined by the solution of the set of equations,

$$\begin{aligned}
-b_1 &= 2\beta_{11}x_1 + \beta_{12}x_2 + \cdots + \beta_{1n}x_n \\
-b_2 &= \beta_{12}x_1 + 2\beta_{22}x_2 + \cdots + \beta_{2n}x_n \\
-b_3 &= \beta_{13}x_1 + \beta_{23}x_2 + \cdots + \beta_{3n}x_n \\
&\;\vdots \\
-b_n &= \beta_{1n}x_1 + \beta_{2n}x_2 + \cdots + 2\beta_{nn}x_n
\end{aligned} \tag{7.14.3}$$

Note that the quadratic function Eq. (7.14.1) may be written in the form

$$\begin{aligned}
y = b_o &+ b_1x_1 + b_2x_2 + b_3x_3 + \cdots + b_nx_2 \\
&+ x_1[\beta_{11}x_1 + \tfrac{1}{2}\beta_{12}x_2 + \tfrac{1}{2}\beta_{13}x_3 + \cdots + \tfrac{1}{2}\beta_{1n}x_n] \\
&+ x_2[\tfrac{1}{2}\beta_{12}x_1 + \beta_{22}x_2 + \tfrac{1}{2}\beta_{32}x_3 + \cdots + \tfrac{1}{2}\beta_{2n}x_n] \\
&\;\vdots \\
&+ x_n[\tfrac{1}{2}\beta_{1n}x_1 + \tfrac{1}{2}\beta_{2n}x_2 + \cdots + \beta_{nn}x_n]
\end{aligned}$$

Substitution of Eq. (7.14.3) for the optimum gives, using the notation $x_j = x_{j0}$, for the optimum point

$$y_{\text{opt}} = b_o + \tfrac{1}{2}b_1x_{10} + \tfrac{1}{2}b_2x_{20} + \cdots + \tfrac{1}{2}b_nx_{no}$$

The optimum for the example above is $y_{\text{opt}} = 68.14$.

We can translate the quadratic, Eq. (7.14.1), to a set of axes about the optimum point by changing the axis so that

$$\begin{aligned}
z_1 &= x_1 - x_{10} \\
z_2 &= x_2 - x_{20} \\
&\;\vdots \\
z_n &= x_n - x_{no}
\end{aligned}$$

giving

$$\begin{aligned}
y = b_o &+ \tfrac{1}{2}b_1x_{10} + \tfrac{1}{2}b_2x_{20} + \cdots + \tfrac{1}{2}b_nx_{no} \\
&+ \beta_{11}z_1^2 + \beta_{12}z_1z_2 + \beta_{13}z_1z_3 + \cdots + \beta_{1n}z_1z_n \\
&\qquad + \beta_{22}z_2^2 + \beta_{32}z_2z_2 + \cdots + \beta_{2n}z_2z_n \\
&\qquad\qquad + \beta_{33}z_3^2 + \cdots + \beta_{3n}z_3z_n \\
&\qquad\qquad\qquad \ddots \\
&\qquad\qquad\qquad\qquad \beta_{nn}z_n^2
\end{aligned}$$

In order to study the character of the quadratic in more detail, we shall make the change of variable to principal axes. The second-degree term cited may be written as

$$\mathbf{z}^T \boldsymbol{\beta} \mathbf{z}$$

where

$$\boldsymbol{\beta} = \begin{bmatrix} \beta_{11} & \frac{1}{2}\beta_{12} & \frac{1}{2}\beta_{13} & \cdots & \frac{1}{2}\beta_{1n} \\ \frac{1}{2}\beta_{12} & \beta_{22} & \frac{1}{2}\beta_{23} & \cdots & \frac{1}{2}\beta_{2n} \\ \cdot & & & & \\ \cdot & & & & \\ \cdot & & & & \\ \frac{1}{2}\beta_{1n} & \frac{1}{2}\beta_{2n} & & \cdots & \beta_{nn} \end{bmatrix}$$

Since this is symmetric there is a matrix $\mathbf{X}$ made up of the normalized eigenvectors belonging to the eigenvalues which are solutions of

$$\det [\boldsymbol{\beta} - \lambda \mathbf{I}] = 0$$

so that

$$\mathbf{X}^T \boldsymbol{\beta} \mathbf{X} = \boldsymbol{\Lambda}$$

is the diagonal matrix of the eigenvalues. The orthogonal transformation

$$\mathbf{z} = \mathbf{X}\mathbf{w}, \quad \mathbf{w} = \mathbf{X}^T \mathbf{z}$$

then gives

$$y = b_o + \tfrac{1}{2} b_1 x_{10} + \tfrac{1}{2} b_2 x_{20} + \cdots + \tfrac{1}{2} b_n x_{no} + \mathbf{w}^T \boldsymbol{\Lambda} \mathbf{w}$$

or

$$y = b_o + \tfrac{1}{2} \mathbf{b}^T \mathbf{x}_o + \lambda_1 w_1^2 + \lambda_2 w_2^2 + \cdots + \lambda_n w_n^2$$

In the preceding example

$$\boldsymbol{\beta} = \begin{bmatrix} -1.539 & -1.544 & -1.094 \\ -1.544 & -0.264 & -0.606 \\ -1.094 & -0.606 & -0.676 \end{bmatrix}$$

and the λ's which satisfy the characteristic equation are

$$\lambda_1 = -3.190, \quad \lambda_2 = -0.069, \quad \lambda_3 = 0.780$$

and

$$\mathbf{X} = \begin{bmatrix} 0.7511 & 0.3066 & 0.5848 \\ 0.4884 & 0.3383 & -0.8044 \\ 0.4443 & -0.8897 & -0.1044 \end{bmatrix}$$

so that the relation between the variables $\{x_j\}$ and $\{w_j\}$ is

$$w_1 = 0.7511(x_1 - x_{10}) + 0.4884(x_2 - x_{20}) + 0.4443(x_3 - x_{30})$$
$$w_2 = 0.3066(x_1 - x_{10}) + 0.3383(x_2 - x_{20}) - 0.8897(x_3 - x_{30})$$
$$w_3 = 0.5848(x_1 - x_{10}) - 0.8044(x_2 - x_{20}) - 0.1044(x_3 - x_{30})$$

and
$$\begin{aligned} x_1 - x_{10} &= 0.7511\, w_1 + 0.3066\, w_2 + 0.5848\, w_3 \\ x_2 - x_{20} &= 0.4884\, w_1 + 0.3383\, w_2 - 0.8044\, w_3 \\ x_3 - x_{30} &= 0.4443\, w_1 - 0.8897\, w_2 - 0.1044\, w_3 \end{aligned}$$

Now the canonical form of Eq. (7.13.2) is

$$y - 68.14 = -3.190\, w_1^2 - 0.069\, w_2^2 + 0.780\, w_3^2$$

The form of the transformed response surface verifies the analysis in the early part of this section that there was no maximum since this quadratic form is not definite. Had it been definite, the signs of the squared terms would have been positive. This was obvious, of course, once the eigenvalues of the matrix had been calculated. From the preceding expression, however, it is clear that y can be increased by moving away from the center along the w_3 axis in such a fashion that

$$\begin{aligned} x_1 &= 0.60 + 0.0548\, w_3 \\ x_2 &= 0.215 - 0.8044\, w_3 \\ x_3 &= 0.501 - 0.1044\, w_3 \end{aligned}$$

It is at this point that the estimates of the statistical significance of the coefficients in Eq. (7.14.2) become necessary for it is important to determine whether this effect is real or not. If the effect is real, then a new experimental program may be necessary to obtain new information and a new quadratic function must be obtained which will represent this information and a new canonical analysis instituted.

7.15 Maxima or Minima with Restrictions

Frequently the foregoing analysis is not adequate since there may be restrictions on the values of certain variables or on combinations of these variables. For example, the yield of a certain process may be represented by a quadratic function of the form

$$y = b_0 + \mathbf{b}^T\mathbf{x} + \mathbf{x}^T\mathbf{B}\mathbf{x} \tag{7.15.1}$$

and the quality of a certain product may be represented by a quadratic function

$$q = c_0 + \mathbf{a}^T\mathbf{x} + \mathbf{x}^T\mathbf{A}\mathbf{x}$$

where b_{ij} and a_{ij} are typical elements of **B** and **A**, respectively, and, if it is specified that q must be q_0, then

$$0 = a_0 + \mathbf{a}^T\mathbf{x} + \mathbf{x}^T\mathbf{A}\mathbf{x} \tag{7.15.2}$$

with $a_0 = c_0 - q_0$. It then follows that if the yield is to be a maximum for $q = q_0$, Eq. (7.15.1) must be maximized subject to the restriction given in Eq. (7.15.2). Therefore we consider the function

$$\phi = b_0 + \mathbf{b}^T\mathbf{x} + \mathbf{x}^T\mathbf{B}\mathbf{x} - \lambda(a_0 + \mathbf{a}^T\mathbf{x} + \mathbf{x}^T\mathbf{A}\mathbf{x})$$

and the necessary conditions that ϕ be a maximum are that the partial derivatives of ϕ be zero or

$$\begin{aligned} 0 = b_j &+ 2b_{1j}x_1 + 2b_{2j}x_2 + \cdots + 2b_{jj}x_j + \cdots + 2b_{nj}x_n \\ &- \lambda a_j - \lambda[2a_{1j}x_1 + 2a_{2j}x_2 + \cdots + 2a_{jj}x_j + \cdots + 2a_{nj}x_n] \end{aligned}$$

or

$$0 = \frac{\mathbf{b}}{2} - \frac{\lambda}{2}\mathbf{a} + \mathbf{B}\mathbf{x} - \lambda\mathbf{A}\mathbf{x} = 0 \tag{7.15.3}$$

Note that the matrices **A** and **B** are symmetric.

Equations (7.15.2) and (7.15.3) must be solved simultaneously to give the conditions of the maximum or minimum point. We consider first

$$\mathbf{B}\mathbf{x} = \lambda\mathbf{A}\mathbf{x} \tag{7.15.4}$$

This is a homogeneous set of linear algebraic equations which will have nontrivial solutions if and only if

$$\det[\mathbf{B} - \lambda\mathbf{A}] = 0$$

If **B** and **A** are matrices of order n, then this is a polynomial of degree n and there will be n values $\{\lambda_j\}$ and certainly for each λ_j there will be a solution vector $\mathbf{x}_j$. Now we note that this is a generalization of the eigenvalue problem discussed in so much detail earlier. In the case under consideration, **B** and **A** are symmetric matrices since each belongs to a quadratic form. Let λ_i belong to $\mathbf{x}_i$ and λ_j to $\mathbf{x}_j$, then

$$\mathbf{B}\mathbf{x}_i = \lambda_i\mathbf{A}\mathbf{x}_i$$

$$\mathbf{B}\mathbf{x}_j = \lambda_j\mathbf{A}\mathbf{x}_j$$

and multiplying the first by $\mathbf{x}_j^T$ and the second by $\mathbf{x}_j^T$ and subtracting, one obtains

$$\mathbf{x}_j^T\mathbf{B}\mathbf{x}_i - \mathbf{x}_i^T\mathbf{B}\mathbf{x}_j = \lambda_i\mathbf{x}_j^T\mathbf{A}\mathbf{x}_i - \lambda_j\mathbf{x}_i^T\mathbf{A}\mathbf{x}_j$$

where from the symmetry of the matrices

$$(\lambda_i - \lambda_j)(\mathbf{x}_j^T\mathbf{A}\mathbf{x}_i) = (\lambda_i - \lambda_j)(\mathbf{x}_i^T\mathbf{A}\mathbf{x}_j) = 0$$

so that if $\lambda_i \neq \lambda_j$

$$\mathbf{x}_j^T\mathbf{A}\mathbf{x}_i = \mathbf{x}_i^T\mathbf{A}\mathbf{x}_j = 0$$

$$\mathbf{x}_j^T\mathbf{B}\mathbf{x}_i = \mathbf{x}_i^T\mathbf{B}\mathbf{x}_j = 0$$

and we have a generalized orthogonality property for the vectors $\{\mathbf{x}_j\}$. Furthermore, by a procedure exactly analogous to that used earlier it can be shown that the eigenvalues, if **A** and **B** are symmetric and either **A** or **B** is positive definite, are real. Therefore, in order to find a solution of Eq. (7.15.3) we assume that **x** may be expressed as

$$\mathbf{x} = \sum_j c_j\mathbf{x}_j$$

Substitution into Eq. (7.15.3), after assuming that $\mathbf{b}$ and $\mathbf{a}$ may be expressed as

$$\mathbf{b} = \sum_j \beta_j \mathbf{x}_j$$
$$\mathbf{a} = \sum_j \alpha_j \mathbf{x}_j$$

gives, if λ is not an eigenvalue,

$$0 = \frac{1}{2} \sum_j \beta_j \mathbf{x}_j - \frac{\lambda}{2} \sum_j \alpha_j \mathbf{x}_j + \sum_j c_j \mathbf{B}\mathbf{x}_j - \lambda \sum_j c_j \mathbf{A}\mathbf{x}_j$$
$$= \frac{1}{2} \sum_j \beta_j \mathbf{x}_j - \frac{\lambda}{2} \sum_j \alpha_j \mathbf{x}_j + \sum_j c_j \lambda_j \mathbf{A}\mathbf{x}_j - \lambda \sum_j c_j \mathbf{A}\mathbf{x}_j$$

Multiplication by $\mathbf{x}_i^T$ will give

$$0 = \frac{1}{2} \sum_j \beta_j \mathbf{x}_i^T \mathbf{x}_j - \frac{\lambda}{2} \sum_j \alpha_j \mathbf{x}_i^T \mathbf{x}_j + c_i \lambda_i \mathbf{x}_i^T \mathbf{A}\mathbf{x}_i - \lambda c_i \mathbf{x}_i^T \mathbf{A}\mathbf{x}_i$$

or

$$c_i = \frac{B_i - \lambda \mathbf{A}_i}{(\lambda - \lambda_i)\mathbf{x}_i^T \mathbf{A}\mathbf{x}_i}$$

where

$$B_i = \frac{1}{2} \sum_j \beta_j \mathbf{x}_i^T \mathbf{x}_j = \frac{1}{2} \mathbf{x}_i^T \mathbf{b}$$
$$A_i = \frac{1}{2} \sum_j \alpha_j \mathbf{x}_i^T \mathbf{x}_j = \frac{1}{2} \mathbf{x}_i^T \mathbf{a}$$

and therefore the solution of Eq. (7.15.3) in terms of λ is

$$\mathbf{x} = \sum_i \frac{B_i - \lambda A_i}{(\lambda - \lambda_i)\mathbf{x}_i^T \mathbf{A}\mathbf{x}_i} \mathbf{x}_i, \quad \lambda \neq \lambda_i$$

In order to find the value of λ we substitute this into Eq. (7.15.2),

$$0 = a_o + \sum_i \frac{B_i - \lambda A_i}{(\lambda - \lambda_i)} \frac{\mathbf{a}^T \mathbf{x}_i}{\mathbf{x}_i^T \mathbf{A}\mathbf{x}_i} + \sum_i \frac{(B_i - \lambda A_i)^2}{(\lambda - \lambda_i)^2} \frac{1}{\mathbf{x}_i^T \mathbf{A}\mathbf{x}_i}$$
$$= a_o + 2 \sum_i \frac{A_i(B_i - \lambda A_i)}{(\lambda - \lambda_i)\mathbf{x}_i^T \mathbf{A}\mathbf{x}_i} + \sum_i \frac{(B_i - \lambda A_i)^2}{(\lambda - \lambda_i)^2} \frac{1}{\mathbf{x}_i^T \mathbf{A}\mathbf{x}_i}$$
$$= a_o - \sum_i A_i^2 \frac{1}{\mathbf{x}_i^T \mathbf{A}\mathbf{x}_i} + \sum_i \frac{[B_i - \lambda_i A_i]^2}{(\lambda - \lambda_i)^2} \frac{1}{\mathbf{x}_i^T \mathbf{A}\mathbf{x}_i}$$

In general this will be a polynomial of degree $2n$ and there will be $2n$ values of λ which will produce extrema, either maximum, minimum, or inflection points. Once these λ's are obtained, then the vectors $\mathbf{x}$ corresponding to real λ's may be computed to give the points at which these extreme values occur and finally the extreme values themselves.

7.16 Simultaneous Maxima

In this section we consider a problem somewhat different from the preceding but related to it. Suppose we have p quantities y_j, such as yield, quality of several products, etc., each of which may be related to a number of variables by means of a second-order function of the form

$$y_j = b_j + \mathbf{b}_j^T\mathbf{x} + \mathbf{x}^T\mathbf{B}_j\mathbf{x} \quad j = 1, 2, 3, \ldots, p \tag{7.16.1}$$

where b_j, $\mathbf{b}_j$, and $\mathbf{B}$ are generalizations of the quantities b, $\mathbf{b}$, and $\mathbf{B}$ in the previous example but which are different for each quantity y_j. Now suppose in addition that p_j is the profit associated with a unit of y_j so that the total profit will be

$$P = \sum_j p_j y_j \tag{7.16.2}$$

Suppose in addition that we wish to study how the maximum value of P varies as the region defined by

$$\mathbf{x}^T\mathbf{x} = R^2 \tag{7.16.3}$$

changes. Hence we wish to maximize P subject to the restriction of Eq. (7.16.3). Then let

$$\phi = \sum_j p_j y_j - \lambda(\mathbf{x}^T\mathbf{x} - R^2)$$

and by the usual procedure we calculate the partial derivatives and set them equal to zero. First consider P,

$$\begin{aligned} P &= \sum_j p_j b_j + \sum_j p_j b_j^T\mathbf{x} + \sum_j p_j \mathbf{x}^T\mathbf{B}_j\mathbf{x} \\ &= \sum_j p_j b_j + (\sum p_j b_j^T)\mathbf{x} + \mathbf{x}^T(\sum p_j\mathbf{B}_j)\mathbf{x} \\ &= c_0 + \mathbf{c}^T\mathbf{x} + \mathbf{x}^T\mathbf{C}\mathbf{x} \end{aligned}$$

so that P is a quadratic function of the same kind as considered previously, and we could make the same kind of response surface canonical reduction as was made in Section 7.14. The extreme point could be computed and the behavior in the vicinity of the extreme point ascertained. Here we shall consider

$$\phi = c_0 + \mathbf{c}^T\mathbf{x} + \mathbf{x}^T\mathbf{C}\mathbf{x} - \lambda(\mathbf{x}^T\mathbf{x} - R^2) \tag{7.16.4}$$

and it follows that the necessary conditions for an extreme point will be

$$0 = \frac{\mathbf{c}}{2} + \mathbf{C}\mathbf{x} - \lambda\mathbf{x} \tag{7.16.5}$$

Since

$$\mathbf{C}\mathbf{x} = \lambda\mathbf{x}$$

has n real eigenvalues and corresponding eigenvectors we know that $\mathbf{x}$ the solution of Eq. (7.16.5) may be expressed as

$$\mathbf{x} = \sum_j \alpha_j \mathbf{x}_j$$

and if $\mathbf{c} = \sum_j \gamma_j \mathbf{x}_j$, then

$$\mathbf{x} = \frac{1}{2}\sum_j \frac{\gamma_j}{\lambda - \lambda_j}\mathbf{x}_j \tag{7.16.6}$$

and the extreme value of P is

$$P_{\text{ext}} = c_0 + \frac{1}{2}\mathbf{c}^T \sum_j \frac{\gamma_j}{\lambda - \lambda_j}\mathbf{x}_j + \frac{1}{4}\sum_j \frac{\gamma_j \mathbf{x}_j^T}{\lambda - \lambda_j}\mathbf{C}\sum_j \frac{\gamma_j \mathbf{x}_j}{\lambda - \lambda_j}$$

$$= c_0 + \frac{1}{2}\sum_j \frac{\gamma_j^2 \mathbf{x}_j^T \mathbf{x}_j}{\lambda - \lambda_j} + \frac{1}{4}\sum_j \frac{\lambda_j \gamma_j^2 \mathbf{x}_j^T \mathbf{x}_j}{(\lambda - \lambda_j)^2}$$

Let $\mathbf{x}^T\mathbf{x}_j$, which is positive, be l_j^2

$$P_{\text{ext}} = c_0 + \frac{1}{2}\sum_j \frac{\gamma_j^2 l_j^2}{\lambda - \lambda_j} + \frac{1}{4}\sum_j \frac{\gamma_j^2 l_j^2 \lambda_j}{(\lambda - \lambda_j)^2}$$

and

$$\frac{1}{4}\sum_j \frac{\gamma_j^2 l_j^2}{(\lambda - \lambda_j)^2} = R^2$$

so that P_{ext} is a function of λ and therefore also of R. Hence we have $2n$ extreme values of P on a circle of radius R, for in general there will be $2n$ λ's for a given value of R. The detailed pathology of the various possible forms of the extrema will not be examined further.

7.17 Transformations

We now generalize some results of three-dimensional analytic geometry. Suppose in a set of rectangular cartesian axes we have a vector $\mathbf{x}$ with components $\{x_j\}$ along the n axes, $x_1, x_2, \ldots x_n$, then the length of the vector will be

$$(\sum_j x_j^2)^{1/2}$$

and the angle that the vector makes with the jth axis is α_j, so that

$$\frac{x_j}{\sqrt{\sum_j x_j^2}} = \cos \alpha_j = v_j$$

where v_j is the direction cosine. Clearly then

$$\sum_j v_j^2 = 1$$

and if there is a second vector $\{\mathbf{x}_j'\}$ with direction consines $\{v_j'\}$, orthogonality implies

$$\sum_j v_j v_j' = 0$$

Consider now two sets of axes, each rectangular cartesian, with the same origin and call one the $x_1, x_2, x_3, \ldots, x_n$ set and the other $x_1', x_2', x_3', \ldots, x_n'$. Then any point or vector in space has two representations, one with coordinates or components $(x_1, x_2, \ldots, x_n)'$ and one with coordinates $(x_1', x_2', \ldots, x_n',)$ each with respect to the appropriate set of axes. In order to find the relation between these coordinates, it is necessary that the relative orientations of the two sets of axes be known. Let the axis x_1' make angles with the $x_1, x_2, \ldots, x_n$ axes so that the cosines are

$$v_{11}, v_{12}, v_{13}, \ldots, v_{1n}$$

Then the orientation of the primed axes with respect to the unprimed is determined by the matrix

$$\mathbf{Q} = \begin{bmatrix} v_{11} & v_{12} & \cdots & v_{1n} \\ v_{21} & v_{22} & \cdots & v_{2n} \\ \cdot & & & \\ \cdot & & & \\ \cdot & & & \\ v_{n1} & v_{n2} & \cdots & v_{nn} \end{bmatrix}$$

It follows then that (see Fig. (7.17.1) for a two-dimensional interpretation) the relation between the coordinates $\{x_j'\}$ and $\{x_j\}$ is

$$\begin{aligned} x_1' &= v_{11}x_1 + v_{12}x_2 + \cdots + v_{1n}x_n \\ x_2' &= v_{21}x_1 + v_{22}x_2 + \cdots + v_{2n}x_n \\ &\cdot \\ &\cdot \\ &\cdot \\ x_n' &= v_{n1}x_1 + v_{n2}x_2 + \cdots + v_{nn}x_n \end{aligned}$$

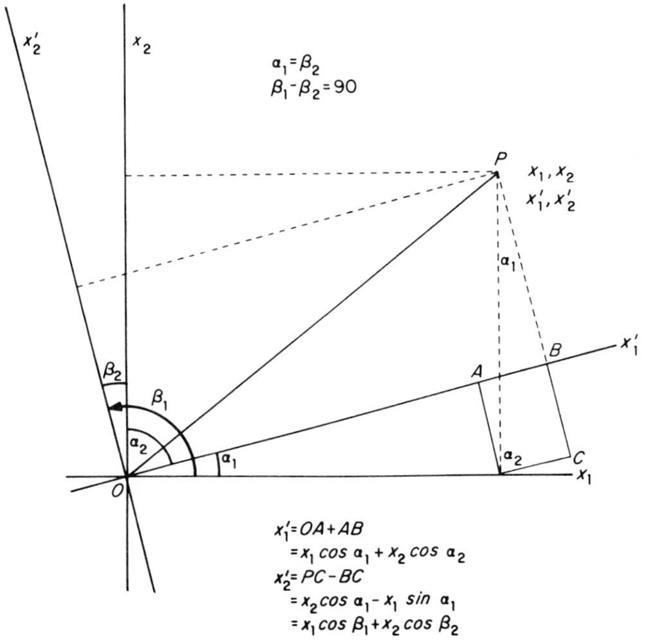

Figure 7.17.1 Two dimensional graphical interpretation of orthogonal transformation of coordinates.

or

$$\mathbf{x}' = \mathbf{Q}\mathbf{x}$$

Now the essential point of the argument is that $\mathbf{x}'$ is a vector in the primed axes and a vector $\mathbf{x}$ in the unprimed axes, but considered as the coordinates of a point we are talking about the *same point* and also the *same vector*. Although it has a different direction with respect to the two sets of axes, its length must be the same in each coordinate system; hence

$$l^2 = \sum_j x_j^2 = \sum_j x_j'^2$$

necessarily. Now

$$l^2 = \sum_j x_j'^2 = \sum_j \sum_i v_{ji} x_i x_j' = \sum_i x_i \sum_j v_{ji} x_j' = \sum_i x_i^2$$

hence it follows that

$$x_i = \sum_j v_{ji} x_j'$$

or

$$\begin{aligned} x_1 &= v_{11}x_1' + v_{21}x_2' + \cdots + v_{n1}x_n' \\ x_2 &= v_{12}x_1' + v_{22}x_2' + \cdots + v_{n2}x_n' \\ &\vdots \\ x_n &= v_{1n}x_1' + v_{2n}x_2' + \cdots + v_{nn}x_n' \end{aligned}$$

and

$$\mathbf{x} = \mathbf{P}\mathbf{x}'$$

so that

$$\mathbf{P} = \mathbf{Q}^{-1}$$

But the matrix $\mathbf{P}$ is just the transpose of Q; hence

$$\mathbf{Q}^T = \mathbf{Q}^{-1}$$

and the change from one set of axes to another is an orthogonal transformation. Such a transformation, since lengths are preserved, must be a rotation of axes alone.

From the condition for orthogonality, $\mathbf{Q}^T = \mathbf{Q}^{-1}$, it follows that the rows of $\mathbf{Q}$ form an orthonormal (orthogonal and normalized) set of vectors

$$\mathbf{v}_j^T \mathbf{v}_k = \delta_{jk}$$

and the columns of the same matrix $\mathbf{w}_j$ are such that

$$\mathbf{w}_j^T \mathbf{w}_k = \delta_{jk}$$

Since the determinant of $\mathbf{Q}$ and the determinant of $\mathbf{Q}^T$ are the same, it follows that the determinant of $\mathbf{Q}^2$ is one and the determinant of $\mathbf{Q}$ is ± 1, where the sign is determined by the directions of the axes. The orthogonality of the transformation is characterized by the orthogonality of the set of row vectors of the matrix $\mathbf{A}$ and the set of column vectors of the inverse transformation, that is, $\mathbf{A}^{-1}$.

7.18 Transformations in Oblique Coordinate Systems

In Section 7.17 we discussed orthogonal coordinate systems and the transformation from one orthogonal coordinate system to another. Although we make no direct use of oblique systems in this work, their treatment is a natural introduction to tensor analysis, a subject long important in some fields and of paramount importance in fluid mechanics and transport theory.

Consider first an n-dimensional orthogonal coordinate system X with n base vectors $\mathbf{e}_j$, $j = 1, 2, 3, \ldots, n$ where $\mathbf{e}_j$ is a vector of unit length in the direction of the axis X_j. Then any vector $\mathbf{x}$ in this coordinate system has a representation

$$\mathbf{x} = x_1\mathbf{e}_1 + x_2\mathbf{e}_2 + x_3\mathbf{e}_3 + \cdots + x_n\mathbf{e}_n \tag{7.18.1}$$

and $x_1, x_2, \ldots, x_n$ are the components of $\mathbf{x}$. Now consider n linearly independent vectors $\mathbf{a}_1, \mathbf{a}_2, \ldots, \mathbf{a}_n$ where

$$\mathbf{a}_j^T = [a_{1j}a_{2j} \ldots a_{nj}]$$

where a_{ij} is the ith component of the vector $\mathbf{a}_j$ along the axis X_i, that is,

$$\mathbf{a}_j^T\mathbf{e}_i = a_{ij}$$

Since the vector set $\{\mathbf{a}_j\}$ is linearly independent, it may be treated as a basic set of vectors in this n-dimensional space, and the vector $\mathbf{x}$ has a representation

$$\mathbf{x} = \xi_1\mathbf{a}_1 + \xi_2\mathbf{a}_2 + \cdots + \xi_n\mathbf{a}_n \tag{7.18.2}$$

where the $\{\xi_j\}$ are the components of the vector $\mathbf{x}$ in a new oblique coordinate system Y. We can find the relation between the components in the orthogonal system X and the oblique system Y by equating Eq. (7.18.1) and Eq. (7.18.2) noting that each of the $\mathbf{a}_j$ has a representation in terms of the basic set of vectors for X, that is,

$$\mathbf{a}_j = a_{1j}\mathbf{e}_1 + a_{2j}\mathbf{e}_2 + a_{3j}\mathbf{e}_3 + \cdots + a_{nj}\mathbf{e}_n$$

and hence

$$\begin{aligned}
x_1 &= a_{11}\xi_1 + a_{12}\xi_2 + \cdots + a_{1n}\xi_n \\
x_2 &= a_{21}\xi_1 + a_{22}\xi_2 + \cdots + a_{2n}\xi_n \\
&\ \ \vdots \\
x_n &= a_{n1}\xi_1 + a_{n2}\xi_2 + \cdots + a_{nn}\xi_n
\end{aligned}$$

or

$$\mathbf{x} = \mathbf{A}\boldsymbol{\xi}$$

This gives the relationship between two representations of the same vector.

The square of the length of the vector $\mathbf{x}$ in the orthogonal system is

$$l^2 = x_1^2 + x_2^2 + \cdots + x_n^2 = \mathbf{x}^T\mathbf{x}$$

which is the generalization to n-dimensional geometry of the familiar formula for two and three dimensions. Therefore

$$\begin{aligned} l^2 &= \mathbf{x}^T\mathbf{x} = \boldsymbol{\xi}^T\mathbf{A}^T\mathbf{A}\boldsymbol{\xi} \\ &= \boldsymbol{\xi}^T\mathbf{G}\boldsymbol{\xi} \end{aligned}$$

where

$$\mathbf{G} = \begin{bmatrix} g_{11} & g_{12} & \cdots & g_{1n} \\ g_{21} & g_{22} & \cdots & g_{2n} \\ g_{n1} & g_{n2} & \cdots & g_{nn} \end{bmatrix}$$

and

$$g_{ij} = \mathbf{a}_i^T\mathbf{a}_j$$

Thus the length of the vector when the vector is expressed in the Y system is

$$\begin{aligned} l^2 = g_{11}\xi_1^2 &+ g_{12}\xi_1\xi_2 + \cdots + g_{1n}\xi_1\xi_n \\ &+ g_{21}\xi_2\xi_1 + g_{22}\xi_2^2 + \cdots + g_{2n}\xi_2\xi_n \\ &\cdot \\ &\cdot \\ &\cdot \\ &+ g_{n1}\xi_n\xi_1 + g_{n2}\xi_n\xi_2 + \cdots + g_{nn}\xi_n^2 \end{aligned}$$

and the result obtained, more or less surprising, that the length depends upon the cross-product terms as well as the squared terms of the coordinates. The vectors $\{\mathbf{a}\}_j$ may be normalized so that

$$\mathbf{a}_i^T\mathbf{a}_i = 1$$

and thus g_{ii} may be taken as unity. The quantity a_{kj} with $\mathbf{a}_j$ normalized to unity, from elementary analytic geometry, is the cosine of the angle which the vector $\mathbf{a}_j$ makes with the axis X_k of X and hence

$$\begin{aligned} \mathbf{a}_i^T\mathbf{a}_j &= \cos\alpha_{1i}\cos\alpha_{1j} + \cos\alpha_{2i}\cos\alpha_{2j} + \cdots + \cos\alpha_{ni}\cos\alpha_{nj} \\ &= \cos\theta_{ij} = g_{ij} \end{aligned}$$

where θ_{ij} is the angle made by the vectors $\mathbf{a}_i$ and $\mathbf{a}_j$.

Suppose now we consider a second oblique system defined by a linearly independent set of vectors $\mathbf{a}_1^*, \mathbf{a}_2^*, \ldots \mathbf{a}_n^*$ defined in such a way that

$$\mathbf{a}^{*T}_i\mathbf{a}_j = 0, \quad i \neq j$$

that is, each vector of the starred set is orthogonal to all but one of the vectors in the unstarred set. The foregoing analysis can be repeated in detail. Let the coordinate system defined by the starred vectors be called Y^* and let the vector $\mathbf{x}$ with respect to this system have components $\{\xi_j^*\}$, that is,

$$\mathbf{x} = \xi_1\mathbf{a}_1^* + \xi_2^*\mathbf{a}_2^* + \cdots + \xi_n^*\mathbf{a}_n^*$$

It follows that

$$\mathbf{x} = \mathbf{A}^*\boldsymbol{\xi}^*$$

and

$$l^2 = \boldsymbol{\xi}^{*T}\mathbf{A}^{*T}\mathbf{A}^*\boldsymbol{\xi}^*$$

so that

$$\mathbf{a}_i^{*T}\mathbf{a}_j^* = g_{ij}^*$$
$$\mathbf{G}^* = [g_{ij}^*]$$

Consider now the matrix product

$$\mathbf{A}^T\mathbf{A}^*$$

From the definition of the coordinate system, this product is a diagonal matrix since the off-diagonal terms are $\mathbf{a}_i^T\mathbf{a}_j^* = 0,\ i \neq j$. The diagonal elements will not, however, be unity in general if the $\{\mathbf{a}_j^*\}$ vectors are also normalized since $\cos\theta_{ii^*}$ is not one. If it is desired to make the diagonal elements in $\mathbf{A}^T\mathbf{A}^*$ unity, then since the unstarred vectors have already been normalized, the starred vectors cannot be normalized to unity (that is, have unit length). Another way of saying the same thing is that though g_{ii} is unity g_{ii}^* is not. This means that each of the Y^* axes has a different scale. Therefore

$$\mathbf{A}^T\mathbf{A}^* = \mathbf{I}$$

hence

$$\mathbf{G}^{-1} = \mathbf{G}^*$$

The quantity $\mathbf{G}$ is called the *fundamental metric tensor* and it determines the length metric in the oblique coordinate system. The system of axes Y^* is called the *reciprocal oblique system* to Y although this is a mutual relationship. The numbers $\{\xi_j\}$ are called the *contravariant* components of the vector $\mathbf{x}$ whereas the $\{\xi_j^*\}$ are the covariant components. From the nature of the two systems it is clear that

$$\mathbf{a}_i^T\mathbf{x} = \xi_i^*$$
$$\mathbf{a}_i^{*T}\mathbf{x} = \xi_i$$

If one were given a second oblique system with matrix $\mathbf{B}$ and its reciprocal $\mathbf{B}^*$ then $\mathbf{x}$ would have contravariant components $\{\eta_j\}$ and covariant components $\{\eta_j^*\}$ and

$$\mathbf{x} = \mathbf{A}^*\xi^* = \mathbf{A}\boldsymbol{\xi} = \mathbf{B}^*\boldsymbol{\eta}^* = \mathbf{B}\boldsymbol{\eta}$$

Various relationships between the contravariant and covariant components in two different coordinate systems can be derived from the equations. One is

$$\boldsymbol{\eta}^* = \mathbf{B}^T\mathbf{A}\boldsymbol{\xi} = \mathbf{C}\boldsymbol{\xi}$$

and the elements of $\mathbf{C}$ are $\mathbf{b}_i^T\mathbf{a}_j = c_{ij}$.

Complex Monomolecular Kinetics and Staged Dynamics

8

8.1 Introduction

In this chapter two main examples will be exploited and in the course of their exploitation some new concepts not hitherto encountered will be introduced. The first problem will be concerned with the analysis of complex reaction systems which are psuedomonomolecular. In this problem one is interested in the structure of the mathematical system and how the change in the form by the reduction of a system of differential equations to normal form simplifies the physical and chemical interpretation of certain experimental facts. This problem will be presented in some detail. The second problem will be the analysis of the dynamics of a distillation column and will be more computational in nature. This problem is important from the standpoint of automatic control; however, the analysis here is for relatively simple models and probably does not have direct application to realistic systems because of the time of computation required.

A. Analysis of Complex Reaction Systems

8.2 Preliminary Remarks

We will consider a system of complex chemical reactions carried out in a batch reactor in which there is coupling between all the species. Suppose, for example, that we have three chemical species, A_1, A_2, and A_3, where these symbols stand for the species as well as their concentration and suppose they are connected by the reaction scheme

$$A_1 \underset{k_{12}}{\overset{k_{21}}{\rightleftarrows}} A_2$$

$$k_{13} \diagdown\ k_{31} \quad k_{32} \diagup\!\diagup k_{23}$$

$$A_3$$

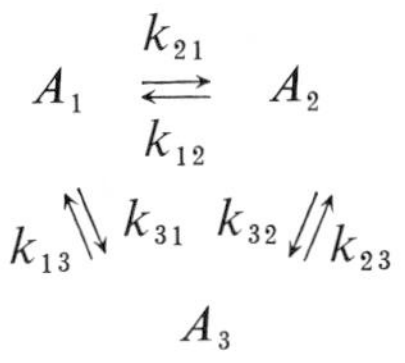

where k_{ij} is the absolute rate constant and $k_{ij}A_j$ is the rate of formation of A_i from A_j in moles per unit volume per unit of time. For the reaction scheme just given, the equations for the batch reactor are

$$\frac{dA_1}{dt} = -(k_{21} + k_{31})A_1 + k_{12}A_2 + k_{13}A_3$$

$$\frac{dA_2}{dt} = k_{21}A_1 - (k_{12} + k_{32})A_2 + k_{23}A_3 \qquad (8.2.1)$$

$$\frac{dA_3}{dt} = k_{31}A_1 + k_{32}A_2 - (k_{13} + k_{23})A_3$$

Such a system may be generalized to n chemical species, each of which is coupled to every other species by chemical reaction to give ($k_{jj} = 0$ for all j)

$$\frac{dA_1}{dt} = -\sum_{j=1}^{n} k_{j1}A_1 + k_{12}A_2 + k_{13}A_3 + \cdots + k_{1n}A_n$$

$$\frac{dA_2}{dt} = k_{21}A_1 - \sum_{j=1}^{n} k_{j2}A_2 + k_{23}A_3 + \cdots + k_{2n}A_n \qquad (8.2.2)$$

$$\vdots$$

$$\frac{dA_n}{dt} = k_{n1}A_1 + k_{n2}A_2 + \cdots + k_{n,n-1}A_{n-1} - \sum_{j=1}^{n} k_{jn}A_n$$

Since these relations are homogeneous in the A_j it is convenient to normalize them by division by $\sum_{j=1}^{n} A_j$. If this is done the resultant variables become mole fractions so that it is convenient to think of the A_j as mole fractions rather than concentrations. We define then the vector

$$\mathbf{A}^T = [A_1 A_2 \ldots A_n]$$

and think of the space in which the $\{A_j\}$ are the coordinates as a *composition space*. **A** is then a vector in this space and because compositions must be real and positive **A** is a vector with real and positive components. Because its components are mole fractions,

$$A_1 + A_2 + A_3 + \ldots + A_n = 1 \tag{8.2.3}$$

and any vector **A** which is a solution of the set of Eq. (8.2.1) must have a terminus which lies on the plane

$$A_1 + A_2 + A_3 = 1$$

an equilateral plane triangle formed in the first octant of this space. The solution of Eq. (8.2.2) must be a vector whose terminus lies in the $(n - 1)$-dimensional subspace defined by Eq. (8.2.3) which is a hyperplane. Associated with the set of Eq. (8.2.2) is a set of initial conditions

$$A_j = A_j^{(i)}$$

and, in general, the method of solution for these equations could be obtained as described earlier. The solution

$$\mathbf{A} = \mathbf{A}(t, \mathbf{A}^{(i)})$$

is a vector whose terminus must stay in that part of the n-dimensional space such that each component is positive and such that the terminus of **A** always lies on the hyperplane $\sum_{j=1}^{n} A_j = 1$. In the limit, as $t \to \infty$ the equilibrium composition $\mathbf{A}^*$ must be attained. The hyperplane is frequently referred to as the *phase plane*, and during the course of the reaction as the vector **A** changes with time to the vector $\mathbf{A}^*$, a *curve* will be traced out on the phase plane; this *curve* is referred to as a *reaction path* or a *trajectory*. The reaction path lies in the generalization of the equilateral triangle and is called a *simplex*.

Now it is relatively simple to determine the reaction paths if the problem is as stated here. Given the initial vector and the pertinent reaction velocity constants, one could write out the solution. The inverse problem, however, is one of even more interest, since the velocity constants must be determined from experimental information on reaction paths. If one proceeds and attempts to determine the velocity constants in this way, enough difficulties arise from a practical and numerical point of view, which need not be dwelled on here, to urge one to find some other technique for solving the problem. Such a technique has been exploited by James Wei and Charles D. Prater* who applied methods discussed earlier in this work and exploited

*The following material is presented with the permission of the authors and publishers: "The Structure and Analysis of Complex Reaction Systems," Chap. 5 in *Advances in Catalysis*, Vol. 13, New York: Academic Press, 1962.

these to give an elegant solution to the problem. We draw heavily on their work first published in *Advances in Catalysis*, and in fact, will follow it closely. (See the source for many of the applications.)

8.3 Reduction to Normal Form

Now as shown in Section 7.12 a set of equations like Eq. (8.2.2) may be reduced to normal form and be otherwise completely uncoupled and the analysis of complex reaction systems depends completely on this fact. If we write Eq. (8.2.2) in the form

$$\frac{d\mathbf{A}}{dt} = \mathbf{KA}$$

and if we define a matrix $\mathbf{X}$

$$\mathbf{A} = \mathbf{XB}$$

then

$$\frac{d\mathbf{B}}{dt} = \mathbf{X}^{-1}\mathbf{KXB}$$

If $\mathbf{X}$ has the property

$$\mathbf{X}^{-1}\mathbf{KX} = \mathbf{\Lambda}$$

where $\mathbf{\Lambda}$ is the matrix of eigenvalues, then

$$\frac{d\mathbf{B}}{dt} = \mathbf{\Lambda B} \tag{8.3.1}$$

and hence the differential equations are uncoupled.

Now in Wei and Prater's article the notation has been changed somewhat from standard notation, and since readers may wish to follow their analysis in more detail we adopt their nomenclature. The matrix $\mathbf{\Lambda}$, following Wei and Prater, is the matrix of negative eigenvalues and the columns of $\mathbf{X}$ are the eigenvectors which satisfy*

$$(\mathbf{K} + \lambda_j\mathbf{I})\mathbf{x}_j = 0$$

with

$$X = [\mathbf{x}_0\mathbf{x}_1\mathbf{x}_2 \ldots \mathbf{x}_{n-1}]$$

and therefore the uncoupled differential equations are

$$\begin{aligned} \frac{dB_0}{dt} &= -\lambda_1 B_0 \\ \frac{dB_1}{dt} &= -\lambda_2 B_1 \\ &\cdot \\ &\cdot \\ &\cdot \\ \frac{dB_{n-1}}{dt} &= -\lambda_{n-1} B_{n-1} \end{aligned} \tag{8.3.2}$$

*We refer to the λ_j which satisfy this equation as the *eigenvalues* although it is clear that they are the negatives of the eigenvalues of $\mathbf{K}$.

Hence, we pass from the **A** to the **B** by the linear transformation

$$\mathbf{A} = \mathbf{X}\mathbf{B}$$

$$\mathbf{B} = \mathbf{X}^{-1}\mathbf{A}$$

and any set of compositions **A** may be transformed to a set of pseudo compositions **B**, the advantage of the pseudo compositions being that the rate expressions are now completely uncoupled, as shown in Eq. (8.3.1) and (8.3.2). The uncoupling will be a true uncoupling in the real sense, provided that the eigenvalues of **K** are real. This will be discussed shortly.

A word about these pseudo compositions is in order. First, one must note that the matrix **K** is a singular matrix, since if one adds the first $n - 1$ rows to the last row, a row of zeroes will be produced in the last row, hence **K** has a zero eigenvalue. This is as it should be since Eq. (8.2.2) from a physical point of view cannot have eigenvalues with positive real parts. If one assumes that the eigenvalues other than the zero one are positive, then those components of the solution belonging to the non-zero eigenvalues must approach zero asymptotically; hence the equilibrium $\mathbf{A}^*$ is determined completely by the zero eigenvalue and its eigenvector and is $\mathbf{x}_0$. This is brought out more clearly in the pseudo compositions, for if we assume that λ_0 is the zero eigenvalue, then, if all the other eigenvalues are positive,

$$\frac{dB_0}{dt} = 0 \quad \text{or} \quad B_0 = \text{constant}$$

and

$$B_j \to 0, \quad j = 1, 2, 3, \ldots, n - 1$$

as $t \to \infty$ and therefore all the mass of the system, when expressed in pseudo compositions, must reside in B_0 and the remaining pseudo compositions contain no mass. Now the vectors $\mathbf{x}_j$ are determined only up to an arbitrary multiplicative constant; hence the B_j's suffer from the same deficiency. Since, however, we are considering that the total molar concentration remains the same during the course of the reaction, $\sum_{j=1}^{n} A_j = 1$. But from $\mathbf{A} = \mathbf{X}\mathbf{B}$ it follows that

$$A_1 = x_{10}B_0 + x_{11}B_1 + x_{12}B_2 + \cdots + x_{1,n-1}B_{n-1}$$
$$A_2 = x_{20}B_0 + x_{21}B_1 + x_{22}B_2 + \cdots + x_{2,n-1}B_{n-1}$$
$$\vdots$$
$$A_n = x_{n0}B_0 + x_{n1}B_1 + x_{n2}B_2 + \cdots + x_{n,n-1}B_{n-1}$$

and adding, one obtains

$$1 = \sum_{j=1}^{n} A_j = B_0 \sum_{i=1}^{n} x_{i0} + B_1 \sum_{i=1}^{n} x_{i1} + B_2 \sum_{i=1}^{n} x_{i2} + \cdots + B_{n-1} \sum_{i=1}^{n} x_{i,n-1}$$

This is a molar balance and since all the mass resides in B_0, $B_0 = 1$. The remaining B_j do not represent mole fractions of any real species or even represent any mass but, on the other hand, $B_j \neq 0, j = 1, 2, 3, \ldots, n-1$ so it follows necessarily that

$$\sum_{i=1}^{n} x_{i0} = 1$$

$$\sum_{i=1}^{n} x_{ij} = 0, \quad j = 1, 2, 3, \ldots, n-1 \tag{8.3.3}$$

It follows therefore that the vectors $\mathbf{x}_j$, $j \neq 0$, may have components which are negative and thus may lie outside of the positive octant in the three-dimensional space or the positive orthant in the n-dimensional space. Obviously these vectors do not necessarily form an orthogonal set.

8.4 The Character of the Eigenvalues

It has been assumed in the foregoing discussion that the non-zero eigenvalues are real and positive; hence we should establish this before we proceed further. Since it is not apparent that the eigenvalues are real, this should be the first order of business. Recall that the equilibrium mole fractions were denoted by A_j^* and let a diagonal matrix $\mathbf{D}$ be defined

$$\mathbf{D} = \begin{bmatrix} A_1^* & 0 & 0 & & \cdots & 0 \\ 0 & A_2^* & 0 & & \cdots & \\ 0 & 0 & A_3^* & 0 & \cdots & 0 \\ \cdot & & & \cdot & & \cdot \\ \cdot & & & \cdot & & \cdot \\ \cdot & & & \cdot & & \cdot \\ 0 & 0 & & 0 & & A_n^* \end{bmatrix}$$

then

$$\mathbf{KD} = \begin{bmatrix} -\sum_{j=1}^{n} k_{j1} A_1^* & k_{12} A_2^* & \cdots & k_{1n} A_n^* \\ k_{21} A_1^* & -\sum_{j=1}^{n} k_{j2} A_2^* & \cdots & k_{2n} A_n^* \\ \cdot & \cdot & & \cdot \\ \cdot & \cdot & & \cdot \\ \cdot & \cdot & & \cdot \\ k_{n1} A_1^* & k_{n2} A_2^* & \cdots & -\sum_{j=1}^{n} k_{jn} A_n^* \end{bmatrix}$$

The principle of microscopic reversibility is now invoked which says that *each* of the individual species in the system is in chemical equilibrium with *each* of the other individual species; that is, there is microscopic equilibrium as well as macroscopic equilibrium. This is called the *principle of detailed balancing*, or

$$k_{ji}A_i^* = k_{ij}A_j^*$$

hence the matrix **KD** is a symmetric matrix.

Now suppose one considers two matrices **U** and **V** connected by the transformation

$$\mathbf{P}^{-1}\mathbf{V}\mathbf{P} = \mathbf{U}$$

where **P** is nonsingular. As shown at the beginning of Chapter 7, the eigenvalues of **U** and **V** are the same.

Consider now the matrix, called for the moment, $\mathbf{D}^{1/2}$

$$\mathbf{D}^{1/2} = \begin{bmatrix} \sqrt{A_1^*} & 0 & \cdots & & 0 \\ 0 & \sqrt{A_2^*} & 0 & & 0 \\ 0 & 0 & & & 0 \\ \cdot & & & & \\ \cdot & & & & \\ \cdot & & & & \\ 0 & 0 & \cdots & 0 & \sqrt{A_n^*} \end{bmatrix}$$

and

$$\mathbf{D}^{-1/2} = \begin{bmatrix} \frac{1}{\sqrt{A_1^*}} & 0 & \cdots & \cdots & 0 \\ 0 & \frac{1}{\sqrt{A_2^*}} & 0 & \cdots & 0 \\ \cdot & & & & \\ \cdot & & & & \\ \cdot & & & & \\ 0 & \cdots & \cdots & 0 & \frac{1}{\sqrt{A_n^*}} \end{bmatrix}$$

Direct computation will show that

$$\mathbf{D}^{-1/2}\mathbf{D}^{1/2} = \mathbf{D}^{1/2}\mathbf{D}^{-1/2} = \mathbf{I}$$

$$\mathbf{D}\mathbf{D}^{-1/2} = \mathbf{D}^{-1/2}\mathbf{D} = \mathbf{D}^{1/2}$$

$$(\mathbf{D}^{1/2})^{-1} = \mathbf{D}^{-1/2}$$

$$(\mathbf{D}^{-1/2})^{-1} = \mathbf{D}^{1/2}$$

If one forms

$$\mathbf{D}^{-1/2}\mathbf{K}\mathbf{D}^{1/2} = \tilde{\mathbf{K}}$$

then

$$\tilde{\mathbf{K}} = \begin{bmatrix} -\sum_{j=1}^{n} k_{j1} & k_{12}\dfrac{\sqrt{A_2^*}}{\sqrt{A_1^*}} & k_{13}\dfrac{\sqrt{A_3^*}}{\sqrt{A_1^*}} & \cdots & k_{1n}\dfrac{\sqrt{A_n^*}}{\sqrt{A_1^*}} \\ k_{21}\dfrac{\sqrt{A_1^*}}{\sqrt{A_2^*}} & -\sum_{j=1}^{n} k_{j2} & k_{23}\dfrac{\sqrt{A_3^*}}{\sqrt{A_2^*}} & \cdots & k_{2n}\dfrac{\sqrt{A_n^*}}{\sqrt{A_2^*}} \\ k_{31}\dfrac{\sqrt{A_1^*}}{\sqrt{A_3^*}} & k_{32}\dfrac{\sqrt{A_2^*}}{\sqrt{A_3^*}} & & \cdots & \\ \vdots & \vdots & & & \vdots \\ k_{n1}\dfrac{\sqrt{A_1^*}}{\sqrt{A_n^*}} & k_{n2}\dfrac{\sqrt{A_2^*}}{\sqrt{A_n^*}} & & \cdots & -\sum_{j=1}^{n} k_{jn} \end{bmatrix}$$

and one sees that $\tilde{\mathbf{K}}$ is symmetric, since

$$k_{ji}\frac{\sqrt{A_i^*}}{\sqrt{A_j^*}} = k_{ij}\frac{\sqrt{A_j^*}}{\sqrt{A_i^*}}$$

from the principle of detailed balancing. Hence, since $\tilde{\mathbf{K}}$ and $\mathbf{D}^{-1/2}\mathbf{K}\mathbf{D}^{1/2}$ are connected by a collineatory transformation and $\tilde{\mathbf{K}}$ is symmetric, the eigenvalues of $\mathbf{K}$ are real, since the symmetric matrix $\tilde{\mathbf{K}}$ has real eigenvalues.

A quadratic form corresponding to the symmetric matrix $\tilde{\mathbf{K}}$ is

$$\boldsymbol{\gamma}^T\tilde{\mathbf{K}}\boldsymbol{\gamma}$$

and it may be reduced to a sum of squares, as shown in Section 7.4 in which the coefficients of the squared terms are the eigenvalues of $\tilde{\mathbf{K}}$. If the eigenvalues are all negative or zero, then certainly $\boldsymbol{\gamma}^T\tilde{\mathbf{K}}\boldsymbol{\gamma}$ will always be negative for any $\boldsymbol{\gamma}$. The converse may also be shown to be true. If $\boldsymbol{\gamma}^T\tilde{\mathbf{K}}\boldsymbol{\gamma}$ is always negative, then the eigenvalues which are not zero must be negative. The matrix $\tilde{\mathbf{K}}$ is*

$$\tilde{\mathbf{K}} = \begin{bmatrix} -\sum_{j=2}^{n} k_{j1} & \sqrt{k_{12}k_{21}} & \sqrt{k_{13}k_{31}} & \cdots & \sqrt{k_{1n}k_{n1}} \\ \sqrt{k_{21}k_{12}} & -\sum_{j=1,3}^{n} k_{j2} & \sqrt{k_{23}k_{32}} & \cdots & \sqrt{k_{2n}k_{n2}} \\ \vdots & \vdots & & & \vdots \\ \sqrt{k_{1n}k_{1n}} & \sqrt{k_{2n}k_{n2}} & \cdots & \cdots & -\sum_{j=1,n}^{n} k_{jn} \end{bmatrix}$$

and $\boldsymbol{\gamma}^T\tilde{\mathbf{K}}\boldsymbol{\gamma}$ may be written

*Once again we use the notation $\sum_{j=1,i}^{n}$ to indicate the $j = i$ is to be omitted from the summation.

$$\begin{aligned}\gamma^T\tilde{\mathbf{K}}\gamma &= -\Big[\gamma_1^2\sum_{j=2}^{n}k_{j1} - \gamma_1\sum_{j=2}^{n}\gamma_j\sqrt{k_{1j}k_{j1}} + \gamma_2^2\sum_{j=1,2}^{n}k_{j2} - \gamma_2\sum_{j=1,2}^{n}\gamma_j\sqrt{k_{2j}k_{j2}} \\ &\quad + \gamma_3^2\sum_{j=1,3}^{n}k_{j3} - \gamma_3\sum_{j=1,3}^{n}\gamma_j\sqrt{k_{3j}k_{j3}} + \cdots + \gamma_n^2\sum_{j=1}^{n-1}k_{jn} - \gamma_n\sum_{j=1}^{n-1}\gamma_j\sqrt{k_{nj}k_{jn}}\Big] \\ &= -\sum_i\Big(\gamma_i^2\sum_{j=1,i}^{n}k_{ji} - \sum_{j=1,i}^{n}\sqrt{k_{ji}k_{ij}}\,\gamma_i\gamma_j\Big) \\ &= -\sum_{i=1}^{n}\sum_{j=1,i}^{n}[\gamma_i^2 k_{ji} - \sqrt{k_{ji}k_{ij}}\,\gamma_i\gamma_j] = -\sum_{i=1}^{n}\sum_{j=1}^{n}\Big[\gamma_i^2 k_{ji} - \sqrt{k_{ji}k_{ij}}\,\gamma_i\gamma_j\Big]\end{aligned}$$

since the omitted term in the first summation is zero.
Then

$$\begin{aligned}\gamma^T\tilde{\mathbf{K}}\gamma &= -\sum_{i=1}^{n}\sum_{j=1}^{n}[\gamma_i^2 k_{ji} - \sqrt{k_{ji}k_{ij}}\,\gamma_i\gamma_j] + \sum_{i=1}^{n}\sum_{j=1}^{n}\Big[\sqrt{k_{ji}k_{ij}}\,\gamma_i\gamma_j - \gamma_j^2 k_{ij}\Big] \\ &\quad - \sum_{i=1}^{n}\sum_{j=1}^{n}[\sqrt{k_{ji}k_{ij}}\,\gamma_i\gamma_j - \gamma_j^2 k_{ij}] \\ &= -\sum_{i=1}^{n}\sum_{j=1}^{n}[\gamma_i^2 k_{ji} - 2\sqrt{k_{ji}k_{ij}}\,\gamma_i\gamma_j + k_{ij}\gamma_j^2] - \gamma^T\tilde{\mathbf{K}}\gamma\end{aligned}$$

so

$$\gamma^T\tilde{\mathbf{K}}\gamma = -\frac{1}{2}\sum_{i=1}^{n}\sum_{j=1}^{n}[\gamma_i\sqrt{k_{ji}} - \gamma_j\sqrt{k_{ij}}]^2 < 0$$

and therefore all the non-zero eigenvalues of $\tilde{\mathbf{K}}$ must be negative, but since $\tilde{\mathbf{K}}$ and $\mathbf{K}$ have the same eigenvalues, the non-zero eigenvalues of $\mathbf{K}$ must be negative.*

8.5 Straight-line Reaction Paths

We know now that the eigenvectors are all real, since the eigenvalues are real, and $\mathbf{X}$ is nonsingular, since we shall assume that the eigenvalues are distinct. Therefore if $\mathbf{A}^{(i)}$, the initial mole fractions at time $t = 0$, are specified, the initial values of the $\mathbf{B}$, $B_0^{(i)}, B_1^{(i)}, B_2^{(i)}, \ldots, B_{n-1}^{(i)}$ may be computed and in fact the solution of Eq. (8.3.2) may be written

$$\begin{aligned}B_0 &= B_0^{(i)} \\ B_1 &= B_1^{(i)}e^{-\lambda_1 t} \\ B_2 &= B_2^{(i)}e^{-\lambda_2 t} \\ &\cdot \\ &\cdot \\ &\cdot \\ B_{n-1} &= B_{n-1}^{(i)}e^{-\lambda_{n-1} t}\end{aligned}$$

*Note that this statement refers to the eigenvalues of $\tilde{\mathbf{K}}$ as defined by the equation

$$|\tilde{\mathbf{K}} - \lambda I| = 0$$

and therefore the λ_j from $|\tilde{\mathbf{K}} + \lambda I| = 0$ are all positive (or zero).

and from the relation

$$\mathbf{A} = \mathbf{XB}$$

it follows that

$$\mathbf{A} = B_0^{(i)}\mathbf{x}_0 + B_1^{(i)}\mathbf{x}_1 e^{-\lambda_1 t} + B_2^{(i)}\mathbf{x}_2 e^{-\lambda_2 t} + \cdots + B_{n-1}^{(i)}\mathbf{x}_{n-1} e^{-\lambda_{n-1} t} \quad (8.5.1)$$

The vectors $\mathbf{x}_1, \mathbf{x}_2, \ldots, \mathbf{x}_{n-1}$ have an interesting property for they have no component perpendicular to the plane $\sum_{j=1}^{n} A_j = 1$. A vector normal to this plane is

$$\mathbf{n}^T = [1 \quad 1 \quad 1 \quad \ldots \quad 1]$$

and

$$\mathbf{x}_j^T\mathbf{n} = \sum_{i=1}^{n} x_{ij} = 0$$

because of Eq. (8.3.1). Therefore the only vector with a component normal to that plane is $\mathbf{x}_0$ and all the other $\mathbf{x}_j$ are parallel to that plane. Let us now refer to the three-component system shown in Fig. 8.5.1, but write our formulae for the n-component system. Consider a vector

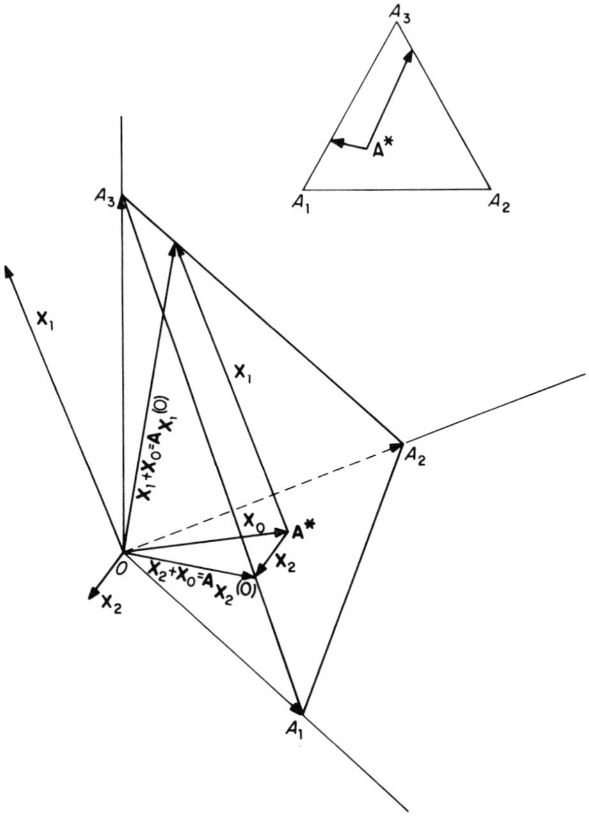

Figure 8.5.1 Schematic diagram showing relation among the vectors $\mathbf{x}_0$, $\mathbf{x}_j$, and $\mathbf{A}_{\mathbf{x}_j}^{(0)}$.

$$\mathbf{x}_0 + \beta_1 \mathbf{x}_1$$

where β_1 is a positive number. This vector has its terminus in the plane of the simplex and also lies in the plane determined by $\mathbf{x}_0$ and $\mathbf{x}_1$ for all β_1 and therefore as β_1 changes from zero and increases (or decreases), it traces a straight line in the plane of the simplex. Now the *length* of $\mathbf{x}_1$ has not been defined up to this point, and we shall define the length of $\mathbf{x}_1$ so that $\mathbf{x}_0 + \beta_1\mathbf{x}_1$ has the value $\mathbf{x}_0 + \mathbf{x}_1$ when its terminus intersects the boundary of the simplex. Therefore, the vector

$$\mathbf{x}_0 + \mathbf{x}_1 = \mathbf{A}_{x_1}(0)$$

defined in this way is a vector in the plane normal to the axis A_1, and its first component must be zero. In the figure it was just a coincidence that the first component was zero, since the naming of the vectors bears no relation to the coordinate axes A_1, A_2, A_3; that is, the subscripts on $\mathbf{x}_1$, $\mathbf{x}_2$, $\mathbf{x}_3$ bear no relation to the subscripts on A_1, A_2, A_3. Now given, in general, any of the vectors $\mathbf{x}_j$, the length of $\mathbf{x}_j$ will be defined so that the terminus of the vector

$$\mathbf{x}_0 + \mathbf{x}_j = \mathbf{A}_{\mathbf{x}_j}(0)$$

lies on the boundary of the simplex and

$$\mathbf{x}_0 + \beta_j \mathbf{x}_j$$

is a vector which for $\beta_j = 0$ is $\mathbf{x}_0$ and for β_j increasing to one traces out a straight line in the simplex. In general, then, each of these vectors for $\beta_j = 1$ will lie in a plane normal to one of the coordinate axes and therefore will have one zero component.

Consider now the initial values of $\mathbf{A}$, $\mathbf{A}^{(i)}$, and the transformation

$$\mathbf{A} = \mathbf{XB}$$

$$\mathbf{B} = \mathbf{X}^{-1}\mathbf{A}$$

The initial values of the composition $\mathbf{B}$ are $\mathbf{B}^{(i)}$ and are related to the $\mathbf{A}^{(i)}$ by

$$\mathbf{B}^{(i)} = \mathbf{X}^{-1}\mathbf{A}^{(i)}$$

so that the coefficients in Eq. (8.5.1) may be computed. On the other hand, if the $\mathbf{B}^{(i)}$ are specified, the $\mathbf{A}^{(i)}$ may be computed. In Eq. (8.5.1) let us specify that $B_0^{(i)} = 1$ (necessarily), $B_1^{(i)} = 1$, and $B_2^{(i)} = B_3^{(i)} = \ldots = B_{n-1}^{(i)} = 0$ so that

$$\mathbf{A}_{\mathbf{x}_1} = \mathbf{x}_0 + \mathbf{x}_1 e^{\lambda_1 t}$$

will give a reaction path in the simplex which is a straight line if the initial composition is as specified. Similarly,

$$\mathbf{A}_{\mathbf{x}_j} = \mathbf{x}_0 + \mathbf{x}_j e^{\lambda_j t}$$

will be straight-line paths for other specified initial compositions. In fact, it is obvious that there are $2(n - 1)$ different initial compositions that will produce $2(n - 1)$ straight-line paths. This follows from the fact that we have arbitrarily chosen a direction of the vector $\mathbf{x}_0 + \beta_j \mathbf{x}_j$ and the reverse or negative direction could have been chosen just as well.

Suppose now that one considers the reaction simplex alone, then the *natural* coordinate system in the simplex is not the set of lines bounding the simplex but rather the straight-line reaction paths. Any point *in the simplex* is given by two of the three coordinates A_1, A_2, A_3 or in general by $(n - 1)$ of the n coordinates $A_1, A_2, A_3, \ldots, A_n$. Any point in the simplex is, however, given just as well by $B_0, B_1, B_2, \ldots, B_{n-1}$ but $B_0 = 1$, so that $B_1, B_2, \ldots, B_{n-1}$ will serve. But from the relations

$$B_j = B_j^{(i)} e^{-\lambda_j t}$$

these are clearly the parametric equations of any reaction path with time as a parameter in the simplex and the B_j are the components of the vector from the equilibrium point to a point on the curve along the natural coordinate axes.

The foregoing relations may be used to compute the ratios of the eigenvalues, for if experiments are performed and the results are translated into the **B** system, then for any two B_j, say, B_j and B_k

$$\log B_j = \log B_j^{(i)} - \lambda_j t$$

$$\log B_k = \log B_k^{(i)} - \lambda_k t$$

and eliminating t

$$\begin{aligned} B_j &= B_k^{\lambda_j/\lambda_k} B_j^{(i)} B_k^{(i)^{-\lambda_j/\lambda_k}} \\ &= \text{constant} \cdot B_k^{\lambda_j/\lambda_k} \end{aligned}$$

Therefore if B_j is plotted versus B_k on log-log paper, a straight line will result, the slope of which is λ_j/λ_k. Thus the $(n - 1)$ ratios of the eigenvalues may be obtained.

It is to be stressed that up to this time no detailed knowledge of the reaction time has been essential for any of the computations described. The computation of the vectors and the ratios of the eigenvalues depend upon a detailed knowledge of the compositions on their path to the equilibrium value. The equilibrium value is also needed. It will now be shown that, with the knowledge so far obtained, the ratios of the reaction velocity constants k_{ij} may be computed.

Suppose that Λ, the matrix of the eigenvalues, is written in the form

$$\Lambda = \lambda_m \begin{bmatrix} 0 & 0 & 0 & & 0 & 0 & \cdots & & 0 \\ 0 & \frac{-\lambda_1}{\lambda_m} & 0 & & & & \cdots & & 0 \\ 0 & 0 & \frac{-\lambda_2}{\lambda_m} & & \vdots & \vdots & & & \\ \vdots & \vdots & & \ddots & & & & & \vdots \\ & & 0 & & 0 & & \cdots & & \\ 0 & & & & -1 & 0 & & & \\ \vdots & \vdots & \vdots & & 0 & \ddots & & & \\ 0 & & & & & \frac{-\lambda_{m+1}}{\lambda_m} & & 0 & \cdots \quad 0 \\ \vdots & \vdots & \vdots & & \vdots & 0 & \ddots & & \\ & & & & & \vdots & & \ddots & \\ 0 & 0 & 0 & \cdots & 0 & 0 & \cdots & 0 & \frac{-\lambda_{n-1}}{\lambda_m} \end{bmatrix}$$

where λ_m is one of the non-zero eigenvalues, then

$$\mathbf{K} = \mathbf{X}\Lambda\mathbf{X}^{-1} = \lambda_m \mathbf{X}\Lambda' X^{-1}$$

Define

$$\mathbf{X}\Lambda'\mathbf{X}^{-1} = \mathbf{K}'$$

so

$$\mathbf{K} = \lambda_m \mathbf{K}'$$

Now the matrix $\mathbf{K}'$ is the same as the matrix $\mathbf{K}$ except that each element of $\mathbf{K}$ is divided by λ_m to get $\mathbf{K}'$. From the analysis which has been presented thus far, the matrix $\mathbf{K}'$ may be computed, since the elements of Λ' are known and the elements of $\mathbf{X}$, although their computation has not yet been made explicit, can be computed. Thus the reaction velocity constants in $\mathbf{K}$ are known up to a single multiplicative constant.

8.6 A New Set of Vectors

We are still not ready to illustrate the computation of the straight-line reaction paths from experimental information since one further calculation is required. In Section 8.4 a new matrix $\mathbf{D}^{1/2}$ was introduced in connection with proving that all the eigenvalues were non-negative. Consider now the transformation

$$y_j = \mathbf{D}^{-1/2}\mathbf{x}_j$$

and its inverse

$$\mathbf{x}_j = \mathbf{D}^{1/2}\mathbf{y}_j$$

The eigenvalue problem defining the $\mathbf{x}_j$ is

$$\mathbf{K}\mathbf{x}_j = -\lambda_j\mathbf{x}_j$$

and, therefore, from the foregoing,

$$\mathbf{K}\mathbf{D}^{1/2}\mathbf{y}_j = -\lambda_j\mathbf{D}^{1/2}\mathbf{y}_j$$

and multiplying by $\mathbf{D}^{-1/2}$

$$\mathbf{D}^{-1/2}\mathbf{K}\mathbf{D}^{1/2}\mathbf{y}_j = -\lambda_j\mathbf{y}_j$$

or

$$\tilde{\mathbf{K}}\mathbf{y}_j = -\lambda_j\mathbf{y}_j$$

But $\tilde{\mathbf{K}}$ is a symmetric matrix, and the $\mathbf{y}_j$ are its eigenvectors and hence form an orthogonal set of vectors

$$\mathbf{y}_j^T\mathbf{y}_k = 0. \quad k \neq j$$

or

$$\mathbf{x}_j^T\mathbf{D}^{-1}\mathbf{x}_k = 0, \quad k \neq j$$

so that the original eigenvectors are orthogonal but with a weighting matrix. These vectors $\{\mathbf{y}_j\}$ are used to predict the region in which to look for initial compositions in order to compute the eigenvectors $\mathbf{x}_j$. It is clear that

$$\mathbf{x}_j^T\mathbf{D}^{-1}\mathbf{x}_j = l_j$$

where

$$l_j = \sum_{k=1}^{n} = \frac{x_{kj}^2}{A_k^*} \quad \text{and} \quad \mathbf{X}^T\mathbf{D}^{-1}\mathbf{X} = \mathbf{L}$$

where $\mathbf{L}$ is a diagonal matrix with l_j as the diagonal elements. The vectors $\{\mathbf{y}_j\}$ may also be transformed to unit vectors by division by $l_j^{1/2}$. If $\sqrt{l_j}\ \mathbf{z}_j = \mathbf{y}_j$, then

$$\mathbf{z}_j^T\mathbf{z}_k = \begin{cases} 0, & j \neq k \\ 1, & j = k \end{cases}$$

From the formula $\mathbf{X}^T\mathbf{D}^{-1}\,\mathbf{X} = \mathbf{L}$, the inverse matrix $\mathbf{X}^{-1}$ may be easily computed from

$$\mathbf{X}^{-1} = \mathbf{L}^{-1}\mathbf{X}^T\mathbf{D}^{-1}$$

8.7 The Treatment of Experimental Data

Although such a discussion may seem inappropriate in this book, in order to apply the methods previously developed the general problem must be discussed here. Our problem is to determine the reaction velocity constants k_{ij} for a complex system of chemical reactions. The methods used for a single reaction are generally not useful and even when used give results of doubtful accuracy. The main object of the foregoing analysis was to give a clue about *what* experiments should be made in order that straight-line reaction paths could be obtained and, once these paths had been obtained, how they could be used to extract the desired information. We shall describe the method of calculation and then, omitting some of the detail, use some of Wei and Prater's examples.

A transient experiment is performed starting with some convenient initial composition $A_1^{(i)}, A_2^{(i)}, \ldots, A_n^{(i)}$, and the composition is followed as a function of the time and tabulated. A satisfactory one is that of a pure component, say A_1. The reaction path is plotted in the simplex if there are three components but if there are more than three the following procedure will be used. We assume that we, as a result of the first experiment, know what the equilibrium composition $\mathbf{A}^*$ is (this fixes $\mathbf{x}_o$) and we want to compute the straight-line reaction path corresponding to $\mathbf{x}_1$. Since in the neighborhood of the equilibrium point the reaction path will be determined largely by the B species with smallest eigenvalue (or slowest decay time), we write

$$
\begin{aligned}
A_1 &= x_{10} + e^{-\lambda_1 t} x_{11} = A_1^* + e^{-\lambda_1 t} x_{11} \\
A_2 &= x_{20} + e^{-\lambda_1 t} x_{21} = A_2^* + e^{-\lambda_1 t} x_{21} \\
&\;\;\vdots \\
A_n &= x_{no} + e^{-\lambda_1 t} x_{n1} = A_n^* + e^{-\lambda_1 t} x_{n1}
\end{aligned}
$$

where we assume that the equilibrium point is in the interior of the simplex and therefore $x_{j0} > 0$ for all j. If we choose A_k and A_m as two typical components, then, upon eliminating the time,

$$A_m = \frac{x_{m1}}{x_{k1}} A_k + \left(A_m^* - A_k^* \frac{x_{m1}}{x_{k1}}\right)$$

so that if A_k is chosen as a base A_m may be plotted versus A_k, for all $m \neq k$. These, in general, will be curves, but in the neighborhood of $\mathbf{A}^*$ the lines should be relatively straight and one can extrapolate back to the boundary of the simplex. The boundary will have been reached when any A_m becomes zero. All A_m must be greater than zero until the boundary is reached. Now this extrapolation back to zero will not be accurate in the sense that this is now a straight-line reaction path. In order to test whether it is a straight-line reaction path, the new composition determined at the boundary by the extrapolation is used as a new initial composition and a second experiment is run, the composition being followed as a function of the time and tabulated. The extrapolation of the tangent line at the equilibrium point back to the boundary is repeated and a new initial composition determined. This is then used in a new experiment as an initial composition to determine whether a straight-line reaction path results. The procedure is repeated until a straight-line path is obtained to within the experimental error. We call the values on the boundary which serve as the initial composition for a straight-line reaction path $\mathbf{A}_{\mathbf{x}_1}^{(i)}$ and therefore the first eigenvector is

$$\mathbf{x}_1 = \mathbf{A}_{\mathbf{x}_1}^{(i)} - \mathbf{x}_0$$

We now have one eigenvector, and the first impulse would be to choose a radically different initial composition in the hope of arriving at a new straight-line reaction path; this would, however, be a coincidence since in all probability this initial composition would have enough of the B_1 component in it so that the extrapolation iteration procedure would bring one back to the eigenvector just computed, since the behavior of the transient reaction paths in the neighborhood of the equilibrium point is dominated by the component with the slowest decay time. This being the case, a new procedure must be instituted to guarantee that the new initial composition contains no B_1 component.

As previously let

$$\mathbf{y}_1 = \mathbf{D}^{-1/2}\mathbf{x}_1$$

and this may be computed, since the equilibrium compositions are known, and

$$\mathbf{z}_1 = \frac{1}{\sqrt{l_1}}\mathbf{y}_1$$

Also, we must remember that

$$\mathbf{y}_0 = \mathbf{D}^{-1/2}\mathbf{x}_0$$

$$\mathbf{z}_0 = \frac{1}{\sqrt{l_0}}\mathbf{y}_0$$

Instead of assuming a new initial composition, assume a vector $\boldsymbol{\gamma}$ and form from it a vector $\boldsymbol{\delta}$ which is orthogonal to $\mathbf{z}_0$ and $\mathbf{z}_1$ having the form

$$\boldsymbol{\delta} = \boldsymbol{\gamma} - a_1\mathbf{z}_0 - a_2\mathbf{z}_1$$

then it is necessary that

$$a_1 = \mathbf{z}_0^T\boldsymbol{\gamma}$$

$$a_2 = \mathbf{z}_1^T\boldsymbol{\gamma}$$

Now $\boldsymbol{\delta}$ is not a unit vector but may be transformed to a unit vector

$$\boldsymbol{\rho} = \frac{\boldsymbol{\delta}}{|\boldsymbol{\delta}|}$$

where $|\boldsymbol{\delta}|$ stands for the length of $\boldsymbol{\delta}$ and so $|\boldsymbol{\rho}| = 1$. Therefore $\boldsymbol{\rho}$ is $\mathbf{z}_2$ and

$$\mathbf{y}_2 = \sqrt{l_2}\,\mathbf{z}_2$$

$$\mathbf{x}_2 = \mathbf{D}^{1/2}\mathbf{y}_2$$

From the relation

$$\mathbf{A}_{\mathbf{x}_2}^{(i)} = \mathbf{x}_2 + \mathbf{x}_0$$

it follows that we should have an initial composition $\mathbf{A}_{\mathbf{x}_2}^{(i)}$ which contains no B_1 component. This is now used as an initial composition for an experimental determination of whether it does in fact give a straight-line reaction path. If it does, then $\mathbf{x}_2$ is the third eigenvector. If it doesn't, then we extrapolate back to a new initial composition $\mathbf{A}_{\mathbf{x}_2}^{(i)}$. Before this initial composition can be used in an experiment, we must be sure that it contains no B_1 component, so we form

$$\mathbf{x}_2 = \mathbf{A}_{\mathbf{x}_2}^{(i)} - \mathbf{x}_0$$

and use this as a $\boldsymbol{\gamma}$ and repeat the foregoing procedure to remove any trace of $\mathbf{x}_0$ and $\mathbf{x}_1$. Finally, after iterating on this procedure until the results are within the experimental accuracy, $\mathbf{x}_2$ is obtained and we have a second straight-line path. Note that when we started the preceding procedure to determine $\mathbf{x}_2$ we could have assumed an initial composition, instead of $\boldsymbol{\gamma}$, and removed any trace of $\mathbf{x}_0$ and $\mathbf{x}_1$ as we did in the successive iterations. Wei and Prater assumed the vector $\boldsymbol{\gamma}$ to have a special form which is automatically orthogonal to $\mathbf{z}_0$, hence the foregoing computations are somewhat simpler.

In order to make the procedure definite let us carry it through one more step. It is clear that, in order to find the third straight-line reaction path, any assumption of a new initial composition must be free of both B_1 and B_2 and from the nature of the process it seems clear that B_1 and B_2 are probably the components of the final solution belonging to the smallest non-zero eigenvalue and the second smallest non-zero eigenvalue. For, if the second reaction path contained no B_1 component, its behavior in the neighborhood of the equilibrium point must be dominated by the second smallest non-zero eigenvalue.

We assume now a composition $\mathbf{A}_{\mathbf{x}_3}^{(i)}$ and form

$$\mathbf{x}_3 = \mathbf{A}_{\mathbf{x}_3}^{(i)} - \mathbf{x}_0$$

This vector will be used as a $\boldsymbol{\gamma}$ and we form

$$\boldsymbol{\delta} = \boldsymbol{\gamma} - a_1\mathbf{z}_0 - a_2\mathbf{z}_1 - a_3\mathbf{z}_2$$

and demand that $\boldsymbol{\delta}$ be orthogonal to $\mathbf{z}_0$, $\mathbf{z}_1$, and $\mathbf{z}_2$. Therefore,

$$a_1 = \mathbf{z}_0^T \boldsymbol{\gamma}$$

$$a_2 = \mathbf{z}_1^T \boldsymbol{\gamma}$$

$$a_3 = \mathbf{z}_2^T \boldsymbol{\gamma}$$

We then normalize $\boldsymbol{\delta}$ and it becomes a new $\mathbf{z}_3$ which can be transformed

$$\mathbf{y}_3 = \sqrt{l_3}\, \mathbf{z}_3$$

$$\mathbf{x}_3 = \mathbf{D}^{1/2} \mathbf{y}_3$$

A new $\mathbf{A}_{\mathbf{x}_3}^{(i)}$ is calculated from

$$\mathbf{A}_{\mathbf{x}_3}^{(i)} = \mathbf{x}_0 + \mathbf{x}_3$$

and an initial composition is available for an experiment. If the reaction path is straight, $\mathbf{x}_3$ is the third eigenvector. If not, the path is extrapolated back to a new initial composition and the whole procedure just described is repeated.

The method used by Wei and Prater to form $\boldsymbol{\gamma}$ is the following: If one writes down the vector for $\mathbf{z}_0$, a vector orthogonal to it may be found in a variety of ways. If all but two components of $\mathbf{z}_0$ are made zero and those two are interchanged with one also of different sign, then the scalar product of these two vectors is zero. For example, if

$$\mathbf{z}^T = [\alpha_1 \alpha_2 \alpha_3 \alpha_4 \ldots \alpha_n]$$

and

$$\boldsymbol{\gamma}^T = [-\alpha_2 \alpha_1 \, 0 \, 0 \ldots 0]$$

then $\mathbf{z}_0^T \boldsymbol{\gamma} = 0$. This is the technique used by Wei and Prater to start the iteration scheme.

Now that all the eigenvectors have been computed, the matrix $\mathbf{X}$ is known and, from the transformation $\mathbf{B} = \mathbf{X}^{-1}\mathbf{A}$, the initial values of $\mathbf{B}$ may be computed. As mentioned in Section 8.5, the ratios of the eigenvalues can be computed from the data on any *single* reaction path. If the time is also used, all the eigenvalues may be computed and therefore

$$\mathbf{K} = \mathbf{X}\boldsymbol{\Lambda}\mathbf{X}^{-1}$$

hence all the velocity constants for the reactions are available.

8.8 An Example of Wei and Prater for a Four-component System

Wei and Prater consider a number of examples, but we give only two in outline form. One is a system of the kind

$$\begin{array}{ccc} A_1 & \rightleftarrows & A_2 \\ \upharpoonleft\downharpoonright & & \upharpoonleft\downharpoonright \\ A_4 & \rightleftarrows & A_3 \end{array}$$

This reaction system shows no direct coupling between A_1 and A_3 or between A_2 and A_4 and if the mechanism suggested in the preceding reaction scheme is correct, then the velocity constants between A_1 and A_3 and between A_2 and A_4 should be very small compared to the others or zero. Table 8.8.1 gives the composition values for various times starting with pure A_1. In order to obtain the first approximation to a straight-line reaction path, the values of the mole fractions of the species must be plotted versus one

TABLE 8.8.1

t	A_1	A_2	A_3	A_4
0	1	0	0	0
t_1	0.6725	0.0189	0.0071	0.3016
t_2	0.5008	0.0306	0.0222	0.4464
t_3	0.3471	0.0468	0.0578	0.5486
t_4	0.2759	0.0782	0.1205	0.5253
t_5	0.2252	0.1325	0.1950	0.4472
t_6	0.1847	0.1872	0.2611	0.3670
t_8	0.0994	0.3036	0.3968	0.2002
∞	0.1000	0.3000	0.4000	0.2000

of them taken as a basis. It is clear from Table 8.8.1 that, as one moves along the reaction path away from the equilibrium point, A_1 increases, A_4 increases and then decreases, and A_2 and A_3 decrease; and it may be shown that A_2 will reach a value of zero on extrapolation of the tangent to the reaction path at the equilibrium point back to the boundary of the simplex before A_3. If A_2, A_3, and A_4 are plotted versus A_1, this indeed proves to be so, and a value of

$$\mathbf{A}_{\mathbf{x}_1}^{(i)} = \begin{bmatrix} 0.330 \\ 0.000 \\ 0.040 \\ 0.638 \end{bmatrix}$$

is obtained. The sum of these elements is not one because of errors, and so the vector is normalized to give

$$\mathbf{A}_{\mathbf{x}_1}^{(i)} = \begin{bmatrix} 0.3274 \\ 0.0000 \\ 0.0397 \\ 0.6329 \end{bmatrix}$$

This is then used as a new experimental initial composition and the process repeated until results within the experimental measurements are obtained. The initial value of a straight-line reaction path is found to be

$$\mathbf{A}_{\mathbf{x}_1}^{(i)} = \begin{bmatrix} 0.3214 \\ 0.0000 \\ 0.0382 \\ 0.6404 \end{bmatrix}$$

Then

$$\mathbf{x}_1 = \mathbf{A}_{\mathbf{x}_1}^{(i)} - \mathbf{x}_0 = \begin{bmatrix} 0.3214 \\ 0.0000 \\ 0.0382 \\ 0.6404 \end{bmatrix} - \begin{bmatrix} 0.1000 \\ 0.3000 \\ 0.4000 \\ 0.2000 \end{bmatrix} = \begin{bmatrix} 0.2214 \\ -0.3000 \\ -0.3618 \\ 0.4404 \end{bmatrix}$$

and

$$\mathbf{D}^{1/2} = \begin{bmatrix} 0.316228 & 0 & 0 & 0 \\ 0 & 0.547723 & 0 & 0 \\ 0 & 0 & 0.632456 & 0 \\ 0 & 0 & 0 & 0.447214 \end{bmatrix}$$

$$\mathbf{D}^{-1/2} = \begin{bmatrix} 3.162275 & 0 & 0 & 0 \\ 0 & 1.825740 & 0 & 0 \\ 0 & 0 & 1.581137 & 0 \\ 0 & 0 & 0 & 2.236066 \end{bmatrix}$$

One then finds

$$\mathbf{z}_0 = \begin{bmatrix} 0.316228 \\ 0.547723 \\ 0.632456 \\ 0.447214 \end{bmatrix}$$

$$\mathbf{z}_1 = \begin{bmatrix} 0.484615 \\ -0.379123 \\ -0.395966 \\ 0.681634 \end{bmatrix}$$

Now for $\boldsymbol{\gamma}$ we choose, following Wei and Prater,

$$\boldsymbol{\gamma} = \begin{bmatrix} 0.000000 \\ -0.632456 \\ 0.547723 \\ 0.000000 \end{bmatrix}$$

and this vector is orthogonal to $\mathbf{z}_0$. Let

$$\boldsymbol{\delta} = \boldsymbol{\gamma} - a_1\mathbf{z}_0 - a_2\mathbf{z}_1$$

and we choose a_1 and a_2 so that $\boldsymbol{\delta}$ is to be orthogonal to $\mathbf{z}_0$ and $\mathbf{z}_1$. Therefore,

$$a_1 = 0 \quad \text{and} \quad a_2 = \mathbf{z}_1^T\boldsymbol{\gamma}$$

and a direct calculation gives $a_2 = 0.022899$, so

$$\boldsymbol{\delta} = \boldsymbol{\gamma} - 0.022899\mathbf{z}_1 = \begin{bmatrix} -0.011097 \\ -0.623774 \\ 0.556790 \\ -0.015609 \end{bmatrix}$$

Now $\boldsymbol{\delta}$ is a member of the orthogonal system of vectors, but we can transform it to the nonorthogonal system by

$$\boldsymbol{\Delta} = \mathbf{D}^{1/2}\boldsymbol{\delta}$$

$$= \begin{bmatrix} -0.003509 \\ -0.341655 \\ 0.352145 \\ -0.006981 \end{bmatrix}$$

It would be an $\mathbf{x}_2$ if $\boldsymbol{\Delta} + \mathbf{x}_0$ were a vector $\mathbf{A}_{\mathbf{x}_2}^{(i)}$ which had the property that one component was zero and all other components were positive. Consider

$$\begin{bmatrix} 0.1000 \\ 0.3000 \\ 0.4000 \\ 0.2000 \end{bmatrix} + \alpha \begin{bmatrix} -0.003509 \\ -0.341655 \\ 0.352145 \\ -0.006981 \end{bmatrix} = \boldsymbol{\epsilon} = \begin{bmatrix} \epsilon_1 \\ \epsilon_2 \\ \epsilon_3 \\ \epsilon_4 \end{bmatrix}$$

then,

$$0.1000 - 0.003509\,\alpha = \epsilon_1$$

$$0.3000 - 0.341655\,\alpha = \epsilon_2$$

$$0.4000 + 0.352145\,\alpha = \epsilon_3$$

$$0.2000 - 0.006981\,\alpha = \epsilon_4$$

and examine the quantities

$$\alpha = \frac{0.1000}{0.003509},\quad \frac{0.3000}{0.341655},\quad \frac{0.2000}{0.006981}$$

If the first or third is chosen for α, then ϵ_2 will be negative. The only way that one component can be chosen to be zero with the others positive is to choose

$$\alpha = \frac{0.3000}{0.341655}$$

Therefore

$$\mathbf{x}_2 = \alpha\boldsymbol{\Delta} = \begin{bmatrix} -0.003081 \\ -0.300000 \\ 0.309211 \\ -0.006130 \end{bmatrix}$$

and

$$\mathbf{A}_{\mathbf{x}_2}^{(i)} = \begin{bmatrix} 0.0969 \\ 0.0000 \\ 0.7092 \\ 0.1939 \end{bmatrix}$$

This is now used as an initial composition for a new experimental run, an extrapolation back to the boundary performed as before, and it turns out that one obtains

$$\mathbf{A}_{\mathbf{x}_2}^{(i)} = \begin{bmatrix} 0.0888 \\ 0.0000 \\ 0.7076 \\ 0.2036 \end{bmatrix}$$

Since this vector may contain some of the B_1 component, one computes $\mathbf{x}_2$ from

$$\mathbf{x}_2 = \mathbf{A}_{\mathbf{x}_2}^{(i)} - \mathbf{x}_0$$

and then removes any trace of $\mathbf{x}_1$ by the preceding procedure of transforming to the orthogonal set of unit vectors. If this is done, then after transforming back, one obtains

$$\mathbf{x}_2 = \begin{bmatrix} -0.011748 \\ -0.300000 \\ 0.309178 \\ 0.002571 \end{bmatrix}; \qquad \mathbf{z}_2 = \begin{bmatrix} -0.050537 \\ -0.745087 \\ 0.665004 \\ 0.007820 \end{bmatrix}$$

We must now find $\mathbf{x}_3$. We choose as

$$\boldsymbol{\gamma} = \begin{bmatrix} 0.447214 \\ 0.000000 \\ 0.000000 \\ -0.316228 \end{bmatrix}$$

and let
$$\boldsymbol{\delta} = \boldsymbol{\gamma} - a_1\mathbf{z}_0 - a_2\mathbf{z}_1 - a_3\mathbf{z}_2$$

where $\boldsymbol{\delta}$ is orthogonal to $\mathbf{z}_0$, $\mathbf{z}_1$, and $\mathbf{z}_2$;

then
$$a_1 = 0$$
$$a_2 = \mathbf{z}_1^T\boldsymbol{\gamma} = 0.0011748$$
$$a_3 = \mathbf{z}_2^T\boldsymbol{\gamma} = -0.025074$$

so

$$\boldsymbol{\delta} = \begin{bmatrix} 0.445378 \\ -0.018237 \\ 0.017239 \\ -0.316833 \end{bmatrix}$$

and

$$\boldsymbol{\Delta} = \mathbf{D}^{1/2}\boldsymbol{\delta} = \begin{bmatrix} 0.140841 \\ -0.009989 \\ 0.010840 \\ -0.141692 \end{bmatrix}$$

As before we must combine this vector with $\mathbf{x}_0$ so that the sum has one zero component and the others positive. A similar analysis will give

$$\mathbf{x}_3 = \begin{bmatrix} 0.198799 \\ -0.014099 \\ 0.015300 \\ -0.20000 \end{bmatrix}, \qquad \mathbf{A}_{\mathbf{x}_3}^{(i)} = \begin{bmatrix} 0.2988 \\ 0.2859 \\ 0.4153 \\ 0.0000 \end{bmatrix}$$

Then

$$\mathbf{X} = \begin{bmatrix} 0.100000 & 0.221400 & -0.011748 & 0.198799 \\ 0.300000 & -0.300000 & -0.300000 & -0.014099 \\ 0.400000 & -0.361800 & 0.309178 & 0.015300 \\ 0.200000 & 0.440400 & 0.002571 & -0.200000 \end{bmatrix}$$

$$\mathbf{D}^{-1} = \begin{bmatrix} 10.000000 & 0 & 0 & 0 \\ 0 & 3.333333 & 0 & 0 \\ 0 & 0 & 2.500000 & 0 \\ 0 & 0 & 0 & 5.000000 \end{bmatrix}$$

$$\mathbf{L} = \begin{bmatrix} 1.000000 & 0.000000 & 0.000000 & 0.000000 \\ 0.000000 & 2.087188 & 0.000000 & 0.000000 \\ 0.000000 & 0.000000 & 0.540390 & 0.000000 \\ 0.000000 & 0.000000 & 0.000000 & 0.596457 \end{bmatrix}$$

$$\mathbf{L}^{-1} = \begin{bmatrix} 1.000000 & 0 & 0 & 0 \\ 0 & 0.479113 & 0 & 0 \\ 0 & 0 & 1.850512 & 0 \\ 0 & 0 & 0 & 1.676563 \end{bmatrix}$$

From

$$\mathbf{X}^{-1} = L^{-1}\mathbf{X}^T\mathbf{D}^{-1}$$

$$\mathbf{X}^{-1} = \begin{bmatrix} 1.000000 & 1.000000 & 1.000000 & 1.000000 \\ 1.060757 & -0.479114 & -0.433357 & 1.055007 \\ -0.217397 & -1.850513 & 1.430344 & 0.023788 \\ 3.332992 & -0.078793 & -0.064133 & -1.676563 \end{bmatrix}$$

Now from a given initial composition $\mathbf{A}^{(i)}$ an experimental run is performed and the time is noted at various compositions. These compositions can then be converted by

$$\mathbf{B}^{(i)} = \mathbf{X}^{-1}\mathbf{A}^{(i)}$$

Suppose that B_1 and B_3 are plotted against B_2 to obtain λ_1/λ_2 and λ_3/λ_2. These are

$$\frac{\lambda_1}{\lambda_2} = 0.151, \qquad \frac{\lambda_3}{\lambda_2} = 2.24$$

then

$$\Lambda' = \begin{bmatrix} 0 & 0 & 0 & 0 \\ 0 & -0.151 & 0 & 0 \\ 0 & 0 & -1 & 0 \\ 0 & 0 & 0 & -2.24 \end{bmatrix}$$

and $\mathbf{K}' = \mathbf{X}\Lambda'\mathbf{X}^{-1}$

$$\mathbf{K}' = \begin{bmatrix} -1.52223 & 0.02936 & 0.00274 & 0.71160 \\ 0.08809 & -0.57935 & 0.41150 & 0.00198 \\ 0.01093 & 0.54866 & -0.46811 & 0.10774 \\ 1.42320 & 0.00132 & 0.05387 & -0.82132 \end{bmatrix}$$

If from the time determination in the experiment one also determines λ_2 itself as

$$\lambda_2 = 36.02$$

then

$$\mathbf{K} = \mathbf{K}'\lambda_2$$

or

$$\mathbf{K} = \begin{bmatrix} -54.82 & 1.06 & 0.10 & 25.63 \\ 3.17 & -20.87 & 14.82 & 0.07 \\ 0.39 & 19.77 & -16.86 & 3.88 \\ 51.26 & 0.04 & 1.94 & -29.58 \end{bmatrix}$$

Thus all the velocity constants are determined and in fact those thought to be zero in the reaction scheme, k_{31}, k_{13}, k_{42}, and k_{24} are indeed small.

8.9 Time Contours in the Reaction Simplex and Their Properties

Although use of the time has been minimized in this work since the purpose of the foregoing analysis was, in many respects, to give as little weight to time measurements as possible, each reaction path in the simplex has time as a parameter along its length from the initial condition to the equilibrium point and in fact the time parameter runs in principle through the whole semi-infinite interval $(0,\infty)$. The vector $\mathbf{B}$ may be written

$$\mathbf{B} = e^{\Lambda t}\mathbf{B}^{(i)}$$

where

$$\mathbf{B}^{(i)} = \mathbf{X}^{-1}\mathbf{A}^{(i)}$$

so

$$\mathbf{B} = e^{\Lambda t}\mathbf{X}^{-1}\mathbf{A}^{(i)}$$

and

$$\mathbf{A}(t) = \mathbf{X}e^{\Lambda t}\mathbf{X}^{-1}\mathbf{A}^{(i)} = \mathbf{T}(t)\mathbf{A}^{(i)} \tag{8.9.1}$$

Then,

$$\mathbf{A}(t) = \mathbf{T}(t)\mathbf{A}^{(i)}$$

$$\mathbf{A}(\tau) = \mathbf{T}(\tau)\mathbf{A}^{(i)}$$

and

$$\begin{aligned}\mathbf{T}(\tau)\mathbf{A}(t) &= \mathbf{T}(\tau)\mathbf{T}(t)\mathbf{A}^{(i)}\\ &= \mathbf{X}e^{\Lambda\tau}\mathbf{X}^{-1}\mathbf{X}e^{\Lambda(t)}\mathbf{X}^{-1}\mathbf{A}^{(t)}\\ &= \mathbf{X}e^{\Lambda(t+\tau)}\mathbf{X}^{-1}\mathbf{A}^{(i)}\\ &= \mathbf{T}(t+\tau)\mathbf{A}^{(i)} = \mathbf{A}(t+\tau)\end{aligned} \tag{8.9.2}$$

Equation (8.9.1) transforms a particular composition $A^{(i)}$ into a composition at time t, and if one thinks of $\mathbf{A}^{(i)}$ as being a set of initial compositions or a curve in the simplex at $t = 0$, then Eq. (8.9.1) is the transformation which produces the curve of compositions at time t, or Eq. (8.9.1) represents a constant time contour as a function of the initial composition.

From Eq. (8.9.2), it follows that

$$\mathbf{A}(\Delta t) = \mathbf{T}(\Delta t)\mathbf{A}^{(i)}$$

$$\mathbf{A}(2\,\Delta t) = \mathbf{T}^2(\Delta t)\mathbf{A}^{(i)}$$

or

$$\mathbf{A}[(n+1)\Delta t] = \mathbf{T}(\Delta t)\mathbf{A}(n\,\Delta t)$$

and this equation may be used to make computations on families of reaction paths in a stepwise fashion.

Given any two initial compositions $\mathbf{A}_1^{(i)}$ and $\mathbf{A}_2^{(i)}$ then

$$\mathbf{A}^{(i)} = (1-r)\mathbf{A}_1^{(i)} + r\mathbf{A}_2^{(i)}$$

is a composition between $\mathbf{A}_1^{(i)}$ and $\mathbf{A}_2^{(i)}$ and in fact is a weighted composition with the weighting factors $(1 - r)$ and r. If $r = 0$, $\mathbf{A}^{(i)} = \mathbf{A}_1^{(i)}$ and if $r = 1$, $\mathbf{A}^{(i)} = \mathbf{A}_2^{(i)}$. Then,

$$\begin{aligned}\mathbf{T}(t)\mathbf{A}^{(i)} &= (1-r)\mathbf{T}(t)\mathbf{A}_1^{(i)} + r\mathbf{T}(t)\mathbf{A}_2^{(i)} = \mathbf{A}\\ &= (1-r)\mathbf{A}_1 + r\mathbf{A}_2 = \mathbf{A}\end{aligned}$$

and hence if a curve of initial compositions is a straight line it remains a straight line at subsequent times. The straight lines will not be parallel, but their orientation will change and the relative distances on the straight line remain the same. If one thinks of a triangular simplex, the boundaries are straight lines of initial composition. As the reaction proceeds through times $\Delta t, 2\,\Delta t, 3\,\Delta t, \ldots, n\,\Delta t, \ldots$ a sequence of triangles must be formed each one of which is inside the previous one until finally the triangle must shrink on the equilibrium point. Each triangle will be a constant time contour.

8.10 Remarks on Applicability of First-order Systems

The preceding rather complex analysis was developed not for simple homogeneous chemical reactions but rather for the elucidation of the mechanism of complex solid catalyzed reactions. Now in the latter case it is clear from a superficial knowledge of the field that the equations written to describe the kinetic behavior of the system are not the equations that a catalytic chemist would write. Wei and Prater show that under rather broad conditions the kinetic equation for A_i may be written

$$\frac{dA_i}{dt} = \phi\left[k_{i1}A_1 + k_{i2}A_2 \ldots - \sum_{j=1,\,i}^{n} k_{ji}A_i + \cdots + k_{in}A_n\right]$$

where the function ϕ is the same for all i, and is dependent upon compositions. If one defines

$$\phi\, dt = d\tau$$

then
$$\frac{dA_i}{d\tau} = k_{i1}A_1 + k_{i2}A_2 + \cdots - \sum_{j=1,\,i}^{n} k_{ji}A_i + \cdots + k_{in}A_n$$

τ is a pseudo time variable and its relation to real time would demand the integration of the set of full nonlinear differential equations where the form of ϕ would have to be determined in order to agree with experimental information. For any but the most simple systems, the state of the art has not progressed far enough for this to be a feasible task. Moreover, the analysis in this chapter has not been dependent on knowing the time in detail and the matrix $\mathbf{K}'$, the relative rate matrix, could be computed from composition data alone. From a design point of view, the time variable must be known to size the reactor, but the distribution of products from a given initial composition can be obtained without it.

8.11 A Simple Irreversible Consecutive Reaction System

We have considered only cases in which all reactions were reversible and each species would be coupled by reaction with each of the other species. If there is no coupling between some species as in the preceding example, then presumably the corresponding reaction velocity constants should be very small or zero. If, in fact, some of the reactions' velocity constants are small, then the schematic reaction mechanism may be idealized in another direction. For example, the triangular system

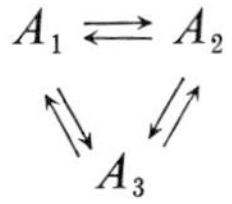

is certainly a generalization of

$$A_1 \xrightarrow{k_1} A_2 \xrightarrow{k_2} A_3$$

in which certain velocity constants in the triangular system are assumed to be zero. The kinetic equations are

$$\frac{dA_1}{d\theta} = -k_1 A_1$$

$$\frac{dA_2}{d\theta} = k_1 A_1 - k_2 A_2$$

$$\frac{dA_3}{d\theta} = k_2 A_2$$

and

$$\frac{d\mathbf{A}}{d\theta} = \mathbf{KA}$$

$$\mathbf{K} = \begin{bmatrix} -k_1 & 0 & 0 \\ k_1 & -k_2 & 0 \\ 0 & k_2 & 0 \end{bmatrix}$$

Thus if

$$\det(\mathbf{K} + \lambda \mathbf{I}) = (-k_1 + \lambda)(-k_2 + \lambda)(\lambda) = 0$$

then the eigenvalues are

$$\lambda_0 = 0, \quad \lambda_1 = k_1, \quad \lambda_2 = k_2$$

Thus

$$\mathbf{x}_0 = \begin{bmatrix} 0 \\ 0 \\ 1 \end{bmatrix}; \qquad \mathbf{x}_1 = \begin{bmatrix} 1 - \dfrac{k_1}{k_2} \\ \dfrac{k_1}{k_2} \\ -1 \end{bmatrix}; \qquad \mathbf{x}_2 \begin{bmatrix} 0 \\ 1 \\ -1 \end{bmatrix}$$

Now

$$\mathbf{A}^{(i)}_{\mathbf{x}_1} = \mathbf{x}_0 + \mathbf{x}_1 = \begin{bmatrix} 1 - \dfrac{k_1}{k_2} \\ \dfrac{k_1}{k_2} \\ 0 \end{bmatrix}$$

$$\mathbf{A}^{(i)}_{\mathbf{x}_2} = \mathbf{x}_0 + \mathbf{x}_2 = \begin{bmatrix} 0 \\ 1 \\ 0 \end{bmatrix}$$

Now we should consider two cases, $k_1/k_2 < 1$, $k_1/k_2 > 1$. In the first case, the straight-line reaction paths are as shown in Fig. 8.10.1, one is definitely inside the simplex, whereas the other is on the boundary, and reaction paths all lead to the vertex A_3. In the second case, the first component of $\mathbf{A}^{(i)}_{\mathbf{x}_1}$ is negative, as shown in Fig. 8.10.2, one of the paths being *outside* the triangle and one on the boundary. In each case, the path on the boundary is physically obvious, since, if one starts with pure A_2, the path must be directly from A_2 to A_3 on the boundary. If $k_1/k_2 < 1$ there is a feasible initial composition which will give a straight-line reaction path. In fact, it is a mixture

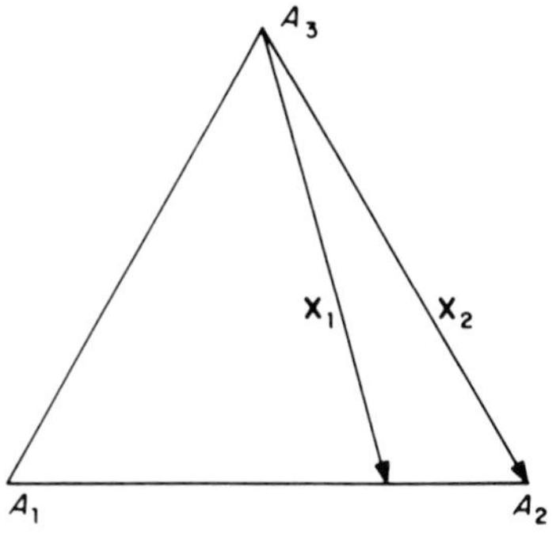

Figure 8.10.1

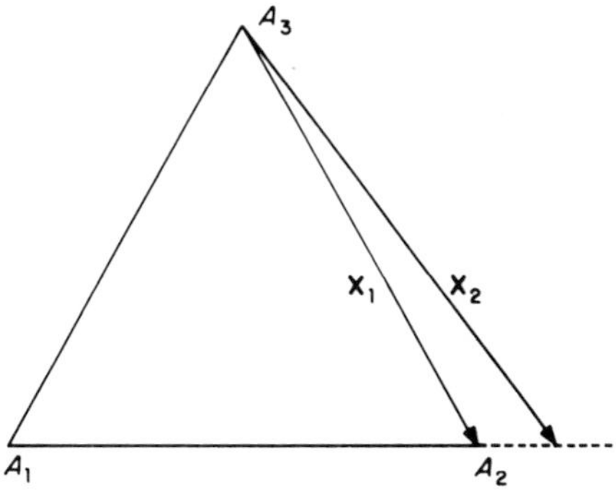

Figure 8.10.2

Schematic diagram of simplex for the chemical reaction system $A_1 \to A_2 \to A_3$.

with a mole fraction of A_2 of k_1/k_2 and a mole fraction of A_1 of $(1 - k_1/k_2)$. On the other hand, if $k_1/k_2 > 1$, there is no laboratory initial composition which will give a straight-line path, since negative amounts of materials are usually not available.

We shall not pursue this subject further, but readers should consult Wei and Prater.

B. Distillation Column Dynamics

8.12 Formulation of the Equations

In Chapter 6, we considered an idealized problem in steady state distillation as well as a somewhat more realistic problem for multicomponent distillation in which the equations describing the system were cast in matrix-vector form. It turns out that the latter approach is the one which may be generalized in the formulation of the equations for the transient behavior of such a column. These equations, in general, will be nonlinear first-order differential equations whose solution may be obtained by straightforward numerical methods; for our purposes, however, the methods recently published by Mah, Michaelson, and Sargent* and by Sargent† are more typical of the approaches suggested by this book so that we shall describe these methods alone. Their approach employs the fact that the matrices involved are always of Jacobi type and involves successive linearizations of the equations over short time steps followed by analytical solutions of the linearized equations. Work on transient behavior of such large systems is still in an early stage of development and it is not clear what techniques will be the most efficient in computer time.

In addition to the notation given in Section 6.7, some new notation must be defined partially because of the transient nature of the problem and partially because this treatment will be somewhat more general.

w_j = molar holdup of liquid in the jth stage

W_j = molar holdup of vapor in the jth stage

P_j = vapor product withdrawal from jth stage

t = time

We shall assume that each stage is a theoretical equilibrium stage with vapor and liquid in equilibrium and with the composition of the effluent streams from each stage the same as the holdup compositions. The latter assumption simplifies the writing of the equations appreciably. It is not difficult then to write the mass balance on each stage, overall and for each component, as

$$\frac{d}{dt}(w_j + W_j) = L_{j-1} + V_{j+1} - L_j - V_j - P_j - S_j + F_j \quad (8.12.1)$$

$$\frac{d}{dt}(x_{ji}w_j + y_{ji}W_j) = L_{j-1}x_{j-1,i} + V_{j+1}y_{j+1,i} - L_jx_{ji} - V_jy_{ji} - P_jy_{ji} - S_jx_{ji} + F_jx_{F_{ji}} \quad (8.12.2)$$

If we call the reboiler stage the Nth stage and the condenser the zeroth stage, then Eq. (8.12.1) and (8.12.2) will be correct if we choose

$$\begin{aligned} V_{N+1} &= 0 \qquad & y_{N+1,i} &= 0 = x_{N+1,i} \\ S_N &= 0 & x_{-1,i} &= 0 \\ L_{-1} &= 0 \\ P_0 &= 0 \end{aligned}$$

Note here that the condenser may be a partial condenser or a total condenser, where the vapor drawoff from the partial condenser is V_0 with $P_0 = 0$. For a total condenser $P_0 = V_0 = 0$, S_0 is the liquid drawoff, and L_0 is the liquid reflux in both cases.

It is convenient to define terms

Chem. Eng. Sci., **17** (1962), 619–39.

†*Trans. Inst. Chem. Eng.*, **41** (1963), 51–60. Presented with permission of the publisher and author.

$$\phi_{ji} = w_j x_{ji} + W_j y_{ji} = w_j x_{ji} + W_j K_{ji} x_{ji}$$
$$= w_j x_{ji} \xi_{ji}$$

where ϕ_{ji} is the partial holdup for each component.

$$\xi_{ji} = 1 + \frac{W_j}{w_j} K_{ji} \tag{8.12.3}$$

and where K_{ji} is the equilibrium constant, a function of temperature, pressure, and composition. With the preceding notation, Eq. (8.12.2) may be written in the form

$$\frac{d\boldsymbol{\phi}_i}{dt} = \mathbf{G}_i \boldsymbol{\phi}_i + \mathbf{f}_i \tag{8.12.4}$$

where $\mathbf{G}_i$ is a square matrix of order $N + 1$ with elements g_{kl} and where the first row and first column is called the *zeroth*,

$$g_{jj} = -[L_j + (V_j + P_j)K_{ji} + S_j] \frac{1}{w_j \xi_{ji}}, \quad 0 \le j \le N$$

$$g_{j,j-1} = L_{j-1} \frac{1}{w_{j-1} \xi_{j-1,i}}, \quad 1 \le j \le N$$

$$g_{j,j+1} = V_{j+1} K_{j+1} \frac{1}{w_{j+1} \xi_{j+1,i}}, \quad 0 \le j \le N - 1$$

and

$$\boldsymbol{\phi}_i = \begin{bmatrix} \phi_{0i} \\ \phi_{1i} \\ \cdot \\ \cdot \\ \cdot \\ \phi_{Ni} \end{bmatrix}; \qquad \mathbf{f}_i = \begin{bmatrix} F_0 x_{F0i} \\ F_2 x_{F1i} \\ F_2 x_{F2i} \\ \cdot \\ \cdot \\ \cdot \\ F_N x_{FNi} \end{bmatrix}$$

Thus there is a matrix equation like Eq. (8.12.4) for each of the species in the system. Obviously, there is one more equation here than necessary since $\sum_i x_{ji} = 1$, $\sum_i y_{ji} = 1$.

A heat balance over each stage can also be written, and this is

$$\frac{d}{dt}(H_{jc} + h_j w_j + H_j W_j) = -(L_j + S_j)h_j - (P_j + V_j)H_j$$
$$+ L_{j-1} h_{j-1} + V_{j+1} H_{j+1} + Q_j + F_j H_{Fj} \tag{8.12.4}$$

where H_{jc} is the heat capacity per stage and includes the metal and lagging associated with each stage and Q_j is any heat absorbed or rejected by heat sinks or sources or heat leaks.

Now the relations $\sum_i x_{ji} = 1$ and $\sum_i K_{ji} x_{ji} = 1$ may be written in the form

$$\sum_i x_{ji} = \sum_i \frac{\phi_{ji}}{w_j \xi_{ji}} = 1$$

$$\sum_i K_{ji}x_{ji} = \sum \frac{K_{ji}\phi_{ji}}{w_j\xi_{ji}} = 1$$

or

$$\sum_i \frac{\phi_{ji}}{\xi_{ji}} = w_j$$

$$\sum_i \frac{K_{ji}\phi_{ji}}{\xi_{ji}} = w_j$$

and Sargent claims that an iteration procedure for determining the temperature on each stage obtains from the formula

$$K_{jr}(T_j^{(n+1)}) = K_{jr}(T_j^{(n)})\left[\frac{\sum_i (\phi_{ji}/\xi_{ji})}{\sum_i (K_{ji}\phi_{ji}/\xi_{ji})}\right]_{T_j=T_j^{(n)}} \tag{8.12.5}$$

where K_{jr} is a particular reference quantity and where the large brace on the right has all quantities evaluated at $T_j = T_j^{(n)}$ and $T_j^{(n)}$ stands for the nth iteration in the use of this formula. Clearly, if the process does converge, then one has shown that

$$\sum_i x_{ji} = \sum_i K_{ji}x_{ji}$$

on each stage and for the *steady state* it may be shown that this condition implies that each summation is one. Note that x_{ji} and y_{ji} may be written

$$x_{ji} = \frac{\phi_{ji}}{\xi_{ji}\sum_i (\phi_{ji}/\xi_{ji})}$$
$$y_{ji} = \frac{K_{ji}\phi_{ji}}{\xi_{ji}\sum_i K_{ji}(\phi_{ji}/\xi_{ji})} \tag{8.12.6}$$

The formulation of the problem still requires one further bit of information which is not quite so easy to specify. We have used, as the holdup in the foregoing equations, molar holdups because these are the most convenient when one uses concentrations in mole fractions. The value of the holdup itself is, however, conveniently measured in volumetric terms and it is obvious that the liquid flow rates $\{L_j\}$ are determined by the liquid volumetric holdup and either the height over a wier or the height above slots or some other hydraulic variable. In short, it is a problem of hydraulics to determine the relationship between h_j and L_j.

8.13 On the Character of the Eigenvalues of $\mathbf{G}_i$

Suppose for the moment that the elements of $\mathbf{G}_i$ are known. Now we showed in Section 5.17 that the eigenvalues of such a matrix had to be real. We showed that if the diagonal terms were negative and the elements on the two adjacent diagonals were positive the eigenvalues were real and this is the case for the elements of $\mathbf{G}_i$. We shall now show that the eigenvalues of $\mathbf{G}_i$ are nonpositive. In order to do this, we employ some results already obtained. In Section 5.17 we discussed Sturmian sequences of polynomials and showed that for a Jacobi matrix with negative elements on the main diagonal and positive elements on the two adjacent diagonals the eigenvalues were all real. We also showed that the sequence of polynomials for large positive λ alternated in sign. Now suppose we consider instead of the matrix $\mathbf{G}_i$ the matrix $-\mathbf{G}_i$, that is, the matrix whose elements are the negatives of the elements of the matrix of $\mathbf{G}_i$. Then the eigenvalues of $-\mathbf{G}_i$ are determined by the zeroes of

$$\det\,[-\mathbf{G}_i - \lambda\mathbf{I}] = 0$$

These eigenvalues are the same as the eigenvalues of

$$\det\,[\mathbf{G}_i + \lambda\mathbf{I}] = 0$$

which are the negatives of the eigenvalues of

$$\det\,[\mathbf{G}_i - \lambda\mathbf{I}] = 0$$

Hence if we can show that $-\mathbf{G}_i$ has all positive eigenvalues, then $\mathbf{G}_i$ must have all negative eigenvalues. We consider now the determinant of $(-\mathbf{G}_i - \lambda\mathbf{I})$ and call it $P_{N+1}(\lambda)$ and let $P_N(\lambda), \ldots, P_j(\lambda), \ldots, P_0(\lambda)$ be formed as in Section 5.17.

$$\begin{aligned}
P_{N+1}(\lambda) &= (-g_{NN} - \lambda)P_N(\lambda) - g_{N-1,N}g_{N,N-1}P_{N-1}(\lambda) \\
P_N(\lambda) &= (-g_{N-1,N-1} - \lambda)P_{N-1}(\lambda) - g_{N-2,N-1}g_{N-1,N-2}P_{N-2}(\lambda) \\
P_{j+1}(\lambda) &= (-g_{jj} - \lambda)P_j(\lambda) - g_{j-1,j}g_{j,j-1}P_{j-1}(\lambda) \\
&\;\;\vdots \\
P_1(\lambda) &= (-g_{00} - \lambda)P_0(\lambda) \\
P_0(\lambda) &= 1
\end{aligned}$$

Clearly, for large positive λ, the signs of $P_{j+1}(\lambda)$ will alternate as j increases from zero, since $P_{j+1}(\lambda)$ is a polynomial of degree $j + 1$ in λ and the coefficient of λ^{j+1} is alternately positive and negative as j increases from zero. Therefore there are $N + 1$ sign changes. Consider now the same sequence of polynomials evaluated at $\lambda = 0$. Then let $P_{j+1} = P_{j+1}(0)$

$$P_{j+1} = -g_{jj}P_j - g_{j-1,j}g_{j,j-1}P_{j-1}$$

$$P_{j+1} = [L_j + (V_j + P_j)K_{ji} + S_j]\frac{1}{w_j\xi_{ji}}P_j - \frac{L_{j-1}}{w_{j-1}\xi_{j-1,i}}\frac{V_jK_{ji}}{w_j\xi_{ji}}P_{j-1}$$

Now this may be written in the form

$$\left\{P_j - \left[\frac{V_j}{(V_j + P_j)}\right]\frac{L_{j-1}}{w_{j-1}\xi_{j-1,i}}P_{j-1}\right\}\frac{(V_j + P_j)K_{ji}}{w_{ji}\xi_{ji}} = P_{j+1} - \left[\frac{L_j + S_j}{L_j}\right]\frac{L_j}{w_j\xi_{ji}}P_j$$

Note that

$$\frac{V_j}{(V_j + P_j)} \leqslant 1; \qquad \frac{L_j + S_j}{L_j} \geqslant 1$$

and therefore if

$$P_j \geqslant \frac{L_{j-1}}{w_{j-1}\xi_{j-1,i}} P_{j-1}$$

then

$$P_{j+1} \geqslant \frac{L_j}{w_j \xi_{ji}} P_j$$

Hence if we can show that

$$P_1 \geqslant \frac{L_0}{w_0 \xi_{0i}} P_0$$

then all the P_j are positive. Since P_0 is positive, P_1 is positive, and by induction, all the P_j are positive. Now ($P_0 = 1$),

$$P_1 = [L_0 + (V_0 + P_0)K_{0i} + S_0]\frac{1}{w_0 \xi_{0i}} P_0$$

or

$$P_1 = \frac{L_0 P_0}{w_0 \xi_{0i}} + \frac{(V_0 + P_0)K_{0i} + S_0}{w_0 \xi_{0i}}$$

or

$$P_1 > \frac{L_0}{w_0 \xi_{0i}} P_0$$

Therefore since all the P_j are positive there are no sign changes in the Sturmian sequence and the net number of sign changes between 0 and very large λ is $N + 1$. Therefore $-\mathbf{G}_i$ has $N + 1$ positive eigenvalues or $\mathbf{G}_i$ has $N + 1$ negative eigenvalues. One special case is of interest theoretically but has little practical interest. If one examines the matrix $\mathbf{G}_i$ for the case of total reflux with no vapor or liquid products withdrawn, then the sum of the elements in any column is zero and therefore there is a zero eigenvalue. The remainder of the analysis remains except it follows easily that there are N negative eigenvalues.

8.14 The Mathematical Solution

Our problem is to find solutions of Eq. (8.12.3) recognizing that the matrices $\mathbf{G}_i$ are not constant matrices but contain elements which are functions of the compositions, temperatures, vapor and liquid flow rates, etc. Suppose we think of the elements of $\mathbf{G}_i$ as being functions of the time alone for the moment and write

$$\dot{\boldsymbol{\phi}} = \mathbf{G}(t)\boldsymbol{\phi} + \mathbf{f}(t)$$

The general solution of this system of equations is given according to Eq. (7.9.2)

$$\boldsymbol{\phi} = \boldsymbol{\Omega}_0^t(\mathbf{G})\boldsymbol{\phi}(0) + \boldsymbol{\Omega}_0^t(\mathbf{G})\int_0^t \boldsymbol{\Omega}_0^{t^{-1}}(\mathbf{G})\mathbf{f}(\tau)\,d\tau$$

For the sake of illustration, and following Sargent, let

$$\mathbf{G}(t) = \mathbf{A} + \mathbf{B}t + \mathbf{C}t^2$$

Using Eq. (7.9.1),

$$\boldsymbol{\Omega}_0^t(\mathbf{X} + \mathbf{Y}) = \boldsymbol{\Omega}_0^t(\mathbf{X})\boldsymbol{\Omega}_0^t(\mathbf{Z})$$

where $$\mathbf{Z} = \boldsymbol{\Omega}_0^{t^{-1}}(\mathbf{X})\mathbf{Y}\boldsymbol{\Omega}_0^t(\mathbf{X})$$

Let $\mathbf{X} = \mathbf{A}$, $\mathbf{Y} = \mathbf{B}t + \mathbf{C}t^2$, then

$$\mathbf{Z} = \exp(-\mathbf{A}t)(\mathbf{B}t + \mathbf{C}t^2)\exp(\mathbf{A}t)$$

Each of the exponentials has a series expansion so that $\mathbf{Z}(t)$ may be expressed as

$$\mathbf{Z} = \mathbf{B}t + [\mathbf{C} + \mathbf{BA} - \mathbf{AB}]t^2 + \text{higher-order terms}$$

Now we must find the matrizant of $\mathbf{Z}$, and for this it will be necessary to find terms of the form $\boldsymbol{\Omega}_0^t(\mathbf{B}t^n)$. As shown earlier, a matrizant is a solution of a differential equation and in particular this one is the solution of

$$\frac{dW}{dt} = \mathbf{B}t^n\mathbf{W}$$

which also has the solution

$$\mathbf{W} = e^{\mathbf{B}t^{n+1}/(n+1)}$$

since the differential equation may be written

$$\frac{d\mathbf{W}}{d(t^{n+1/(n+1)})} = \mathbf{BW}$$

so $$\boldsymbol{\Omega}_0^t(Bt^n) = \exp(\mathbf{B}t^{n+1/(n+1)})$$

Then

$$\boldsymbol{\Omega}_0^t(\mathbf{Z}) = \exp(\tfrac{1}{2}\mathbf{B}t^2)\boldsymbol{\Omega}_0^t[\exp(-\tfrac{1}{2}\mathbf{B}t^2)\cdot(\mathbf{C} + \mathbf{BA} - \mathbf{AB})t^2\cdot\exp(\tfrac{1}{2}\mathbf{B}t^2)]$$
$$+ \text{higher-order terms}$$
$$= \mathbf{I} + \tfrac{1}{2}\mathbf{B}t^2 + \tfrac{1}{3}(\mathbf{C} + \mathbf{BA} - \mathbf{AB})t^3 + \text{higher-order terms}$$

and finally

$$\boldsymbol{\Omega}_0^t(\mathbf{G}) = [\exp\mathbf{A}t][\mathbf{I} + \tfrac{1}{2}\mathbf{B}t^2 + \tfrac{1}{3}(\mathbf{C} + \mathbf{BA} - \mathbf{AB})t^3]$$

If one supposes that the feed vector is of the form

$$\mathbf{f}(t) = \boldsymbol{\alpha} + \boldsymbol{\beta}t + \boldsymbol{\gamma}t^2$$

then the solution $\phi(t)$ may be shown to be of the form

$$\boldsymbol{\phi}(t) = \boldsymbol{\Omega}_0^t(\mathbf{G})\cdot\boldsymbol{\phi}(0) + [\exp(\mathbf{A}t) - \mathbf{I}]\mathbf{A}^{-1}\mathbf{f}(t) - \tfrac{1}{2}\boldsymbol{\beta}t^2$$
$$+ \tfrac{1}{3}\boldsymbol{\beta\alpha} - \mathbf{AB} - 2\boldsymbol{\gamma})t^3 + \text{higher-order terms}$$

8.15 The Numerical Procedure

In order to simplify the discussion, we shall follow Sargent and assume that

$$G_i(t) = \mathbf{A}_i + \mathbf{B}_i t$$

$$\mathbf{f}_i(t) = \boldsymbol{\alpha}_i + \boldsymbol{\beta}_i t$$

The solution for this case is

$$\boldsymbol{\phi}_i(t) = [\exp \mathbf{A}_i t]\,\{[\mathbf{I} + \tfrac{1}{2}\mathbf{B}_i t^2]\boldsymbol{\phi}_i(0) + \mathbf{A}_i^{-1}\mathbf{f}_i(t)\} - \{\mathbf{A}_i^{-1}\mathbf{f}_i(t) + \tfrac{1}{2}\boldsymbol{\beta}_i t^2\}$$

The general procedure is an iterative one. We shall assume that we have chosen a computation interval Δt and that we know the condition of the column at time $t = 0$. The steps in the computation can then be summarized as follows:

1. Compute $\mathbf{G}_i(0) = \mathbf{A}_i$ for all components.
2. Calculate the eigenvalues and eigenvectors of $\mathbf{A}_i$ as described in the next section.
3. Use the solution of the next section, that is, one for which $\mathbf{G}_i(0) = \mathbf{G}_i$, to obtain $\boldsymbol{\phi}_i$ after time Δt.
4. Determine the boiling points at the end of time Δt for all stages using Eq. (8.12.5).
5. Calculate the liquid and vapor compositions from Eq. (8.12.6).
6. Calculate liquid and vapor enthalpies and the heat capacity per stage of metal and lagging.
7. Obtain L_j, V_j, and the holdups at the end of time Δt from Eq. (8.12.1) and (8.12.4) and the tray hydraulics.
8. With the information now available recalculate $\mathbf{G}_i(\Delta t)$ and estimate $\mathbf{B}_i$. This may be done, since $\mathbf{G}_i(0)$ is known.
9. Compute a new $\boldsymbol{\phi}_i(\Delta t)$ from Eq. (8.12.7).
10. Iterate on this time step to improve the solution. When the preassigned accuracy has been attained pass to the next time interval.

The method suggested for Eq. (8.12.1) and (8.12.4) is replacing the derivatives by the simple difference quotient over the time interval and replacing the quantities on the right-hand side by their arithmetic average over the time interval. If one assumes that the holdup w_j is a function of $L_j + S_j$ and $V_j + P_j$, then if

$$L_j + S_j = u_j(t)$$

$$V_j + P_j = v_j(t)$$

with

$$w_j = w_j(u_j, v_j) = w_j(t)$$

then over a small interval t

$$w_j(t) = w_j(0) + \left(\frac{\partial w_j}{\partial u_j}\right)_0 [u_j(t) - u_j(0)] + \left(\frac{\partial w_j}{\partial v_j}\right)_0 [v_j(t) - v_j(0)]$$

and the holdup may also be estimated in an iterative manner.

8.16 The Computation of Eigenvectors and Eigenvalues

At each stage of the computation previously described, one must compute the eigenvalues and eigenvectors of $\mathbf{A}_i$ and solve a differential equation of the form

$$\frac{d\mathbf{x}}{dt} = \mathbf{Ax} + \boldsymbol{\phi} \tag{8.13.1}$$

where $\mathbf{A}$ is a constant matrix and $\boldsymbol{\phi}$ is a constant vector. If

$$\mathbf{x} = \mathbf{Cy}$$

then, if one considers the equation,

$$\mathbf{C}\frac{d\mathbf{y}}{dt} = \mathbf{AC}y$$

or

$$\frac{d\mathbf{y}}{dt} = \mathbf{C}^{-1}\mathbf{AC}y$$

and if $\mathbf{C}$ is the matrix made up of the eigenvectors of $\mathbf{A}$

$$\boldsymbol{\Lambda} = \mathbf{C}^{-1}\mathbf{AC}$$

and

$$\frac{d\mathbf{y}}{dt} = \boldsymbol{\Lambda}\mathbf{y}$$

or

$$\mathbf{y} = e^{\Lambda t}\mathbf{y}_i$$

where $\mathbf{y}_i$ is the vector of initial values of $\mathbf{y}$ and related to $\mathbf{x}_i$ by

$$\mathbf{x}_i = \mathbf{Cy}_i$$

Therefore,

$$\mathbf{x} = \mathbf{C}e^{\Lambda t}\mathbf{C}^{-1}\mathbf{x}_i$$

In order to find the solution of Eq. (8.13.1), we assume a solution of the form

$$\mathbf{x} = \mathbf{C}e^{\Lambda t}\mathbf{C}^{-1}\mathbf{z} + \mathbf{K}$$

where $\mathbf{K}$ is a vector of constants to be determined so that it is a particular solution. There results

$$\mathbf{K} = -\mathbf{A}^{-1}\boldsymbol{\phi},$$

hence

$$\mathbf{x} = \mathbf{C}e^{\Lambda t}\mathbf{C}^{-1}[\mathbf{x}_i + \mathbf{A}^{-1}\boldsymbol{\phi}] - \mathbf{A}^{-1}\boldsymbol{\phi}$$

is the solution of Eq. (8.13.1), satisfying the condition

$$\mathbf{x} = \mathbf{x}_i, \quad t = 0$$

Now in order to solve the problem posed in this section it is necessary to compute the eigenvalues and the eigenvectors of **A**.

The computation of the eigenvalues for a Jacobi matrix may be obtained without too much difficulty as suggested in Section 5.16. Calculating the eigenvectors is another matter, however, and, as Wilkinson* shows, some care must be exercised in their computation. Wilkinson suggests the following scheme. Suppose λ'_j is an approximate eigenvalue, then to obtain $\mathbf{x}'_j$ corresponding, assume *any* vector **b** and compute the solution $\mathbf{x}'_j$ by

$$(\mathbf{A} - \lambda'_j\mathbf{I})\mathbf{x}'_j = \mathbf{b}$$

where the solution is obtained from

$$\mathbf{x}'_j = (\mathbf{A} - \lambda'_j\mathbf{I})^{-1}\mathbf{b}$$

This is solved by successive substitutions with

$$\mathbf{x}''_j = (\mathbf{A} - \lambda'_j\mathbf{I})^{-1}\mathbf{x}'_j$$

and with the obvious continuing procedure, computing $\mathbf{x}'''_j$, $\mathbf{x}''''_j$, etc. This solution will produce $\mathbf{x}'_j$ with large components which can be scaled down by division since the vectors are determined only up to an arbitrary multiplicative constant. Wilkinson suggests that the system $(\mathbf{A} - \lambda'_j\mathbf{I})\,\mathbf{x}'_j$ be reduced to upper triangular form by pivotal condensation, producing zeroes in all elements below the main diagonal and then using the vector for the right-hand side with each element equal to one.

It is necessary to compute $\mathbf{X}^{-1}$ and two methods are suggested by Mah, Michaelson, and Sargent. Since the eigenvectors of **A** must be obtained one can also obtain the eigenrows **Y**. Then

$$YX = \Gamma$$

where Γ is a diagonal matrix and therefore

$$\mathbf{X}^{-1} = \Gamma^{-1}\mathbf{Y}$$

where the inverse of the diagonal matrix is trivial to find. As a second method consider the diagonal matrix **W** with elements $\xi_1, \xi_2, \ldots, \xi_n$ where

$$\xi_1 = 1$$

$$\xi_j = \sqrt{\frac{a_{j-1,j}}{a_{j,j-1}}}\,\xi_{j-1}, \quad j = 2, 3, \ldots, n$$

and a_{kj} are the elements of a tridiagonal matrix **A**. Then

$$\mathbf{WAW}^{-1}$$

is a symmetric matrix. Now let **X** be the matrix of eigenvectors of **A** and Λ the diagonal matrix of eigenvalues of **A**, then

$$\mathbf{AX} = \mathbf{X}\Lambda$$

*J. H. Wilkinson, *Computer J.*, **1** (1958), 90.

Multiply by $\mathbf{W}$ on the left

$$\mathbf{WAX} = \mathbf{WX\Lambda}$$

then

$$\mathbf{WA} = \mathbf{WX\Lambda X}^{-1}$$

and

$$\mathbf{WAW}^{-1} = \mathbf{WX\Lambda X}^{-1}\mathbf{W}^{-1}$$

or

$$(\mathbf{WAW}^{-1})(\mathbf{WX}) = \mathbf{WX\Lambda}$$

and therefore $\mathbf{WX}$ is the matrix of eigenvectors of a symmetric matrix $\mathbf{WAW}^{-1}$. Hence,

$$(\mathbf{WX})^T(\mathbf{WX}) = \mathbf{H}$$

where $\mathbf{H}$ is a diagonal matrix. Therefore

$$\mathbf{H} = \mathbf{X}^T\mathbf{W}^2\mathbf{X} \quad \text{or} \quad \mathbf{X}^{-1} = \mathbf{H}^{-1}\mathbf{X}^T\mathbf{W}^2$$

and this gives the inverse matrix of $\mathbf{X}$ easily.

References

(Those marked with an asterisk are particularly useful.)

***1.** Aitken, A. C., *Determinants and Matrices*. Edinburgh: Oliver and Boyd, Ltd., 1959.

2. Allen, D. N. de G., *Relaxation Methods*. New York: McGraw-Hill Book Company, 1954.

3. Allen, R. G. D., *Mathematical Economics*. London: Macmillan & Co., Ltd., 1957.

4. Beckenbach, E. F., *Modern Mathematics for the Engineer*. New York: McGraw-Hill Book Company, 1959.

5. Bellman, R., *Introduction to Matrix Analysis*, New York: McGraw-Hill Book Company, 1960.

6. Bôcher, M., *Introduction to Higher Algebra*. New York: The Macmillan Company, 1938.

7. Bodewig, E., *Matrix Calculus*. Amsterdam: North Holland Publishing Co., 1959.

8. Booth, A. D., *Numerical Methods*. London: Butterworth Scientific Publications, 1955.

*9. Courant, R., and D. Hilbert, *Methods of Mathematical Physics*, Vol. I. New York: Interscience Publishers, 1953.

10. Crandall, S. H., *Engineering Analysis*. New York: McGraw-Hill Book Company, 1959.

11. Dorfman, R., P. A. Samuelson, and R. M. Solow, *Linear Programming and Economic Analysis*. New York: McGraw-Hill Book Company, 1958.

12. Dwyer, P. S., *Linear Computations*. New York: John Wiley & Sons, Inc., 1951.

*13. Fadeeva, V. N., *Computational Methods of Linear Algebra.* New York: Dover Publications, Inc., 1959.

*14. Ferrar, W. L., *Finite Matrices.* Oxford: Oxford University Press, 1951.

*15. Frazer, R. A., W. J. Duncan, and A. R. Collar, *Elementary Matrices.* Cambridge: Cambridge University Press, 1952.

16. Friedman, Bernard, *Principles and Techniques of Applied Mathematics.* New York: John Wiley & Sons, Inc., 1956.

17. Gale, David, *The Theory of Linear Economic Models.* New York: McGraw-Hill Book Company, 1960.

*18. Gantmacher, F. R., *The Theory of Matrices.* New York: Chelsea Publishing Co., 1959.

*19. ———, *Applications of the Theory of Matrices.* New York: Interscience Publishers, 1959.

20. Guillemin, E. A., *The Mathematics of Circuit Analysis.* New York: John Wiley & Sons, Inc., 1949.

21. Halmos, P. R., *Finite Dimensional Vector Spaces.* Princeton, N.J.: Princeton University Press, 1942.

*22. Hildebrand, F. B., *Methods of Applied Mathematics*, 2nd ed. Englewood Cliffs, N.J.: Prentice-Hall, Inc., 1965.

*23. Hoffman, K., and R. Kunze, *Linear Algebra.* Englewood Cliffs, N.J.: Prentice-Hall, Inc., 1961.

24. Householder, A. S., *The Theory of Matrices and Numerical Analysis.* New York: Blaisdell Publishing Co., 1964.

*25. ———, *Principles of Numerical Analysis.* New York: McGraw-Hill Book Company, 1953.

26. Indritz, J., *Methods in Analysis.* New York: The Macmillan Company, 1963.

27. Jaeger, Arno, *Introduction to Analytic Geometry and Linear Algebra.* New York: Holt, Rinehart and Winston, Inc., 1960.

28. Jenson V. G., and G. V. Jeffreys, *Mathematical Methods in Chemical Engineering.* London & New York: Academic Press, Inc., 1963.

29. Lanczos, C., *Linear Differential Operators.* Princeton, N.J., D. Van Nostrand Co., Inc., 1961.

30. Pipes, L. A., *Applied Mathematics for Engineers and Physicists.* New York: McGraw-Hill Book Company, 1958.

*31. ———, *Matrix Methods for Engineers.* Englewood Cliffs, N.J.: Prentice-Hall, Inc., 1963.

32. Schreier O., and E. Sperner, *An Introduction to Modern Algebra and Matrix Theory.* New York: Chelsea Publishing Co., 1952.

33. Varga, R. S., *Matrix Iterative Analysis.* Englewood Cliffs, N.J.: Prentice-Hall, Inc., 1962.